THE KINGS OF
BIG SPRING

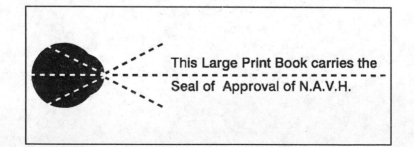

This Large Print Book carries the
Seal of Approval of N.A.V.H.

THE KINGS OF
BIG SPRING

GOD, OIL, AND ONE FAMILY'S SEARCH FOR
THE AMERICAN DREAM

BRYAN MEALER

THORNDIKE PRESS
A part of Gale, a Cengage Company

Farmington Hills, Mich • San Francisco • New York • Waterville, Maine
Meriden, Conn • Mason, Ohio • Chicago

Copyright © 2018 by Bryan Mealer.
Map by Virginia Norey.
Thorndike Press, a part of Gale, a Cengage Company.

**LIBRARY OF CONGRESS CIP DATA ON FILE.
CATALOGUING IN PUBLICATION FOR THIS BOOK
IS AVAILABLE FROM THE LIBRARY OF CONGRESS.**

ISBN-13: 978-1-4328-4782-1 (hardcover)

Published in 2018 by arrangement with Macmillan Publishing Group, LLC/Flatiron Books

Printed in the United States of America
1 2 3 4 5 6 7 22 21 20 19 18

To my children, and their children . . .

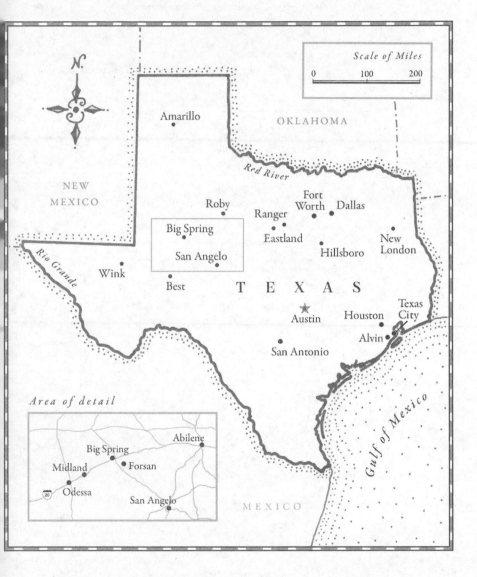

N.

Scale of Miles

0   100   200

OKLAHOMA

Red River

NEW
MEXICO

Amarillo

Roby   Fort
          Ranger   Worth   Dallas

Big Spring   Eastland   New
                              Hillsboro   London

Rio Grande

Wink

San Angelo

Best   T E X A S

          ★
          Austin   Houston   Texas
                              City
San Antonio   Alvin

Area of detail

                    Abilene
     Big Spring
Midland   Forsan
  Odessa
20   San Angelo   MEXICO

Gulf of Mexico

THE MEALERS

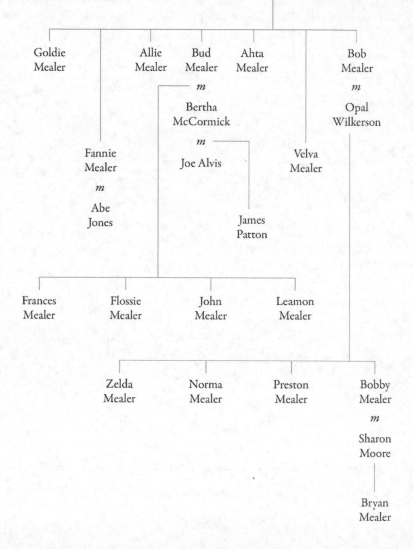

Catherine Cowart *m* Robert Mealer

John Lewis Mealer *m* Julia Bateson

Goldie Mealer

Allie Mealer

Bud Mealer

Ahta Mealer

Bob Mealer

m

Bertha McCormick

m

Opal Wilkerson

Fannie Mealer

m

Joe Alvis

Velva Mealer

m

Abe Jones

James Patton

Frances Mealer

Flossie Mealer

John Mealer

Leamon Mealer

Zelda Mealer

Norma Mealer

Preston Mealer

Bobby Mealer

m

Sharon Moore

Bryan Mealer

THE WILKERSONS

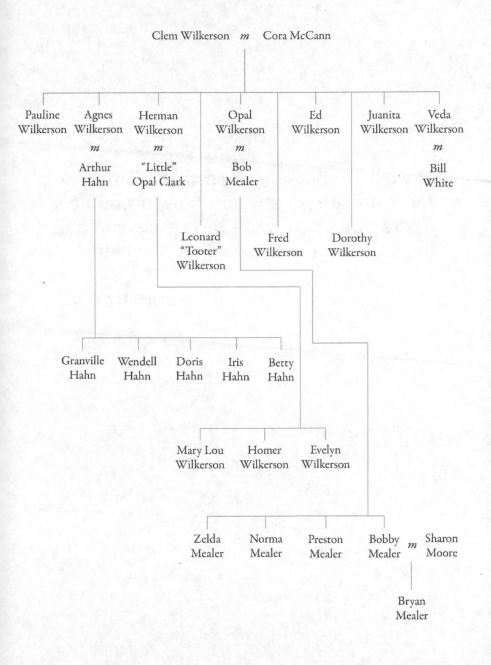

How well I have learned that there is no fence to sit on between heaven and hell. There is a deep, wide gulf, a chasm, and in that chasm is no place for any man.

— JOHNNY CASH

PROLOGUE

In January 1981 my father was given a choice to make.

He was twenty-seven years old, with a wife and three children, and for most of his adult life he'd struggled to find his niche. From one year to another, he'd bounced between jobs in the oil fields, painting houses, and selling used cars. By the time I was six we'd lived in three different towns in two states, leaving whenever Dad found better work or when his lifestyle became too fast or frightening. Mostly we ended up in places where we had family, since family provided sanctuary and a spiritual zip line into the abiding arms of Christ, whom Dad raged against and returned to for most of his life.

But by the early eighties, he seemed to have achieved a balance. We were living in a small town south of Houston surrounded by relatives. He had a career-track job at a nearby chemical plant, we belonged to a good church, and Dad and his brother were making plans to start a business together. Life for our family was not only stable, but the future had promise. Then one evening, as Dad was ready to walk out the door for a graveyard shift, his childhood friend Grady called.

"Bobby Gaylon! How'd you like to be a millionaire?"

Back in their hometown of Big Spring, another oil boom was kicking off and Grady had it by the horns. He was looking for a partner. They were going to get rich.

For Dad, it was the biggest decision he ever had to make, but it was an easy one. Although he never put much stock in the notion, somewhere in his mind he was trying to shake what my family had always called the Mealer Luck. We could trace it back to Ireland, where it rose from the sacked estates and spilled blood of our ancestors, then followed us across the ocean and down through the generations, whispering its name whenever forces greater than us, ones we didn't see coming or fully understand, left us busted and picking up the pieces. And in better times, and there were many, it reminded us what not to take for granted.

Nearly a century had passed since we'd

struck west from Appalachia to settle the raw country, in a time when America was still young. Like others around us, we were eager to put down roots and start something better, to help build this nation during its greatest century. We planted its cotton and drilled for oil, left our mothers and wives to fight its wars. We prayed for peace and rain and thanked God when the streets filled with trucks and men and a sour smell on the wind promised meat on the table. And when our sons didn't come home we endured it. When the oil and cotton went away we moved and started again. Only in Texas was there enough space for so many second acts.

Along these roads, of course, there was life: love and heartbreak, sin and redemption, small victories and unbearable tragedy, and laughter when little else could save us. We drew our strength from the enduring power of our own flesh and blood. My family's story is like the stories of so many others who came looking for their own square of soil and promise of America. It is the story both of Texas and of how this country came to be. And for us, it begins in a Georgia hollow after the Civil War, with a man facing his own fateful decision.

PART 1

1

John Lewis leaves the hollow and heads
west . . . a child is born . . . the boll weevil
comes to Texas . . .

In the spring of 1892, my great-grandfather
John Lewis Mealer left his home near Sharp
Top Mountain, in the foothills of the Georgia Blue Ridge, and headed west in the
direction of his brother. He was in his early
twenties, unmarried, and had begun to feel
the closeness of the hollow in new and
unsettling ways.

Lately, there'd been trouble between the
moonshiners and revenuers. Some of the
boys who kept their stills near the creek had
organized for vengeance, donning black
hoods and setting fire to homes of suspected
informers. The Honest Man's Friend and
Protector, they called themselves. The
sheriff and deputy had given chase, and the
boys had met them with gunfire and taken

to the woods. The lawmen wasted no time destroying their stills, and ever since the presence of the feds had been felt in the valley.

With so few ways to earn money in the hollow, John Lewis and his brothers, along with their father, had found easy work with the moonshiners, providing them firewood, along with apples and corn for their mash. But few men wanted a war. The revenuers were a ruthless bunch who practiced a kind of terror justice that brought back memories of the Confederate Home Guard. During the Civil War, the militia — whose mission was to protect the families of fighting men — had instead pillaged their way across Gilmer and Pickens counties. Two of them had murdered John Lewis's uncle Peter Cantrell in the summer of 1864 after accusing him of desertion. Uncle Peter lay buried in the family plot near Burnt Mountain, his stone proclaiming for the ages: KILLED BY THE JORDAN GANG.

At its best, the hollow was as peaceful as the first breaths of creation, crowded with pine and yellow poplar and broad Spanish oak. The Mealer house stood almost hidden in a grove of cottonwoods, save for clusters of daffodils that served as landscaping. John Lewis's father, Robert, had cleared enough

18

trees to allow a few vaults of sunlight for raising food and animals.

The family had lost their mother when John Lewis was four years old, leaving Robert to raise four kids on his own. His second wife had given him ten more mouths to feed in as many years, and for the most part, the forest had provided. Within a ten-minute walk they found wild strawberries, blackberries, honey hives, and a grove of apple trees. They'd learned to harvest and prepare pokeweed and chinquapin so they could eat it without being poisoned, and how to brew sassafras root into tea. The water that bubbled cold from the spring tasted like the iron and copper that lined the valley floor, something the Cherokee believed carried healthful properties. A dairy cow grazed amid the trees and the garden provided herbs and vegetables. As for meat, there were hogs and pullets and a forest full of squirrels, which they parboiled with cayenne to vanquish the gaminess, then pan-fried with milk gravy.

But short of moonshine, the forest offered meager paying work for Robert and his boys. A pair of men with a crosscut saw could make something from the rough timber, and if you were handy with a froe, you could split white oak into shingles to sell in town. But that way of life didn't hold young men the way it used to, not since the war, especially when half the South, it seemed, was bounding westward.

By 1892, it was possible to take a train clear to California, leaving from Jasper or Ellijay on the Marietta Line, as long as you had the money. Even those with empty pockets were stowing away on freighters and heading in that direction. Out west, the nation was still busy expanding, annexing, trying to fill its new borders. And what the West needed most were men to plow the soil and populate the towns, to work the mines, railroads, and factories that were fortifying this new world.

For the first time, thanks to the U.S. Army, the Indian no longer posed a threat to settlers crossing the coverless plains. The great warrior tribes that had repelled Manifest Destiny from the Powder River to the Rio Grande had been broken and contained, and their buffalo slaughtered.

Gold miners now blasted the Black Hills of the Lakota Sioux, while cattlemen drove their herds atop the lush bluestem where Comanche once trailed the buffalo. Behind them came farmers from the crowded East and the busted plantations of the Confederacy. Each year, tens of thousands were rushing into the Oklahoma Territory, where the government ceded Indian land to a stampede of covered wagons. Even more came from around the globe: from Germany, Bohemia, Scandinavia, and beyond, a great army of tomorrow men seeking cheap land, unobstructed views, and less government.

Many were going to Texas. The advancing railroad had opened farmland in the eastern part of the state, while the army's defeat of the Comanche had freed the western frontier. Railroad agents lured homesteaders with pamphlets and newspaper ads promising cheap and abundant land, an agrarian paradise unmolested by ice and snow, blessed with abundant rain and a kind of miracle soil that would grow any kind of crop. Since the end of the war, the population of Texas had nearly tripled.

Hundreds of thousands had gone there from John Lewis's home state of Georgia — so many that in 1879 an Atlanta newspaper bemoaned the impact of "Texas fever." "As long as the idea prevails that Texas is a very much better state than Georgia, the people who share this delusion will be discontented, shiftless, and inefficient."

Georgians, along with Southerners from Alabama and Tennessee, rushed first into East Texas, settling as tenant farmers and sharecroppers on the large plantations that had gone bankrupt after the war. And after the railroads made headway into the western range, they came to plow up the grassland and pushed the stockmen aside, since the land was more valuable under cotton than beef. In 1886, a New Orleans paper wrote that farmers were moving into western Texas at such a rate "that ranchers have just enough time to move their cattle out and prevent their tails being chopped off by the advancing hoe."

John Lewis's brother Newt had gone to Texas a few years earlier, chasing the new railroad and whatever fortune he could pull down from its trail of smoke. The oldest brother, Thomas, had left at the same time, but never made it out of Georgia. When he reached Bartow County, forty-six miles away, the flat green river bottom enchanted him enough to stay. He eventually opened a general store in Adairsville and counted among his neighbors the Floyd family, whose son Charles later became the beloved outlaw known as Pretty Boy.

Newt had landed in Hillsboro, in north central Texas, where the soil was dark and rich and cotton wagons jammed the streets four and five deep. The immigration flyers touted the potential for corn, cotton, and wheat, so easy to grow that even a mountain dweller like John Lewis could tame the land and prosper.

With the moonshiners now gone from the hollow, John Lewis's prospects paled against the bright Texas dream. He was young and strong and possessed something that was desired in the promised land — he was restless. And while I don't know the details of how he left Georgia, I can imagine the excitement as he bid his family farewell, walked to the nearest depot, then gave himself to that great wave rolling west.

The train journey likely took him from Jasper to Atlanta, through Alabama and across Mississippi and on to Fort Worth, then down into Hillsboro. The town sat along the blackland prairie, which stretched from Oklahoma down to Austin. It contained some of the most fertile soil in the state, and the Missouri, Kansas, and Texas line, known as the Katy, had connected the markets and allowed the cotton trade to boom. The Katy now hauled cotton by the ton down to Galveston, while coming the other way were New York financiers and shrewd German farmers, who'd scoop a handful of the waxy soil and feel the money in their fingers.

But the promises made by the pamphlets and newspaper ads came to an end shortly after John Lewis arrived. In 1893 the country entered a prolongued depression triggered by the collapse of the northern railroads. Banks folded by the hundreds, and over fifteen thousand businesses closed. In Texas, the value of the blackland plummeted, and when it did, the eastern holding companies gobbled it up. The self-contained farmer couldn't afford the mortgages, so he was forced to rent his fields, customarily paying the companies a fourth of his cotton and a third of other crops. But the companies had no use for vegetables, since they weren't commodities that could be sold commercially. If you wanted a mortgage, or even to rent a farm, you had to grow cotton.

It's unclear how John Lewis's brother Newt fared in Hillsboro, but in 1895, records show that he returned to Georgia with a Texas bride, and in that same year, John Lewis married Julia Bateson.

The Batesons had arrived from Arkansas in 1875 and settled in the town of Cleburne, some thirty miles north of Hillsboro. Over time, Julia's father came to own several large farms that brought the family wealth and prominence. Her brother John would later become a celebrated stockman, known throughout the country for his blue-ribbon Jersey cattle. His sons were builders and developers.

In the only photo that exists of Julia, what stands out besides her swirls of dark hair and smoky eyes is the expensive jewelry she's wearing: a pair of pearl earrings and a necklace with an ivory-colored pendant in the shape of a heart. Most likely she was educated in the Cleburne schools and, as the oldest girl, instilled with gentility and standards when it came to choosing a husband. John Lewis was tall, powerfully built, and known to clear the cane-bottom chairs from a room and dance an Irish reel. He could read and write and tell a good yarn. But unlike the Bateson men, he was poor and acreless, a lower hillbilly from the East.

The story of how they met did not survive them. The marriage certificate from Hill County is dated May 31, 1895, the Reverend T. N. James officiating. Julia was twenty years old and John Lewis just shy of his twenty-fifth birthday. But other records reveal a surprising twist — a daughter, Goldie, had been born March 9, nearly three months earlier. One could assume there was a delay in record keeping, or that someone in the clerk's office made a mistake when filing the documents. But if the records are true, Goldie's birth constituted a scandal in their time. And for a prominent family like the Batesons, this was a mark on their name, and could help explain their absence later when the couple needed them most.

By 1902, two more daughters had arrived, Fannie and Allie. Somehow that same year, John Lewis managed to rise above his tenancy and achieve the dream of ownership. He purchased thirteen acres south of Hillsboro, which he flipped two years later for forty-two acres hemmed with live oaks. A son was born, John Jr., whom everyone called Bud, followed by another daughter, Ahta.

That was 1907, when steady showers fell throughout the spring and summer. The cotton emerged like ropes of green pearls, and in their usual procession, flowered white before shedding their petals as the boll readied to bloom. But starting that year, the rain could not be trusted, because traveling with it came the boll weevil.

For more than a decade, the tiny gray beetles had made their terrifying advance from Mexico, ravaging crops and livelihoods with their long snouts that penetrated the young boll and severed its heart. The local gins were full of horror stories of bright, vibrant buds that flared one morning and were dead the next. Farmers spoke of the boll weevil in reverent tones, while bluesmen honored them with ballads the way they did the Devil and loose women.

For more than a decade, entomologists had rallied both chemistry and biology against it, unleashing poisonous clouds of powdered sulfur, Paris green, London purple, and lead arsenate, in addition to employing armies of parasite wasps and Guatemalan ants. But the boll weevil was able to adapt and change its habits to withstand any attack.

By 1907, Louisiana reported total infestation; already the swarms had stretched into Mississippi, Arkansas, and Oklahoma, and they would keep devouring until they reached the beaches of Georgia and could find no more cotton to destroy.

Once the weevils ravaged a farmer's crop and left his cupboards bare for the winter, they retired to the surrounding brush and waited for him to replant. In the pine forests east of Hillsboro, the damage was so complete that many farmers hadn't bothered to harvest at all. Some families sold land for pennies on the dollar, while others simply walked away and left their farms to the plague.

A great many of them headed to West Texas, where the boll weevil hadn't learned to survive atop the cold, dry plains. From 1903 to 1910, thousands of families loaded into wagons and Pullman cars and headed for higher ground, scuttling westward like the bugs who'd put them on the run. They left behind the pine forests and black river bottoms, pulled their teams across the ninety-eighth meridian, and entered the American West.

There's no account of John Lewis's own struggles with the boll weevil. But it's likely the pests discovered his little place in the groves, because by 1909 the family had abandoned their farm and followed the migration, traveling 250 miles by horse and wagon to work another man's fields.

The tenant farm was outside of Roby, where thousands of boll weevil refugees came to settle. The geography was ideal. Roby had sweeping green hills, loamy soil, and thickly wooded riverbanks. Steady rains fell during the early years of the migration, and under so many hands, there came bumper harvests of cotton, corn, and wheat.

The refugees wanted land of their own, but their very presence had pushed prices to unreachable heights. Unable to afford their own farms, many decided to turn back eastward and face down the weevil, whose advance was unstoppable. They returned to the blacklands and pinewoods to combat the plague in new ways, to diversify crops, or to quit farming altogether and find careers in town.

One hundred miles east of Roby was the town of Eastland, whose officials were attempting to spin the weevils' destruction in its favor. Perhaps John Lewis saw one of the many immigration pamphlets advertising Eastland as "the ideal place for a farmer to make money and school his children without raising cotton." They even went so far as to promise no crop failures. "Here is found an ideal place for the man with the push and energy to make good," and if John Lewis had anything, it was that.

Best of all, the land was cheap. In 1913, records show that John Lewis purchased 171 acres south of Eastland from a local lawman named George Bedford. (Years later, Bedford served as chief of police in nearby Cisco, where on Christmas Eve 1927 he was killed while exchanging gunfire with a bank robber dressed as Santa Claus.)

When I walked this land over a hundred years later, little seemed to have changed. A wide slice of green pasture sloped upward from the dirt road and solitary oaks bloomed like mushrooms on the expanse. And just south of the pasture, sitting on a hill overlooking the whole scene, sat the farmhouse, shaded by another large oak. The rest of the property was ringed with juniper and pecan and open for grazing cattle. The Leon River flowed a mile up the road, yet the ground held plenty of water.

It was November when John Lewis and Julia arrived with their children, now totaling six. Another daughter, Velva, had been born in 1910. I assume John Lewis looked ahead to the spring, when he could plant peanuts, corn, and peas — cash crops of the day that were thriving in the sandy soil. And now with so much land of his own, I'm sure he wasted little time carving out his patch of paradise. A vegetable garden needed tilling for Julia to tend with the girls. And when Bud wasn't in school, he could help raise the barn with proper corncrib, hayloft, and stable for a milk cow and team. There had to be a henhouse, of course, and a pigpen, and a smokehouse for when the shoats got big enough to butcher. And if they were lucky, the root cellar could go at the base of the windmill, where they could keep milk or watermelons cool.

By now, John Lewis was forty-four and

Julia thirty-nine. Goldie, their oldest daughter, was already seventeen. Shortly after they came to Eastland, she married a widower named Lem Wilson and took to raising his two girls. They lived together on a farm in tiny Desdemona, forty miles away, and the family no longer saw her much. With Goldie gone, Julia was left to care for five kids while trying to make a new home. She also had extra incentive for getting settled: she was pregnant again, and three months after arriving, she gave birth to their seventh child — a boy named Robert Odell, who would become my grandfather.

The older girls coddled the baby and called him Little Bob, and their help came as a relief to their mother. For John Lewis, the new child spurred him into action, and for a while, the weather cooperated. The rains fell steadily throughout the spring of 1914 and the crops were strong in the three surrounding counties, with corn producing at twenty bushels an acre. In 1915, records show that John Lewis walked into the bank and paid off his farm.

A farmer is never truly at ease, but I wonder if after twenty years of roaming, John Lewis finally allowed himself to imagine growing old on the land and seeing his two boys running their own plows over the fields. I like to picture him on the porch after a long day's work, surrounded by children as he pulls off muddy boots, the fields changing colors under the setting sun. I want him to cherish these moments, to understand that a man grows richer with every child, and a houseful of children makes for a citadel against the darkness.

I want to leave them all in happy twilight. But I can't.

2

Oil, a brief history . . . double heartbreak
. . . Bud on the edge of the dream . . .

A drought arrived in Eastland in 1915. On the rainfall charts, it's almost disguised behind scattered showers that fell from January through March. But the driest period occurred at the worst possible time, when fields were to be plowed and seeds planted.

By June, the crops had failed and John Lewis was in trouble. Worse, he was in debt. Although he owned his farm outright, he'd assumed a pair of loans belonging to George Bedford when he bought the place, which were held by deed of trust. He was paying them off in twice-yearly installments of $9.62, and after paying off the farm, he'd managed to whittle down the balance to $48.10.

But with his fields barren, he missed a

payment in July, followed by another in January 1916. Knowing little else to do, John Lewis drove to Eastland to see his banker. Frank Day was the First National Bank clerk who'd loaned him the money for the farm. Day was twenty-seven years old, handsome, with the brawny physique and confidence of a football star — a man whose ambitions far outsized the small-town bank.

Day listened to John Lewis explain how the drought had caused him to fall behind, how he needed to find a job in order to feed his family and make good with his debtor. He heard they needed laborers to work the fields down in Burnet County and wanted to go there and try his luck. But of course, he said, that would mean leaving his land and his outstanding loans. Could the banker help in any way? What did he advise?

Day assured John Lewis that everything would be fine. Go to Burnet County, he said, do what's best for your family. In fact, if the deed holder starts asking about his money and threatening trouble, I'll settle the difference myself and keep the land from foreclosure. After all, no hardworking man should worry about losing his homestead over $48.10.

So with that assurance, in the spring of

1916, John Lewis gathered Julia and their six children — including my grandfather, Bob, who was two years old — packed their wagon and set off to Burnet County, 150 miles south. But as he embarked on yet another slow retreat from his dead fields, others were headed in the opposite direction, seeing only money. Because what the land could no longer provide in food and living things, it would soon give in oil.

In the history of oil discovery, Texas arrived relatively late. Men had been drilling for oil across Pennsylvania and Appalachia since before the Civil War, mainly to produce kerosene. By the 1890s, this popular fuel made from petroleum had largely surpassed coal, and companies like Standard Oil were searching for bigger, more lucrative reserves.

The Mid-Continent field in Kansas, discovered in 1892, pushed exploration westward into Wyoming and along the Pacific. By the turn of the century, California produced more oil than anywhere in the country — particularly the San Joaquin Valley and Los Angeles, where hundreds of companies sank pipe beneath the modern-day city. But the West Coast fields were too far away from where most Americans lived, and without adequate pipelines, much of that oil was exported to Asia on boats. By themselves, the bubbling wells of Pennsylvania and Kansas weren't enough to drive an industry forward, much less ignite the imagination.

Until then, few people in America had ever laid eyes on a bona fide gusher — there'd been a few in California, wells that had shot between one thousand and fifteen hundred barrels of oil a day and were given names like Wild Bill and Blue Goose, yet news of their discoveries hardly made it east. Then came Spindletop.

On January 10, 1901, while John Lewis was still laboring on a tenant farm back in Hillsboro, the biggest gusher the country had ever seen erupted from a salt dome on the outskirts of Beaumont, Texas. The roar was so deafening that it terrified people in town, until they saw the green-black fountain rising hundreds of feet against the clear blue sky. Spindletop brought in seventy-five thousand barrels of oil per day and sparked a fever that spread throughout the state more fiercely and with greater speed than the boll weevil ever could. Within months, lines of wooden derricks sprouted along the Gulf of Mexico like a forest of naked timber. Tens of thousands rushed into Sour Lake, Humble, and Batson before the fever spread north to the Oklahoma line.

Farmers sold or leased their fields to companies looking to drill, hoping to reap the royalties when the oil sold for top dollar on the market. Agriculture gave way to energy, the barren land be damned.

Three years after Spindletop, the Texas and Pacific Coal Company struck oil by accident — just twenty miles from where John Lewis would buy his farm — while drilling shallow holes for coal. Quietly, the company began exploring in earnest, encouraged by geology reports that suggested vast reserves of crude beneath the ground. After a string of dry holes, a decent producer finally came in near Strawn, north of Eastland, followed by an even bigger one in October 1916 near Breckenridge. That well kicked off at two hundred barrels a day, just enough to start a commotion. It was right around the time that John Lewis and his family began their long journey home.

By then, the drought had reached Burnet County, where they'd gone to find work. And with most of the state now suffering, I assume John Lewis — having saved some money — decided it was better to wait for the rain on his own front porch. That, or a neighbor must have noticed the alarming sight at their farm and sent word to hurry back.

Although the Texas and Pacific Coal Company had practiced discretion while drilling around Eastland, there's no indication their work was kept secret. And any number of people in town could have gained access to the geologist's reports, understanding right away the potential bonanza that sat below the parched and useless fields — especially a loan officer at the local bank, someone like Frank Day.

A week after assuring John Lewis that he'd look after his farm, Day purchased the outstanding note for $48.10 and sold the land from under him. The buyer was John Lewis's neighbor — a man named C. M. Murphy — who John Lewis would later accuse in a lawsuit of conspiring with the banker. When the family arrived home that October, they discovered the gate was locked. Murphy had possession of their house and was farming their fields.

A ferocious confrontation with Day certainly followed. But all the banker had to do was explain to the simple farmer the terms under deed of trust: the farmer was entirely at the mercy of the bank. Day merely had to produce the damning documents and ask, *Didn't you understand what you were signing?* To which the farmer could only counter with flimsy, nonbinding emotion: *But you gave me your word!*

To this day, the 171 acres in Eastland is the most property that anyone in my family has ever owned.

Without a home, the family drifted. According to the lawsuit John Lewis later filed, he took his family south into Comanche County to look for a place to live, but then the record goes blank. There's no trace of them going to Cleburne, where Julia's family could have easily taken them in. The only other mention of the farm comes in April 1917, when Murphy leased it to a drilling company — just in time for one of the biggest oil booms in American history.

A week after Murphy leased the farm, the United States entered World War I. Already the British Royal Navy had made the revolutionary decision to switch from coal-driven vessels to faster ones powered by oil. The United States produced nearly 70 percent of the world's crude, and in the first years of the war, about a quarter of it was going to fuel the British trucks, armored tanks, and airplanes being used to fight the Germans. In this new era of combustible-engine warfare, it was clear that victory would come to whatever side possessed the most fuel. For its part, Germany dispatched submarines into the Atlantic to sink American tankers, a move that had ultimately pushed the U.S. to enter the conflict. Aside from America, there were few places to get oil. German forces already controlled strategic fields in Romania, a major European supplier, while in Russia, the Bolshevik revolution was paralyzing the vast reserves in Baku.

By July 1917, Germany's assault on oil-carrying vessels had become so effective that America's ambassador to England declared that "the Germans are succeeding" and that the Royal Navy was in danger of collapse for lack of fuel. That fall, Walter Long, the British Secretary of State for the Colonies, told the House of Commons that "oil is probably more important at this moment than anything else."

America's entrance into the war only increased the Allied demand for fuel, while at home, the need was also dire. For one, the nation was running out of coal, due in part to the never-ending procession of coal-fired ships being dispatched to the conflict. A brutal winter in 1917–18 diminished reserves even more as homes and businesses devoured coal for heat. But nothing drove up the demand for oil — and its price — more than the Model T, the new gasoline-powered car that was rolling off Henry Ford's assembly line to great demand. Between 1916 and 1918, the number of automobiles on American roads nearly doubled.

While a global fuel shortage threatened to tip the war to the Germans and stall a bustling auto industry, it seems remarkable that the rescue would eventually come ten miles from John Lewis's farm, in a region undergoing one of the bleakest economic periods in its history.

Ranger was a small hamlet northeast of Eastland, with little more than a depot and a few shops, its name derived from the Texas Rangers who'd made camp there fifty years earlier to fight the Comanche. Beef and cotton had once moved its trains and built a community. But the boll weevil had since closed its cotton gin, and drought was starving the cattle and pushing ranchers from their homes.

So desperate, the town's businessmen turned to oil, hoping to profit off the global demand. After hearing of the T&P Coal Company's success at Strawn and Breckenridge, they offered to lease the company twenty-five thousand acres if it would drill

41

around Ranger. The first hole shot only natural gas, which was useless without pipelines to bring it anywhere. Then, on October 17, 1917, the T&P hit its pay.

The well on J. H. McCleskey's farm blew in at a thousand barrels of oil a day, so close to the farmer's house that it coated his white leghorn chickens in black sludge. An even bigger discovery followed on New Year's Eve, and the boom was officially on.

Oilmen from Fort Worth were the first to arrive on the trains. Unlike the farmers undone by the land, the drillers in Witch-Elk boots seemed to harness the power of God, directing armies of men to tunnel into the earth. And with every gusher, from Spindletop to Sour Lake, Goose Creek to Batson, they'd learned a little more about how to tame it.

The major oil companies — Humble, Magnolia, Texas Company — moved in like an occupying force and laid down the apparatus of extraction. Teams of oxen and dapple-gray Percherons appeared on the roads, their tassels swinging as they pulled steam-powered boilers, welded tanks, and premeasured lumber to build the derricks. Men with pulleys and gin poles raised them from the ground, then outfitted the joints of pipe according to how deep the oil was. The drillers spoke of doubles, thribbles, and fourbles, however many lengths it took to punch through the lime and sand and bring up a gusher. Against the driller's technology, the land could no longer resist. Eventually it would break, shoot its mud, then give the men what they wanted.

By spring, the population of Ranger had swelled sixfold and exhausted the town, spilling into Eastland and surrounding hamlets. With no empty hotel rooms, cheap row houses and boom shacks appeared overnight, their walls made from thin beaverboard and tarpaper. Men rented beds and slept in shifts, sharing the same dirty sheets, while others rented chairs in hotel lobbies. The conditions became so crowded and dangerous that in April a fire erupted in the tenements and burned down two city blocks. Then the water ran dry, since the oilmen took what they wanted to drill their wells. Sanitation and hygiene became luxuries. Men wanting a bath had to travel thirty miles by train, and those with automobiles drained their radiators each night for fear of thieves.

By now the drought had spread across the entire state, turning the land pale and brown. In August 1918, D. J. Neill, the state representative to Gorman, just south of

Eastland, told the Legislature that every crop in his district had failed. "Many thousands have turned their faces eastward, homeless, friendless, moneyless. Those who cannot move will die."

Not long after Neill's address, a man traveling to Fort Worth reported seeing five hundred families camped along the roadside near Eastland. I can't help but fear that John Lewis and his family were among those people, that whatever had happened between Julia and her parents after Goldie's birth kept them from seeking help. The crowds were made up mostly of tenants and sharecroppers whose landowners had lost everything, for not every farm was lucky enough to sit over a fortune in oil.

The state responded by passing the Drought Relief Law, which gave emergency loans to farmers in several counties. The commissioners in Comanche County called this "an inconvenience" and likened it to welfare, even as their neighbors starved in the weeds. John Lewis's name does not appear on these lists.

The drought was finally broken in September, when storm clouds appeared and unleashed heavy rains. But the dry, parched ground couldn't soak up the water quickly enough and the land flooded. Roads filled with mud, so deep in Ranger that a horse was said to have slipped and drowned. At the train depot, some enterprising boomers rigged boards into sleds and sold rides down Main Street. Heavy traffic caused sinkholes on the roads that stranded trucks and teams. Farmers living near these traps, broke and desperate, started charging fifteen dollars to pull people out. And just as the rains had brought the boll weevil in earlier years, now they spread typhoid and other diseases in the cramped and filthy boom shacks.

The most deadly was Spanish flu. The great influenza pandemic first reached El Paso in September 1918, killing more than five hundred people within the first month. Elsewhere across the nation, as many as eight hundred people were dying each day in New York City, and in Philadelphia, news accounts described bodies stacked and rotting in the city morgue, so gruesome a scene that embalmers refused to enter. In Texas, the pandemic rode the railroads and summer winds, and by the end of September, it reached the oil patch. In less than a month, over twenty-five hundred cases were reported in Ranger.

It was around that time that Julia died.

According to my grandfather Bob, her death occurred sometime that summer. Bob, who was four at the time, claimed his mother had fallen sick shortly after his birth, which suggests something degenerative such as tuberculosis, which was prevalent in the area. Influenza probably compounded the TB and accelerated her death. But whatever her sickness, she likely carried it with her on the road and was sick when Frank Day took their home and flung them into uncertainty.

I found no record of Julia's death, which tells me they were estranged from her wealthy family, outside the realm of hospitals and coroners, just drifting among the castaways. "She died while they were traveling," one of my uncles later told me, after I'd spent months searching courthouses and weed-strewn graveyards. "They just buried her somewhere on the road. That's what they did back then."

I suspect John Lewis and the children were with her in the final moments, and others, too — a preacher to say a prayer, hopefully even a doctor. One of my aunts remembered hearing that they laid her body in a cheap box in whatever place they'd found as shelter. It sat on a table, its lid closed, and beneath the table lay my grandfather and his sister Velva, curled up on the floor and crying for their mother.

I'm sure John Lewis arranged her burial — perhaps in a cemetery that I've visited, wondering who lay beneath the flat, unmarked stones. Or perhaps, as my uncle suggested, they simply dug a hole on the side of the road, since that's what people did back then. Whatever the case, the trucks and mule teams plodded past under heavy loads of pipe and tools. Coming behind them with blaring horns were the shining new Fords of the drillers and the newly rich farmers, their dead fields painted black, their wheels bouncing onward into the dream.

By now, I'm sure that John Lewis had heard about Frank Day — how he'd quit the bank and entered the oil game, and how he was on his way to making a fortune. Day's charisma was such that no one bothered to question how he'd arrived, never bothered to look for his boot prints on the backs of working men.

Boyce House, the editor of the Eastland newspaper and one of Day's close friends, later told a story about an elderly farmer who entered Day's bank one afternoon and, "out of friendship," House wrote, "offered to give the banker his farm as he was no longer able to pay the taxes." When the boom hit, Day leased the land for forty thousand dollars — a windfall that House chalked up to winner's luck.

When the war ended in November, the Pullmans arrived full of soldiers looking for work, and the boom grew even bigger. Behind them came farmers who'd reversed their retreat to get some payback from the land. They rode into town dragging "half-dead milk cows," House wrote, and found easy work as menial laborers on the rigs. Most of their families lived in tent cities by the derricks, in a slurry of mud and oil and sewage. The boomers quickly exhausted all resources. The price of food soared into the stratosphere. A man lucky enough to get a seat in a café could not dine in peace, for another man was soon hovering over him, hand on his chair, waiting for him to finish. And yet still more came — from Canada, Mexico, Russia, and Egypt — and in such great number that five trains a day were not enough. The Pullmans arrived at the depot with men clinging from the roofs and windows.

Billy Sunday, the most popular evangelist in the country, heard about Ranger while preaching a monthlong revival in Fort

Worth and came down to survey the action. At one of the rigs, Sunday dipped his hands into a puddle of oil and let it run down his arms. Other men smeared their hair and faces with it, removed their shirts, and bathed themselves in its stink. The drillers called this a "Roman orgy." After Sunday took his own bath, he stood ankle-deep in the muck and preached to five thousand men on the pitfalls of demon rum.

More celebrities arrived looking for a piece of the bonanza. One of them was Tex Rickard, the famous boxing promoter, who began leasing land from Ranger to San Angelo. He arrived in town with the heavyweight champion of the world, Jess Willard, who was looking for his own investments. The champ soon found a partner in Frank Day.

Together, Day and Willard ate steaks in the McCleskey Hotel. They pushed through the crowded streets as the player piano chimed from behind the mud-splattered doors of the Blue Mouse Cabaret. Willard hailed from Kansas and was known as the Pottawatomie Giant. He was "much of a man," according to House, six feet four and 240 pounds, while Day was "220 pounds of brawn and sinew." The two men drove around Eastland County looking for land to lease and for oil to drill. One afternoon, a mule team wouldn't get out of the road, so Day clipped it with his car. When the teamsters raised hell, Day and Willard jumped out with fists raised. "Good heavens, boys," the men shouted. "It's the world's champion!"

Day and the champ bounded toward Desdemona, where a new field was wide open. Desdemona, twenty miles from Eastland, was home to John Lewis and Julia's oldest daughter, Goldie, who'd married the widower Lem Wilson and cared for his two children.

Goldie had borne two kids of her own by the time the gas gusher blew on Joe Duke's farm, just down the road. It mixed with some embers from the tool dresser's forge, exploding into a column of fire that swirled like a vision of St. John. It took three days for the steam boilers to cap the runaway well, its bright corona visible for twenty-five miles. And when it began producing a thousand barrels a day, more boomers descended like a swarm.

The boom landed right where the people lived, since most of the oil lay concentrated under the center of town. Drillers spudded in every five hundred feet on leases of one-hundredth of an acre, their derricks like a cluttered board game when viewed from the top of a flow tank. The oil was wrapped up in the rocks, so drillers dropped torpedoes of nitroglycerine down the wells and the ground vibrated until dark, when the glow of gas flares brought forth an odd chemical daylight. Gushers were left spewing to attract crowds and investors. So much oil poured from the ground that drillers dammed the creeks and gullies and filled them to the banks. One morning, a tidal wave of oil rushed three feet deep over the road to De Leon. Oil and mud stranded the mule teams and buried the Model Ts up to their doors. Gas wells blew untethered and sent men running through town shouting, "Shut everything down, nobody light a match!"

Twenty thousand workers crowded in, many of them war vets straight from the trenches. A deputy sheriff pistol-whipped one and knocked his eye loose from its socket, so a mob destroyed the jail and the café where he was arrested. Then they sacked the nearby clothing store, whooping as they stomped bolts of silk into the mud. Twenty thousand men lived on the outskirts of town in a ring of tents, where disease festered and spread.

This was Desdemona during its short-lived boom. And it's where John Lewis was summoned in January 1919 — less than a year after losing Julia — by a message that his daughter Goldie was gravely ill. It's unclear where he and the children were living at the time, but it must have been far, because by the time they arrived and inquired of Goldie's whereabouts, they were told she'd been dead for two weeks. She was twenty-four years old.

The cause was pneumonia, most likely in conjunction with Spanish flu. Unlike John Lewis, Lem Wilson had managed to give his wife a decent burial in the Desdemona Cemetery, its air choked by a ring of belching rigs. Nearly one hundred years later, I visited her grave and wondered how John Lewis must have felt as he stood in the same spot, and if he regretted making the trip out west, leaving behind the peace of the Georgia hollow and all its familiar comforts.

The old homestead must've weighed heavily on John Lewis's mind, because four months after losing his daughter he received news that his little brother, Elijah, having contracted empyema during the war, had died in France. We don't hear from him again until late December, when the hearing was scheduled in response to his lawsuit against Frank Day and C. M. Murphy.

Probably knowing his chances were slim, John Lewis failed to appear in court and the case was dismissed. His farm went to Murphy, and Day got off clean. The judge ordered John Lewis to pay their legal fees.

If there was any consolation, it came on July 4 of 1919 — in Toledo, Ohio, of all places. There, an undersized contender named Jack Dempsey beat Jess Willard in one of the most lopsided victories in boxing title history. By the end of the third round, Dempsey had broken several of Willard's ribs, shattered his cheekbone, and dispatched six of his teeth. Unable to continue, the Pottawatomie Giant threw in the towel. As he stumbled back to his dressing room, a reporter heard him mumbling, "I have a hundred thousand dollars and a farm in Kansas. I have a hundred thousand dollars and a farm in Kansas. . . ."

Frank Day watched it all from his ringside seat, dumbfounded. He had spent the entire week in Willard's training suite, schmoozing with race-car drivers and other celebrities, and emerged so confident that he wagered ten thousand dollars on Willard's chances. Dempsey ended not only Willard's boxing career but also his relationship with Day. From what I could find, Willard never set foot in Eastland County again.

It hardly mattered anyway. Within two years, the boom collapsed. All the gushers from Ranger to Desdemona gassed the pressure out of the ground and the oil became too expensive to reach. Frank Day was caught overspeculating and lost everything. The boomers and wool-suited moneymen simply stepped over him as they moved to another play, this one in Burkburnett, on the Oklahoma line, where the gushers were popping like champagne corks on the red Rolling Plains.

In March 1920, nearly two years after losing Julia, John Lewis responded to a U.S. census taker. He was living back in Eastland County as a tenant farmer. The previous year, his second-oldest daughter, Fannie, had married a gray-eyed pipeliner named Abe Jones, who swept her off her feet and into the North Texas boom. This left John Lewis to look after five children on his own — including my grandfather, Bob, who was now six years old.

John Lewis often traveled to find work, and when this happened, he left the young children under the care of his two oldest, Allie and Bud, who were nineteen and fifteen years old. Despite their supervision, the children ran wild.

One Christmas while their father was away, they sold his egg-laying hens to buy roman candles, then waged a war in the kitchen. They stuck cotton soaked in coal oil to the ends of cane poles, set them ablaze, and chased one another through the house. They cut holes in tow sacks and wore them as bathing suits. The nearest cattle tank was two miles away, where Allie and Bud taught the littlest ones — Ahta, Velva, and Bob — to swim in the deep red water.

When money ran out, Bud shot jackrabbits and squirrels for their supper, but often they went hungry. They rarely attended school and couldn't afford a doctor, not even when Bob nearly took off his own foot with a hatchet. They simply poulticed the wound with coal oil and bound it with shredded rags.

There were times when John Lewis surprised the children with his tenderness. Although he was a hard man, and rightly so, each Christmas, he cleared the cane-bottom chairs and danced a jig, to the delight of his daughters. The girls loved their father fiercely and felt likewise about Bob. They doted on the boy constantly and regarded him with pity, feeling sorry that at such a young age he'd been deprived of his mother's affection. They remembered with sorrow the way he and Velva had cried while curled up beneath her casket.

But none were closer than Allie and Bud. Allie called Bud her dark-eyed boy, Bud called his older sister Skinny Legs, and he worried when she ran off and got married and came home for visits with the spark missing from her eye. Yet by that time Bud was no longer a boy. He was growing into a man, and a handsome one, too — tall and wiry, with thick brown hair that he kept trimmed and greased. On one side of his nose was a faint beauty mark.

The most remarkable thing about Bud was that he seemed genuinely happy. Trauma and privation hadn't soured him — unlike Bob, who even at a tender age was prone to rage and general meanness. In fact, Bud maintained a buoyant disposition that earned him many friends. And like his father at the same age, he was eager to travel the big country. As soon as he turned seventeen and working age, he sniffed the air for the nearest boom and raced headlong into its crazy arms.

Bud wound up in Reagan County, two hundred miles west, working as a mechanic in a derelict town on the wild edge of the pay zone. The town was called Best, "the town with the best name and the worst reputation," and its motto was not untrue. After oil was found four miles from the tiny rail depot, the town had materialized in a matter of months, thrown up near the tracks like a Hollywood back lot. As in Ranger, housing came at a premium, but at seventeen years old, Bud lived happily in a tent, relishing the filth and danger.

Best had all the action of Ranger but very little class, a last resort for prostitutes and bootleggers who'd burned through eight lives in other booms. The town floated on an ocean of Choc beer and moonshine whiskey. Blood covered the floors in the back rooms of saloons, where dogfights were wagered on by drunk, violent men — men whose wives appeared in the post office with black eyes and swollen lips.

Best was surrounded by flat, lonesome country, so dry and sand-choked that farmers wouldn't go near it. There was barely enough rain to sustain the native grasses that over millennia had toughed it out and adapted. It sat in the center of a vast region known as the Permian Basin, some 250 miles wide and 300 miles long, which stretched from the Edwards Plateau west up the Caprock Escarpment, where the Staked Plains yawned into New Mexico. To the south, it covered the Pecos River valley and up the bone-colored Guadalupe range.

The region takes its name from its thick layers of rock, dating back to the Permian age, more than 250 million years ago. A pair of deep inland seas covered the land, fed periodically by a channel from the western ocean and fringed by great coral reefs. Over millions of years of subsidence and uplift, of land masses colliding and shifting apart, the seas became isolated. Their floors collected sediment, such as limestone, and as the sea finally evaporated and the reefs perished, massive deposits of marine life filled the basin and turned it rich with hydrocarbons.

But up until the 1920s, the Permian Basin was considered an oilman's graveyard. Not that there wasn't oil to be found — it had long bubbled up in ranchers' water wells, and just the previous year wildcatters had discovered a respectable field a hundred miles south of Best. But it was just that — respectable, yielding only twenty barrels a day.

Elsewhere in the Permian Basin, the oil was elusive, a fortune just out of reach — yet it was close enough to keep the wildcatters crazy with fever, drilling test well after test well, only to come up brokenhearted.

Rusted boilers and coils of cable littered the roadsides, a testament to the dry holes and saltwater at the end of every gamble.

Reagan County was part of two million acres that the state legislature had endowed to the University of Texas. Cattlemen rented large swaths of the county for grazing, but aside from that, the land held zero commercial value. Then, in 1919, a group of wildcatters requested 430,000 acres to drill for oil. The university leased it to them for a dime an acre.

El Paso businessman Frank Pickrell eventually came to lead the effort. Pickrell was working off an obscure survey that suggested a formation called the Marathon Fold ran beneath a narrow strip in the scrub. He knew little about the oil business, and his hunch was so farfetched that he spent two years trying to attract investors — who typically fronted a percentage of the drilling cost in exchange for royalties once the well came in. The drilling tools they found were secondhand, bought for ten cents on the dollar in Ranger, which by then was nothing but a salvage yard blooming rust.

The well was spudded in August 1921 using a cable-tool rig that pounded deep into the earth like a pick and hammer. The drilling dragged on for twenty-one long months, the work often interrupted due to lack of funds to pay workers and buy supplies. During this time, the driller, Carl Cromwell, lived with his wife and daughter in a wooden shack near the well, their loneliness broken only by passing trains carrying sheep. Sometimes a rancher felt sorry for them and left a quarter of beef swinging from the derrick.

While Cromwell pounded the hole, Pickrell hustled money to keep the operation going. In New York City, a group of nuns agreed to invest in the well, but did so with trepidation. Before Pickrell left, one of the sisters handed him an envelope full of rose petals and issued specific instructions. When he returned to the rig, he climbed to the top and sprinkled the petals onto the derrick floor, then declared, "I hereby christen thee Santa Rita," who was the patron saint of impossible causes.

On the morning of May 28, 1923, Cromwell was having his breakfast when he heard an unmistakable roar. The Santa Rita No. 1 had not only struck oil, but shot a head of crude that blew for a week, coating Cromwell's shack in a gooey sludge. A few days later, over a thousand people crowded the well site and another boom was born — this one different from all the rest.

The Reagan County field was only four miles long, but its reserves would prove vast and steady-flowing. Better, it triggered a surge of exploration that unlocked the rest of the Permian Basin. In a few short years, the providence of the Santa Rita not only made the University of Texas wealthy but introduced the Permian — the erstwhile wasteland — as the largest and most lucrative oil field in the United States, one that would power a growing nation and sustain our family for decades to come.

As an oil field mechanic in a boomtown, Bud most likely serviced trucks and heavy engines that powered the drilling rigs. Despite the bootleg dens and vice all around, he married Bertha McCormick and settled down.

Bertha's people were from northern Alabama, up in the Cumberland Plateau, where her father had worked as an itinerant coal miner. Out of James and Rilla McCormick's six children, three of their four boys had died — all separately and all at the age of ten. Only two daughters had come to them, and to compound the family's trauma, James violated them repeatedly.

For whatever reason, the McCormicks

were living on a farm in Snyder, 130 miles north of Best, when the oil boom hit. Bertha was fifteen years old when she met Bud, smitten by the dark-eyed boy whose past rolled off his back, whereas hers could not. In Bud, she found safety and comfort. They married that year, then welcomed two daughters into the maelstrom of the Reagan County boom — Frances Murl Dean, born in 1925, followed by Flossie Mae two years later.

By then, most of the action in Best was heading out of town — to Big Lake, ten miles east, which was bigger and better suited to absorb the surge of newcomers. Bud's job eventually left, too, but instead of following the boom, he pulled up stakes and found another one.

His older sister Fannie and her husband, Abe Jones, were now living near a town called Big Spring, a hundred miles north, on a cotton farm that Abe and his brothers inherited from their father. Once they put down roots, the rest of the family followed: John Lewis and my grandfather Bob, plus his sisters Allie, Velva, and Ahta, all of whom were married now and raising their own children. Bud and Bertha arrived and rented a house close to town, where they welcomed another child, a boy they named

John Odell.

Oil had been discovered on the big ranches around Big Spring, and lots of it. The boom it triggered was less frenzied and better managed than the others, yet promised the same opportunities: trucks needed driving, rigs needed building, pipeline still had to be sunk. Several refineries were in the works and tall buildings were being raised, including a fifteen-story hotel like one you'd find in Fort Worth or Chicago. The surrounding farmland was rich and loamy and seasoned for cotton, and yields had been high. Come picking season, if the boll weevil kept its distance, there'd be enough work for every man. Most of all, though, the citizens in Big Spring seemed possessed of a vision, and it was clear that people were moving there to stay.

Bones, con men, and oil . . . a town is born . . . fortunes rise and fall . . . the Mealers find their home . . .

The town of Big Spring sits in the eastern Permian Basin and straddles two of the state's most distinct geographical regions. The craggy red hills of the Edwards Plateau, which slope southeast for three hundred miles into the Texas Hill Country, dominate the town itself, which sits at twenty-four hundred feet above sea level. Two of those hills — which residents call "mountains" — rise along its southern edge and afford sweeping views to the north. Just a short drive in that direction one encounters the "breaks," where the land rises dramatically up a jagged shelf, called the Caprock Escarpment, and rolls like a carpet all the way to Colorado. This marks the beginning of the *Llano Estacado,* or Staked Plain, that

Spanish explorer Francisco Vázquez de Coronado crossed in 1541 while searching for the Seven Cities of Cibola.

The town gets its name from a deep, clear pool that once sprang from the limestone. For centuries the spring was a prime watering stop along the Comanche War Trail when the tribe's empire stretched from modern-day Kansas down past the Rio Grande. It sustained warriors on their yearly raids into Mexico for cattle, slaves, and horses, and again when they returned with buffalo, hundreds of thousands of which came to graze on the grama and bluestem before loping north in the spring. The remainder of the year, antelope were as thick as jackrabbits, gray wolves stalked the canyon bottoms, and nightfall brought a strange thunder as vast herds of mustangs crossed the plain.

The pool remained the quiet domain of the Comanche until 1848, when gold was discovered in California's Sacramento Valley. The initial discovery back in January had come less than two weeks after the signing of the Treaty of Guadalupe Hidalgo, which ended the Mexican-American War and gave the United States the California territory and much of the American West. In one stroke of the pen, the nation grew by

66 percent. Only three years prior, Texas had joined the Union as the twenty-eighth state. The new parts of the country were young and raw and wholly perilous, particularly the vast middle section of the Great Plains and Llano Estacado.

The sea of grass that blanketed the western frontier constituted a chasm in the American mind. By the 1840s, most maps of the United States identified the region as the Great American Desert, described by military expeditions, travelers, and journalists as a wasteland void of water, shade, and timber. It was prone to grass fires, tornadoes, rattlesnakes, and blizzards. And of course, there were the Comanche, whose murderous mounted assaults had not only turned back the Spanish from the Llano Estacado, but practically ground American westward expansion to a halt.

But California gold instilled enough courage for thousands to cross anyway in search of their fortune. Up north, they struck out along the established emigrant trails, mainly the Santa Fe and Oregon. But southern routes through the Comancheria proved more treacherous, and in 1849, the government began looking for ways to protect them.

The job of surveying one of these "California roads" through Texas was given to Captain Randolph Marcy, then stationed at Fort Smith, Arkansas. In April, he and his men led nearly five hundred emigrants up the Llano Estacado and over to Santa Fe and El Paso, where the travelers broke off to California.

On his trek back to Fort Smith, Marcy searched for a secondary route. He hired a Comanche guide who led them through the Guadalupe Mountains and across the Pecos River until they reached a branch of the Comanche War Trail that took them up the plains. On October 3, while still on the trail, they encountered "a spring . . . flowing from a deep chasm in the limestone into an immense reservoir some 50 feet in depth."

The "big spring" was added to their maps, and later to Marcy's hugely popular *Prairie Traveler: A Hand-Book for Overland Expeditions,* which was published in 1859. Ragged caravans of emigrants — mostly men — began pushing toward California as the danger of the Great Desert ceded to gold fever. Thus Big Spring had its origins in the first boom of the American West. But it would take another bonanza — a dark and shameful one — to place it permanently on the map.

The great bison slaughter that began in

Kansas and the middle plains after the Civil War reached Big Spring in the mid-1870s. Hunting teams fanned out from Fort Griffin, just north of Eastland, where the U.S. Army was making a last stand against the Comanche. Around Big Spring, groups of hunters hid in trees near watering holes, then opened up with Sharps rifles like carnival shooters. A crew of four men could kill a hundred buffalo in one morning and several thousand in a three-month season, especially when using chains and a team of horses to skin out the carcasses. The hides fetched two dollars apiece from buyers in the East, who sent freight wagons into the putrid camps to haul the pelts to Fort Worth and Dodge City. The remains were left to rot in the tall grass and churn a septic breeze.

By 1875, the killing had spiraled into a full-scale massacre, so out of control that the Texas Legislature debated a bill to protect the remaining herds. But General Philip Sheridan, who'd just broken the last Comanche resistance at the Battle of Red River, rebuffed lawmakers. Still bristling from his victory in the Panhandle, he pressed for the expedited elimination of "the Indian's commissary."

"Send [the hunters] powder and lead, if

you will," he told the Legislature. "But for a lasting peace, let them kill, skin, and sell until the buffaloes are exterminated. Then your prairies will be covered with speckled cattle and the festive cowboy."

By 1878, both the Comanche and their buffalo had mostly disappeared from the Llano Estacado. When the cattlemen did arrive, as Sheridan predicted, they found a prairie covered in bones. One of the first homesteaders in Big Spring, William Roberts, described so many skeletons on his land that a person could walk atop them without ever touching the ground. It wasn't until an Englishman named Jimmy Killfall appeared at Roberts's door asking to collect the bones that Roberts discovered that a railroad was on its way.

The financier and rail magnate Jay Gould had just taken the reins of the failing Texas and Pacific Railway Company, which Congress had chartered in 1871 to build a line from Texas to San Diego using Marcy's California road as its guide. In January 1880, a construction crew of five thousand men and three thousand mules finally broke ground in Fort Worth and began hacking their way west.

Killfall, the man who appeared at Robert's door, was soon joined by a wave of fellow scavengers, couriers of a new industry that rode on the back of the hide trade. Companies in St. Louis, Detroit, Philadelphia, and New York were paying as much as fifteen dollars a ton for the bleached buffalo bones. Sugar companies used the calcium phosphate in the bones to neutralize acid in cane juice. Fertilizer plants ground them into meal. Bones undamaged by the sun went to make buttons.

The pickers traveled on buffalo roads from Fort Dodge through Kansas, where they'd scoured the prairies before moving south. Many were the same hunters who'd helped lay waste to the herds, while others were homesteaders who'd come west seeking fortune, only to be thwarted by drought, grasshoppers, and debt. Together with other families, they traveled in convoys of oxen-pulled wagons loaded down with bones. They combed creek bottoms and ravines, sent kids rooting under bushes and trees, and set fire to the prairies so they would reveal their prize.

The more industrious scavengers operated large, long-haul freight wagons that traveled in columns across the range, their sides "picketed," as one historian noted, in a macabre array of pelvic and thigh bones. The freight was driven and stacked along the planned T&P route — in what would become the towns of Big Spring, Abilene, and Sweetwater — where they were loaded onto "bone trains" and carried to market. From West Texas up into the plains, ghastly white mountains swelled along the horizon.

On average, the bone mounds were the shape of hayricks and stretched as long as a city block, as high as the haulers could pitch them. Others covered a half mile, grand installations of death for men and boys to clamber atop and pose for passing photographers.

At Big Spring, the scavengers made their camps near the water's edge, next to crews of Chinese and Irish laborers who'd come to build the railroad. The daily rumble of their dynamite as they blasted the roadbed came as a welcome distraction to the monotony of bone picking. Eager cowhands hoping to prove themselves on the ever-expanding ranches also made their camps alongside the spring. In the evenings, they gathered to drink red whiskey in a single-tent saloon owned by an Indian scout, and at closing time raced their horses through the tent rows, firing pistols into the lantern lights. Whenever red whiskey ran dry, the bartender tricked the Irish paddies by mixing moonshine with chewing tobacco. When the Texas and Pacific Railway arrived in March 1881, the town of Big Spring was born and these were its first citizens.

The semiarid climate meant that cattle needed big spaces to feed — a whole section, 640 acres, just to keep six animals alive — and the hills and pasture around Big Spring provided plenty of room. The first cattlemen claimed vast holdings for their herds, since much of the land was public domain. Within a decade, the bone heaps gave way to a dominion of sweeping ranches.

Biggest of all was Colonel C. C. Slaughter's Long S Ranch, with over a million acres of open range, so vast it was said to take seven days to cross it by horseback. Other ranches included the McDowell, the Roberts, the Otis Chalk, and Settles, to name a few. These men had emigrated from Ohio, Tennessee, Kentucky, and East Texas, moving west once the Comanche were pushed from the plains; Slaughter himself had hunted Indians with the Texas Rangers. Once their herds grazed along the Caprock, they weathered wolves and coyotes, stampedes and vacillating markets, and heaped a thousand curses upon the prairie dog, whose grass-covered holes could snap a horse's leg and pitch a rider to his death.

But their greatest enemy of all was the weather.

The Caprock lay at the convergence of three principal winds: those blowing cold off the Rockies and the northern plains, and those pushing up warm and damp from the Gulf of Mexico. Nowhere else would it rain and hail one minute, then blow dirt the next, and nowhere else would a blizzard mix with sand so thick it would blind ten thousand cattle, sending cowhands wandering lost with chilblains.

And nowhere else, it seemed, could the land dry up so quickly. Dry years were more common than wet ones, and every generation had its defining drought. The drought of 1886–87 nearly bankrupted Slaughter, and families with smaller holdings had to scavenge buffalo bones to stay afloat. The drought of 1916–18 that cost John Lewis his farm nearly devastated the McDowell ranch. In a panic, Lorin McDowell sent Lorin Jr. with most of their herd into Nebraska and South Dakota to save it, a journey his son would later say "made a furrow in my brain." The following winter, in 1918, a freak blizzard killed two thousand of their steers.

Yet the cattlemen stayed, as did the farmers who came later, burrowing their crude dugouts into the hills with only bran sacks for a door. They stayed even as their fields died and their cattle starved and the wind pushed their wives to the margins of madness. And why? Because when the land was green and healthy, they said, it was like staring into heaven. "This country can promise less and deliver more than anywhere on earth," said one rancher. That statement proved particularly true once oil was discovered, for when the land surrendered its oil, it offered a lasting insurance against the fickle skies.

As one popular Christmas card later declared: "No steer is so fat as one which scratches his ribs against the legs of an oil derrick."

Big Spring's first oil boom arrived with a whimper, and a fraud. It was 1919. Ranger was ripping 174 miles to the east, while to the north, Burkburnett was starting to stir. The whole state was like a frenetic landscape of kicked-over ant beds. People still reeled from the drought that had left hundreds of families like mine homeless. And as oil fever spread across Texas, jumping fence lines and county roads, they grew desperate for the salvation oil could provide from the dry, evil weather. This made them vulnerable to a slickster like Alphabet Cox.

S. E. J. (short for Seymour Ernest Jacobson) Cox was, in the words of the Federal Trade Commission, "the most seductive and unreliable promoter in America." Before arriving in Texas, he'd created dozens of companies from Michigan to New York City, peddling everything from "miracle" carburetors to wallpaper and virility pills. But an oil boom presented untold opportunity. In 1917, Cox had moved to Houston and began selling interests in three separate ventures, passing himself off as a savvy dealmaker. He was dark-eyed and charismatic, his wife Nelda was cosmopolitan and gorgeous, and both were accomplished aviators who piloted their own planes, the *Texas Wildcat* and the *Texas Kitten*. The duo routinely competed in cup races.

After starting the General Oil Company in early 1919, Cox drilled several successful wells in Burkburnett, which gave him an air of legitimacy when he rolled into Big Spring that summer looking for a new play.

His arrival came shortly after a group of businessmen had pooled money to drill several test wells, without any luck. So when Cox marched into the local chamber of commerce and demanded to lease two hundred thousand acres, people paid attention. Even more so when Cox told them that his "geologist," an Indian named Geronimo, had already scouted their pastures and discovered a sea of oil.

One of the sites was on Lorin McDowell's place. Since the rancher's herd was still recovering from the drought, he was placing his last hopes in Cox. As crews got busy spudding in during the summer of 1920, the *Big Spring Herald* reported that even fortune tellers had become interested in the oil field. The excitement swelled. "According to their dope an oil well is to be brought in between June 20 and 27," the paper said.

The crystal gazers were correct. Sometime before dawn on June 20, Cox's first well — the No. 1 McDowell — struck pay. "We got her!" the driller screamed into the night. "Honest to God, we got her!"

Oil was seeping into the pit, all right, and by that afternoon over five hundred people had arrived at the ranch to get a closer look. It was no gusher — barely forty barrels a day — but enough to stoke a frenzy, and enough for Cox to stretch the discovery into the realms of fantasy.

"There is no doubt that the McDowell well will be a gusher with a daily capacity of from 2,000 to 5,000 barrels, and possibly more," Cox wrote to potential investors, dictating the letter from the derrick floor. "Here is evidence to prove that we have at Big Spring not only the largest oil field in Texas but one of the largest and best oil fields in the world."

Days later, residents packed the Elks Hall as Cox spread his arms and, with eyes closed, described a vision of oil derricks that stretched seventy miles long. People lined up with cash. According to the *Herald,* "The only man in Big Spring who does not own some General Oil stock or a parcel of soaring leases, lies under the sod and daisies."

Residents were eager to celebrate their maiden well. Together with Cox, they began raising thousands of dollars for a huge party that would toast the coming boom and their salvation from the weather. Cox employed a line of trains, "Investor Specials," to bring in ten thousand guests, according to pamphlets, who traveled from as far as Hawaii, New Zealand, and "the chilly borders of the Arctic circle." Residents offered spare beds, even loaned the visitors their vehicles. To feed everyone, ranchers donated 150 beeves and 60 sheep that were stored in refrigerated cars. The festivities kicked off with calf roping and bronc busting, followed by an airplane race. At one point, a local cowpuncher named Shorty Wells appeared riding Cox's single-engine plane. The *Herald* described him as "perched on a saddle in front of the tail and blazing away with his six shooter."

The following day, Cox led a motorcade to the well site, where crews swabbed the hole with suction to pull the oil into the pit while other accounts have them shooting the hole with nitroglycerine. "Behold," Cox said with a wave of his hand. "This black gold is a messenger of a new day in Big Spring and West Texas."

But despite the presentation, the well turned out to be a low-performing dud, made worse after crews, in their haste, collapsed the casing and filled the hole with saltwater. The rest of General Oil's wells turned up dry, and soon Big Spring got smart to Alphabet Cox. After his company went bust, Cox was indicted for mail fraud. He managed to escape the first set of charges, but not a second. A promotion scheme with Dr. Frederick A. Cook, who dubiously claimed to have been the first explorer to reach the North Pole, finally landed Cox in federal prison.

But rather than scare people away, the McDowell failure had the reverse effect. Now there was real proof that oil could be found in this part of the Permian Basin, it was just a matter of hitting the right seam. Exploration intensified, leading to the discovery of the Santa Rita well seventy-five miles away, and sparking the boom that would lure my uncle Bud to the town of Best. Meanwhile, drillers were punching holes all around Big Spring. Finally, in 1926, crews delivered Otis Chalk from the cattle trade by pulling two hundred barrels a day off his ranch south of town. But it wasn't until the following year, when a gusher on Dora Roberts's place came in at thirty-three hundred barrels a day, that Big Spring entered a new age.

Within three years, the population nearly tripled as the town — like others in the Permian Basin — shifted from agriculture to petroleum. All the major oil companies established permanent offices, while four refineries sprang up to process the crude into gasoline, diesel, and kerosene.

Joshua S. Cosden built the biggest one, on the east side of town. But the Cosden & Company refinery was relatively small compared to the $50 million empire that Cosden had already built and lost in spectacular fashion.

"J. S. is down, but he'll bounce back," friends of the tycoon said. In fact, Big

Spring was where Cosden was plotting his historic rebound and where he would establish, once and for all, his reputation as the "Rubber Ball" of the oil industry. Cosden's comeback would position Big Spring to shine on a global stage, while setting into motion the events that eventually brought my father back to town, hoping to glean just a fraction of the oilman's prevailing luck.

Cosden was born in 1881 and raised on a farm in Kent County, Maryland. At twenty, he took a job as a reporter for the *Public Ledger* in Philadelphia before moving to Baltimore to sell insurance and real estate. In 1908, after reading news of oil discoveries in Oklahoma and of ordinary men making fortunes, Cosden headed west, stopping in the Osage Indian reservation in the town of Bigheart, where a boom was under way.

That same year, the first Model T Ford came off the assembly line in Detroit and pushed the demand for gasoline. Out in Bigheart, there was no refinery to distill the oil into fuel, and Cosden saw his niche. His only problem was that he was broke. So he did what Sid Richardson, Clint Murchison, H. L. Hunt, and other legendary Texas oilmen would do in the coming years when possessed with vision and empty pockets:

he borrowed other people's money.

Back in Baltimore, Cosden assembled a group of wealthy friends he would come to rely on over the course of his career. Their seed money allowed him to buy six acres south of Bigheart, where he set out building a small "teapot" refinery. While it was under construction, Cosden and his wife lived in a tent before building a small cabin.

The Southwestern Refining Company was completed in 1910, but Cosden quickly ran into problems. The nearest wells were several miles away, with no pipeline or road to connect them to the refinery. Rather than wait for these to arrive, Cosden transported the oil in a rolling tank wagon that he'd bought secondhand. The contraption was so rickety it leaked oil as it bounced over the rocks and pastures, forcing Cosden to run behind with a bucket. The little amount of fuel he and his wife managed to produce, they sold themselves door-to-door.

The following year, a giant tornado struck Bigheart and destroyed the refinery, along with their cabin. It also flattened oil rigs in the fields and leveled most of the town. The only place for Cosden and his wife to live was in the hole they'd dug as a tornado shelter, where they stayed for two weeks. As soon as Cosden rebuilt, a fire destroyed his boiler and stills, forcing him to start all over again.

But somehow he remained undeterred, and his backers never lost faith. By 1912, rail tankers bearing his company's name were rolling through Bigheart and he was finally turning a profit. That year, he borrowed more money to purchase a second refinery, in West Tulsa, on the banks of the Arkansas River, and sank everything into expanding it. Wanting to ensure a steady stream of oil, he started his own pipeline company. By 1915, he'd connected the Cushing field, Oklahoma's biggest producer, straight to his refineries, which by then were processing eighteen thousand barrels per day into fuel and other derivatives. The lubricating oil and wax plant he built alongside it was said to be the finest anywhere, producing some twenty thousand pounds of wax daily.

But transporting and refining oil weren't enough for Cosden. He wanted control over the entire process, from the ground to the gas tank. He wanted to *drill* for oil. So in 1916, around the time my family was fleeing the Texas drought, Cosden visited one of the biggest investment banks in New York City and laid out his plan.

"I walked into their offices at two-thirty in

the afternoon and walked out at four-fifteen with the money assured," Cosden told one reporter. With few questions, the bankers had given him $12 million.

Cosden began drilling the Cushing field, then built his West Tulsa plant even bigger. He added a hundred cracking stills, which squeezed more gasoline out of a barrel of oil by "cracking" the heavier components. Only the mighty Standard Oil had been using that kind of technology. Standard also owned most of the patents for cracking oil into gasoline and other products, so Cosden simply hired better engineers and wrote patents of his own.

When finished, his Tulsa refinery was the largest in the world, and as oil continued to flow from his own wells, Cosden became one of Tulsa's richest men. That year, in a bold expression of his value, his company took out a whopping million-dollar insurance policy on his life that required thirteen different firms to broker it — one of the biggest individual policies in the nation.

The press found the thirty-four-year-old Cosden irresistible. He was blond and handsome and spent money with extravagance, once writing a $12 million check for an oil company while sitting in the lobby of a Tulsa hotel. At the same time, friends and associates described him as refreshingly humble and well liked. The papers remarked that he rarely touched booze.

In 1912, Cosden built the most expensive home in Tulsa, a four-bedroom Craftsman known as Mission Manor, then built an even bigger one that featured the city's first indoor swimming pool. In 1917, he began erecting his monument. The sixteen-story Cosden Building, located downtown on the corner of Fourth and Boston, was Tulsa's first skyscraper and the tallest building west of the Mississippi. In the meantime, Cosden divorced his wife, with whom he had three children, and married Nellie Neves Roeser, the wife of Charles Roeser, a less-successful oilman who lived across the street. The tabloids wrote of how Nellie spoke French and called her the most gorgeous woman in Oklahoma.

For a while the couple lived in the penthouse of the skyscraper before absconding to the East Coast. They purchased a sprawling estate in Port Washington, Long Island, and another in Newport, Rhode Island. For the winter they built a spectacular Spanish-style villa in Palm Beach, designed by Addison Mizner, which *Vogue* described as "the finest private residence in America."

The servants' quarters alone had thirty bedrooms.

By 1920, hardly a decade since sleeping in a hole in Bigheart, Cosden was worth between $50 and $75 million. On Long Island, the Cosdens' neighbors included Vincent Astor, the Guggenheims, and the Whitneys, all of whom they regularly entertained. After meeting the Prince of Wales in London, the Cosdens made headlines in September 1924 when they hosted the prince on his visit to America. The party lasted until dawn, when His Royal Highness was seen stumbling to his hired car.

Society writers, perhaps jealous, often dismissed the Cosdens as gold-dipped social climbers. "Men and women whose surnames had been in Blue Book and Social Register since the days of their great-great-grandparents permitted themselves small gasps at the slashing, heedless way the Cosdens spent and spent," wrote one reporter. Cosden's money, said another, "lacked the age that gives wine its bouquet and families their dignity and prestige."

In fact, it was the splashy, new-moneyed millionaires of the Jazz Age like Cosden whom F. Scott Fitzgerald had in mind when he wrote *The Great Gatsby*. Cosden even confessed that the book was one of his favorites. Yet by the time *Gatsby* was published, in April 1925, Cosden was already ruined.

Actually, his self-made empire had been damaged for some time. Reports vary about what caused the crash, but most agree that Cosden bet too aggressively on Wall Street, even going so far as pledging his company stock as security. When the market plunged in 1920–21, he lost heavily, dropping three hundred thousand dollars in one hour, according to one report. Cosden tried buying much of the company stock himself to hold it up, but it was no use. Overproduction and surplus had caused oil prices to dip and stagnate, and profits were on the decline. In January 1925, the board of directors staged a takeover of the company and assumed $100 million in assets, including the sixteen-story monument in downtown Tulsa. The company's name was changed to the Mid-Continent Petroleum Corporation.

To recoup, Cosden was forced to unload his personal empire piece by piece. He sold the Long Island estate to his neighbor, Vincent Astor, while the widow of Horace Dodge, the car magnate, swept up the Palm Beach villa for a cool $4 million. Cosden's "retirement," reported the *New York Times,* "is said by oil men to have removed him from the field of major personalities in that business."

And thus, the first of Joshua S. Cosden's fortunes was gone.

For months, nobody saw or heard from him. At one point, a reporter managed to track down Cosden's first wife, now remarried, and asked what she thought about his turn of luck.

"He'll stage a comeback," she said. "I ought to know. While Joshua and I were married, he went broke twice — but never for long."

Indeed, while the New York society writers were savoring his downfall, poking fun at how other millionaires were snatching up his thoroughbreds for pennies on the dollar, Cosden was already down in Texas making deals.

The low prices that had helped bury his company now provided him a perfect foothold. By the fall of 1925, he'd leased sixteen sections near Odessa, some sixty miles southwest of Big Spring. Then, in February 1927, a local newspaper announced that he'd sold a block of leases in Brown County — some seventeen hundred acres west of Comanche — to the Prairie Oil and Gas Co., which had paid him a million dollars. News of the sale quickly reached Wall

Street, where Cosden had already gone, brimming with confidence and new life. Once again, he convinced wealthier men, such as Standard Oil brokers Pforzheimer & Co., to gamble on his hunches. In just a few days, he returned to Texas with $5 million guaranteed, reorganized his company, and started buying land.

He based the new Cosden Oil Company in a plush office in Fort Worth and started building. He snatched up a string of small oil companies and a spread of leases near Oklahoma City, where he planned a giant tank farm. When his staff warned he was moving too fast, Cosden chided them. "The trouble with you fellows," he said, "is that you have no vision."

In 1928, he set his ambitions on Big Spring and Howard County, where Magnolia had just hit its gusher on Dora Roberts's ranch. He bought leases near the biggest proven wells — paying Roberts a reported one thousand dollars an acre — but he wasn't about to sell his oil to anyone. As crews started production, he set out to build a refinery.

Once again, Cosden's timing was impeccable. The T&P Railway was just in the process of switching from coal oil to diesel and needed a steady supply. John Lancaster, head of the railroad, took Cosden on a personal tour around Big Spring to choose an ideal location. Lancaster not only agreed to buy fuel for the railway, but to ship it anywhere along its route. And with that assurance, Cosden bought two hundred acres on the east side of town, installed new stills and a thermacracker, and started connecting pipe to the newly discovered Howard-Glasscock field, which produced over ten million barrels of oil before the year was over.

By August 1929, just eighteen months after reentering the game, the Prince of Petroleum had spun a profit of $15 million, with company assets valued at $25 million. "Cosden Recovers," the *New York Times* proclaimed, "without investing a dollar himself." His comeback, reported the Associated Press, was being regarded by oilmen as "one of the outstanding achievements in the history of the industry."

With rail cars of oil rumbling into Big Spring, the population tripled to over twelve thousand people and the modern town took shape. By then John Lewis and the family were living on his daughter Fannie's farm north of town, where, like so many farmers in the South, they were trying their hands at growing cotton. But it seems unlikely that farming was profitable enough to keep Bud away from the action in the oil fields. With his experience, he probably had no trouble finding a mechanic's job and the high boom wages would've been plenty to support Bertha and the kids, all of whom were now under the age of five.

I'm sure my family was aware of Josh Cosden, since he was the most famous oilman in the Southwest. There's no evidence of them ever crossing paths, or of any of them working at his refinery. But I'd like to believe they weren't so removed from people like him, so displaced and out of range that they missed the providence that came to men like Cosden and Frank Day. After losing Julia, Goldie, and the farm, and wandering from place to place, I like to think my family had found some purchase in Big Spring, that they'd found a home where the boom could swirl around them with all its befalling luck, and just maybe they could reach out and grab a piece.

Already, Big Spring was shaping up to be a different kind of boomtown, less lawless and depraved. The shrewd merchants, railroad engineers, and cattlemen who founded the town had sensed what was coming and planned ahead. They raised money to pave the streets so the trucks could move freely without mud, installed electric lights along the main thoroughfares to stave off crime, and fortified the county jail so it was impervious to drunken mobs. They built a water system to prevent outbreaks of disease and installed fire hydrants, while construction crews built offices for the influx of new business.

But when office space ran short, rancher Dora Roberts stepped in to help. Roberts was one of the earliest settlers in the county and no doubt its most influential. In 1883, she'd married a young cattle trader and moved to a dugout on the edge of the frontier. Her husband was later killed while breaking a wild bronc, and in 1909, a second husband also died tragically and in similar fashion, crushed by his own horse. Roberts continued to manage the ever-growing ranch on her own, and when oil made her the richest person in West Texas, she dedicated much of her wealth toward the betterment of the town. In 1928, as the population grew, she and Lorin McDowell — whose ranch was finally giving oil — financed the construction of Big Spring's first office tower. The Petroleum Building, as it was called, featured brass ornaments, a marble interior, and a row of concrete Aztec warriors who stood sentry along its rooftop. And while standing only six stories high, it would loom over the future of Big Spring and my family in particular.

The true totem of the oil bonanza, however, was going up a few blocks away — the gift of another come-lately oil millionaire. After a few gushers were tapped on W. R. Settles's ranch, he remained the modest cattleman and devout Presbyterian he'd always been. After donating his first well to his church, Settles was bombarded with myriad suggestions of what to do with his money. Big Spring had only three hotels to accommodate the crush of newcomers, and rooms were in short supply. Town leaders convinced Settles to spend his royalties on building not only a new hotel for the expanding city, but one that would dwarf all others.

They chose a prime location on the corner of Third and Runnels and hired David S. Castle, the architect who'd designed many of Abilene's finest buildings. No expense was spared. The cost of the fifteen-story hotel, plus furnishings, ran to seven hundred thousand dollars, almost equal in value to all other construction in Big Spring that year. A spectacular marble staircase rose from the lobby and split in two directions toward the mezzanine, where the grand ballroom contained one of the largest crystal chandeliers in the world. The floors were polished maple, the paneling was mahogany, and the furnishings were the finest walnut. An eleven-piece orchestra played the first two nights of its opening, with over three hundred people packed into the ballroom.

The *Herald* devoted an entire section to the event. Fred W. Crow, the hotel's manager, told the paper, "I have no hesitancy of comparing the class, service and beauty [of] the Settles Hotel to the New Yorker in New York City. In fact, I think this hotel is superior in many respects."

As tall buildings cut a new skyline in Big Spring, the Cosden Oil Company continued to grow in Fort Worth. When the value of its stock reached $130, Cosden urged investors to let it soar. "It will go to $510," he triumphed. Cosden's second fortune had come in half the time as his first, and the oilman was looking to bounce even higher. He was flying, pushing for orbit, when, in October 1929, the bottom fell out of the world.

The stock market crash didn't penetrate the oil patch's joie de vivre right away. In fact, Cosden voiced confidence the lull would pass, even breaking ground on new construction at the refinery. For nearly a year, his stock price held firm despite the flagging markets. But by June 1930, with crude having plummeted from four dollars a barrel to ten cents, the company's value fluttered to the ground like a punctured balloon. Profits vanished under mounting expenses and employees were cut loose. Before long, the Cosden Oil Company was in receivership and the second of Josh Cosden's fortunes was gone.

As the Great Depression settled over the region, drilling rigs were stacked and hauled away, holes were capped, and oil towns emptied as the boom chasers returned to their farms and cities to wait out the bust. Landowners defaulted on loans as oil royalties disappeared. Banks began to fail along with shops and restaurants in every little town that had grown fat off the boom. It would take several more years for the oil economy to recover, at which point it would thrive in the face of global despair. But until that miracle happened, Josh Cosden returned to New York, where he became terribly ill. My family, meanwhile, returned to the road.

As it turned out, too many farmers across Texas and the South were planting cotton.

In fact, when the family ginned their first harvest back in 1926, no greater amount of cotton had ever been grown in the United States. The land was awash in cotton, and just as with oil, there came a point when there was just too much for prices to remain high. Low prices had hammered the farmers in 1927 and 1928, yet still they planted more. And when it came time to harvest the crop in the fall of 1929, Wall Street was in a panic and people were lining up for food. "Cotton is selling now not only below the cost of production, but below the cost of existence," the *Herald* wrote in December.

Once again, the Mealers were busted. And with nothing going in the oil fields, they resorted to wandering. As a cold winter settled on West Texas, they went home to Georgia.

4

A journey back east . . . then west along the rails . . . the Dust Bowl begins on the plains . . . cotton rebounds . . .

Most likely, it was John Lewis who instigated the trip back to Pickens County. His father, Robert Moore Mealer, had suffered a stroke that rendered him mute and partially paralyzed, confined to a rocking chair. No one knew how long he would live. Nearly forty years had passed since John Lewis had seen him.

They tied a mattress to the roof of Bud's Model T Ford and filled the car with clothes and cookware. John Lewis, Bud, Bertha, and the kids squeezed inside, but there was no room for my grandfather Bob, who wrapped himself in two coats and a blanket and rode a thousand miles hanging off the running board.

The road was full of families fleeing the

dead oil fields, heading back toward kin and familiar country. The family headed north into Arkansas, where icicles hung like stalactites from the roadside shelters where they camped. They crossed Mississippi and Alabama, where Bertha's people were from, then finally entered Georgia. They drove through the city of Atlanta, then up into the Bluc Ridge, where the air grew thin and smelled of sharp, bitter evergreen. Near the town of Jasper, the Ford labored up the mountain and came to the lip of a great basket filled with jagged pine. In the distance, a needlepoint peak jutted from a layer of morning mist.

"They call that Sharp Top," John Lewis said. Beneath its white blanket lay the valley where he was born.

He was four years old when his mother, Catherine Cowart Mealer, died while giving birth, the same age as my grandfather Bob when Julia departed them. After Catherine's death, her sister Drucilla had arrived to help Robert with the four children, but her father, a Baptist minister, forbade her to live in the home as a single woman. So Robert married Drucilla, mainly out of convenience, and she gave him ten more children. But it was clear that Drucilla later resented her station in life, as if she'd woken up from

97

a childhood dream to find herself imprisoned by a washtub and all those little
mouths to feed. Her harsh manner was one
thing John Lewis had been happy to put
behind him when he left for Texas.

Robert and Drucilla left Sharp Top in
1917, just after John Lewis lost the farm in
Texas. They built a home twenty miles
north, near Burnt Mountain, next to John
Lewis's brother, Daniel, then another one
just down the road. It was in the first house,
which sat vacant, where John Lewis and the
family spread their things and took refuge
from a world in crisis.

They spent most of their time at Robert's
home, where he sat on the long front porch
in silence watching the children play. Each
morning Daniel's kids spilled down the
wagon road to get a glimpse of the Texans,
taking them to catch mountain minnows
and horny heads in the stream, wading
barefoot in its sharp clear water until their
skin ached from the cold. Having only
known the flat, dun-colored oil fields, Bud's
daughters, Frances and Flossie, spent the
first week staring into the canopy, searching
for familiar sky.

For five-year-old Frances, her first memories of life were of the Georgia woods, helping Drucilla gather wild berries and poke-

weed and drinking fresh milk plucked from a rock cradle in the stream. Her first impressions of her father were there, too — him stomping out of bed at 3 a.m. and hurling rocks at the jabbering whippoorwill keeping him awake. Or him chasing her mother through the woods carrying a bucket of cold water, the two of them giggling. Frances remembered them holding hands, young and in love.

Eighty-five years later, I would sit with Frances at her dining-room table staring at yellowed photos while her mind skipped back a lifetime. Back to the Georgia woods, where the sounds of her parents' laughter still rang in her memory, along with all the unfortunate things that happened after that fleeting moment of joy.

There was talk about staying in Georgia for good. Bud even found a job as a cook in nearby Ellijay, and Frances remembered the family going on about the cherry pies he baked. But for some reason — she never understood why — the plan to stay was abandoned. After several weeks the family said good-bye to Burnt Mountain and returned home to Texas.

Back in Big Spring, the family slept on the floor at the farm of Bud's sister Fannie and her husband, Abe, and plotted their next move. The town was still recovering from the crash, and there was little work in the oil fields. Mornings, John Lewis, Bob, and Bud drove to the railyard on First Street to look for tools or pipe arriving on the trains, a sign that jobs were imminent. But the station was quiet. The only things coming off the freights, it seemed, were hoboes. As the trains approached the station, dozens of men leapt from the boxcars, their skin and clothes smeared with coal dust, then vanished into the streets to bum a meal.

Big Spring was a rail hub for passenger and freight traffic going to and from the west. And because of this, the town became a way station for those riding the rails to California, where citrus groves, orchards, and cotton fields lined its golden valleys.

Somehow a decision was made that the men would go to California to work and bring home money. It's unclear if they had a particular destination in mind. All Frances remembers is that early one morning her father said good-bye and walked to the T&P tracks. Waiting outside were Bob, John Lewis, and Davey Jones, Abe's younger brother, who'd also left his family at the farm. The men walked to the edge of the depot and stood along the ballast, the crushed stone grinding beneath their feet. A westbound locomotive shrieked in the far darkness, then lurched on its wheels, and within minutes they stared into a black wall of screeching steel. They jumped for an open car.

The trains were still full of old tramps chasing the iron road, but now they shared space with families: exhausted mothers with hungry kids and fathers with bellies full of dread. There were lone breadwinners like Bud and old men like his father. There were blacks, whites, and browns, and there were criminals of every stripe.

But mostly, like Bob, who was fifteen, there were children — kids who'd been turned out of their homes to fend for themselves, or who'd left on their own to ease the burden. At the height of the Depression, as many as a million teenagers traveled the rails looking for work and community, moving in vagabond packs and living in hobo jungles, finding both charity and brutality in the broken-back cities of America. They crowded the cars and hid down in the tenders where the coal and water were stored; they squeezed between cars and clung atop their bucking roofs. In 1932, about 75 percent of the nearly six hundred thousand transients on the Southern Pacific line through Texas, New Mexico, and Arizona were under the age of twenty-five.

The family stayed in the network of jungles along the route, tucked amid the timber and hidden from view. Each camp had its own division of labor — one person went into town to find a potato for soup while another brought salt or an extra spoon. Pots and pans hung from tree branches; crude shelters were made from cardboard or tin scrap and sometimes built up in the trees. Bob and Bud both carried pocketknives, which were like gold in the jungles. They hired their blades in exchange for soap or a bowl of stew. In a pinch, a knife could be traded for a pair of shoes or sold for cash. They cooked possums and jackrabbits caught in snares along the brush lines and traded their hides for food. But mostly, they survived by rolling dice.

Davey Jones, who was twenty-three, was a road gambler who'd taught Bob and Bud to play craps back in Eastland County. His brothers' cotton fields didn't provide enough excitement to hold him to one place, plus he hated to farm. So he'd haunted the boomtowns across the Permian, running tables in the bootleg dens.

There were few men in West Texas who could beat Davey Jones in a game, and because of this, John Lewis and his boys never went long without food. In later years, to the detriment of his wife and children, Davey became one of the most feared gamblers in the state, holing himself up in the Hotel Settles for weeks at a time, one game bleeding into the desperate dawn of another while his sons banged on the door, begging him to come home.

The jungles were full of peril. Thieves and highwaymen, known as yeggs, rode the trains and stalked the camps at night. One morning Bob awoke to find a fellow traveler dead in some nearby weeds, stripped naked and missing his shoes.

Lice and other vermin infested the shanties and bedrolls that hoboes shared. John Lewis contracted crabs and, having no money for a doctor, said to Bob, "Go fetch me a sheet of newspaper," which he twisted tight. The old man dropped his pants and set fire to the paper. Then, with his boys looking on aghast, he pressed it to his genitals.

Jumping trains carried its own dangers. Riders who lost their grip were crushed under the wheels. They froze to death and suffocated when cargo doors slammed shut.

Others fell asleep in coal compartments and were buried alive. Train tunnels took their toll of careless riders, especially in the South, where as many as five hundred people would cling to the roof of each train. Those who rode too close to the chimneys inhaled lungfuls of cinders and spat blood for days.

If a fireman caught a boy riding the blinds between cars, he soaked him with his hose so he'd freeze when the train gathered speed. Worse were the railroad bulls who made up their own law. Boys were beaten and robbed, and others murdered. Some bulls were notorious — men like Texas Slim, who stood six feet seven, wore a suit and white hat, and kept two guns on his hips. He worked the T&P between Fort Worth and El Paso and most certainly stalked the yards in Big Spring. He claimed to have killed seventeen men, most of whom he shot off the tops of moving trains or plucked from open boxcars.

Late one night, two bulls captured John Lewis and the boys after they'd jumped off a train. The bulls pushed the men against a fence and began frisking them for weapons. One of the bulls patted Bud's waist and shouted, "Hey, this one's got a gun." He motioned over his partner, who yanked down Bud's loose trousers to reveal only a sharp, protruding hipbone. "This man's just starved," he said.

The second time they were caught, two bulls were waiting in a boxcar when the group climbed aboard. Everyone managed to escape except Bob, who was trapped. The bull held him at gunpoint until the train bucked with speed, then pushed him out the open door. His body hit the ballast and tumbled down an embankment.

"I must have rolled the length of a city block," he later said.

John Lewis and the others found him in the weeds, bloody and with his clothes in shreds. But luckily, nothing was broken. They carried him to the nearest jungle, where they treated his wounds with coal oil, and he stayed there until well enough to travel. After that, they avoided the boxcars completely and rode the thin brake rods beneath the trains, tying their arms and legs with belts to keep from bouncing off. Hanging two inches above the ballast, they crossed the desert and rode into California.

In the San Joaquin Valley, the men found jobs picking cotton, and when that was over, they followed other harvests, yet it's unclear where they went. The only memory Frances has of this period is receiving a postcard from her father that read, "Greetings from California." On it was a leafy orchard brimming with oranges.

On their way back to Texas, a railroad bull turned up murdered on one of the trains. Word spread through the jungles that Bob had done it, and no matter how many times he denied the charges, they brought him an air of celebrity. The hoboes greeted him with reverence in every camp he entered, and gave the family a wide berth.

Back home in Big Spring, the travelers returned to Fannie and Abe Jones's farm. Abe and his brothers had inherited the 665 acres north of Big Spring from their father Newt, who had purchased it in 1923. The shack where the Jones family lived was only two rooms and a woodstove, with no running water or electricity, not even a proper outhouse.

Bud cleaned up an old shed and moved his family in, while John Lewis and Bob squeezed along the floor with Fannie's boys. Within weeks after Bud's return, Bertha was pregnant with their fourth child. Another son, Leamon, was born in early March 1932. The morning Bertha went into labor, Dr. T. M. Collins arrived from the drugstore carrying his leather satchel and the children were sent outside. When Frances returned and saw her new baby brother, she couldn't understand how the doctor had fit such a thing in his bag.

Fannie had always wanted a girl, but after her sixth son was born, it struck her that bringing a female into that house would be child abuse. The boys were named Troy, Raymond, Earl, Barnie, Bobby, and Curley. They were close in age and tangled like bobcats. Once they got started, it was usually John Lewis who broke them apart, not by force, but by staging his own tantrum. He flipped over the cane-bottom chairs and howled in such a way that the boys froze from sheer discomfort.

Having six sons was Fannie and Abe's divine compensation, or punishment, for such a foolhardy endeavor as dry-land farming on the edge of a desert. Each boy was driving a mule team by the time he was five, disappearing behind a cloud of red dust under the blue sky. Their mother kept them nourished on tea cakes and fried chicken. Milk came from a Jersey cow they rode in like a horse each evening from the pasture.

The law of averages for a dry-land farmer dictated that the cotton would grow only one year in four — never mind the prices. The other three years were either total losses — years when the sky gave nothing, not even enough moisture to plant a seed — or produced stubble that grew runty and backbreakingly close to the ground. The Jones boys had rarely laid their eyes on a field of knee-high cotton. All had experienced their share of blizzards, watched tornadoes drop from a sagging thunderhead, and seen sandstorms so bad they shaved a field like a grater.

And they knew that rain, precious and rare, acted in peculiar ways. A story the Jones boys liked to tell was about the time Old Pinkie, the milk cow, got loose in the cotton field and Earl went to get her. He noticed something strange and called over Raymond, who was in earshot mending fence.

"Come here, Raymond, and look at these cotton bolls."

Raymond came over. "What's wrong with them?"

"What's wrong with them? There's fifteen bolls on this stalk, and only one on this stalk across the row."

Raymond scratched his chin. "I wonder what happened to them?"

"Well, it's easy to see what happened," Earl said. "The rain stopped in the middle of the row."

■ ■ ■ ■

That summer along the High Plains, the first dust storms began to blow. From Nebraska down to the Texas Panhandle, the year had passed largely without rain, and the vast fields of wheat that had been planted during the fever of the First World War began to wither. The dirt that had anchored the grasslands for thousands of years before being planted with crops now turned to fine powder and traveled with the wind. The Dust Bowl had begun.

Yet ironically, farther south on the Llano Estacado, a record rain fell on Big Spring. Starting in August, it came hard and late in the season, so late that farmers began losing sleep dreaming of cotton bolls blooming with fungus. But the skies cleared just before the bolls opened. The weevils and worms stayed away, and prices — at record lows just a year earlier — eased up, as less acreage was under seed. For once, it seemed, God heaped his reward on the soil beneath my family's feet. While the sodbusters up north entered the greatest environmental disaster the country had ever seen, Howard County enjoyed its best cotton crop in a quarter century.

The boom lasted through the fall and winter, with Big Spring at the center. The sudden flood of money and workers provided a respite to the indignities of the Depression. Newspaper ads sang of the brief fortune:

Cotton Pickers Needed! Fifty Cents per
    Hundred Pounds. Meet at Chamber of
    Commerce 6am.
Want 3 Families That Can Pick a Bale a
    Day Each. House Furnished.
A Family of Six Will Gather Your Crop and
    Care for Your Stock. Must Have Good
    Cotton. Apply 407 Gregg.

Up on the Jones farm, everyone dragged a cotton sack and picked their rows. And when the rows had been stripped clean and the cotton hauled to the gin, they moved to other farmers' fields. Then winter arrived, and mornings were unbearably cold. The family huddled in the dark and shivered, waiting for the sun to rise and melt the frost. The kids wore baggy secondhand coats doled out by the Salvation Army, while the men dressed in coveralls stained from the oil fields. They stomped their boots on the soft dirt and rolled cigarettes to stay warm, their fingers as stiff as kindling.

Cotton tramps poured off the morning trains and headed into town, while cotton trucks, overloaded and threatening to tip on the narrow, humpbacked roads, slowed traffic and filled the air with dander.

But by the end of the season, migrants from the Rio Grande Valley started showing up in great numbers offering to pick for half the money — even a quarter — and after that, the picking jobs disappeared. By 1933, the drought on the High Plains finally reached West Texas.

The wind blew hard and sucked away all the moisture, burned the tender shoots and lashed the cotton low against the ground. Winter winds then lifted the topsoil and carried it north, depositing tons of red gypsum dust all across the plains. In April, a Big Spring businessman named Bob Cook sent the state of Nebraska an invoice for all the West Texas "fertilizer" that now covered their fields. He estimated its worth at $12.5 million.

Abe Jones didn't stand a chance. That year, his cotton died before it could bloom, his feed crops shriveled, and by spring his cattle — fewer than a dozen — ran out of grass and began to starve. Starting in July 1934, the Roosevelt administration enacted the Agriculture Adjustment Act in West Texas to lessen the damage. The government started buying up cattle to get prices back on the level and feed the millions on relief. Healthy animals fetched upwards of twenty dollars apiece. By the third day of the program, ranchers around Big Spring had unloaded over three thousand head. Most of which were taken to the cannery on Ninth and Main, where they were rendered into food.

As much as it clawed at his guts, Abe went to see the local agent in charge of the program, who arrived with an inspector and examined Abe's herd. The inspector concluded they were too skinny for mass consumption, and some were diseased. Condemned cattle only warranted the minimum. The agent handed Abe a receipt worth six dollars for each cow, which would be paid in three weeks' time. The healthy ones would be butchered, the meat returned to the family in a box of cans.

As for the sick cows, the government had hired the Grantham brothers to dig a deep trench up the road with their mules and fresno scraper. A few animals already lay tangled up at the bottom, their limbs tightening with rigor mortis. The hired guns were local men who leaned against a trailer and smoked, not saying a word, for it was work no man was proud to do. They waited for the cows to be guided within short range, then planted their feet, took aim, and dispatched them down into the hole. And while these were hard days for Abe Jones, his dark and awesome year was not yet over. On December 7, the *Herald* ran a small item at the bottom of page 1:

"The home of Abe Jones was destroyed by fire Saturday night around 8 o'clock. How the fire started is unknown. The family was eating supper when the fire started in the front room. They did not save anything."

In the face of despair, there were signs of better times. The programs generated by the New Deal and the rebounding markets helped spark a modest recovery across much of the nation. And as the country got back on its feet, it needed oil to keep moving.

In West Texas, new fields were explored, equipment began moving along the roads and rails, and best of all, companies were hiring. The timid economy wouldn't allow for the roaring booms of the previous decade, and it would take more than new oil fields to replace what had been lost. In the short run, the drilling budgets remained tight and the wages terrible. Oftentimes oil-men paid their crews in groceries until a well came in, and a lot of men worked "bean jobs" during these years. But they were jobs all the same.

At the first sign of an uptick in 1934, my grandfather Bob began haunting the drilling company offices looking for work as a roustabout, which was the entry-level job on any rig. Roustabouts were the working cowboys of the oil fields, performing the dumb and dirty tasks for the pumpers and drillers. They had little stake in the outcome but enjoyed all the freedom a boom could offer.

I can imagine my grandfather — nineteen years old, single, and without commitments — living like his brother Bud had in the beaverboard shacks that sprang up near the rig sites, taking meals in flophouse cafés and hitting beer joints at the day's end. By now he'd grown into a strapping young man, his body as taut as a spool of baling wire. He had dark eyes and dark skin like his father and a standing gaze that was both cold and distrustful until a smile softened its edge. He loved cars and trucks, particularly Fords, and liked to get his hands dirty tinkering beneath their hoods, a hobby he'd picked up from Bud. A photo of Bob from this period shows him in a pair of dirty overalls, cowboy hat cocked confidently to one side, posing beside a Ford coupe, his boots caked in white mud. Already he'd roamed the country on freight trains and had a temper so remarkable as to associate him with murder. He was also motherless, volatile, and came with baggage — tailor-made for the Texas oil fields.

5

The dusters continue . . . oil returns to the Permian . . . Bud's fate is sealed . . . Bob comes alive . . .

That same year, 1934, Bud found a job driving a truck for Shell Pipeline Company. The position was in Forsan, a town twenty miles south of Big Spring, which took its name from the four oil-rich sand formations found beneath it, although a fifth was later discovered. By the time Bud and his family arrived, the shacks and bootleg dens of its boom days were gone and a real town had emerged, one with schools, cafés, and an active Main Street.

At first, the family lived in an oil camp out by the highway. Bud drove twelve-hour shifts hauling steel pipe and heavy equipment, and he knew his job well. He could pull apart an engine and reassemble it blindfolded and guide a loaded rig through a sand drift without putting it in the ditch. Because of this, he never lacked for work. After a year of saving money, he and Bertha bought a small lot on the edge of downtown, just walking distance from the elementary school. The vagabond years were finally over.

John Lewis and Bob drove down from Big Spring and helped Bud build a house. Although it was all wood and contained only two rooms, it had running water, as well as lights, a real stove in the kitchen, and an Electrolux refrigerator — all powered by natural gas piped from the fields. Their backyard was a pasture, where Bud built a small smokehouse, a garage, and a pen for their two milk cows.

The new home and Bud's steady schedule introduced routine and a semblance of normalcy. Frances and Flossie enrolled in school, where most of the children were gypsy oil patch kids like themselves. Mornings began as they gathered outside to send their father off to work, waving good-bye as his Ford pulled out of the drive.

When he returned home in the evenings, his coveralls were caked in grease and smelled of tobacco, gasoline, and tire rubber. They waited by the door to greet him, yet there was no embrace. Frances noticed how her father horsed around with her brothers John and Leamon, mussing their hair or wrestling with them while Bertha prepared supper. But with the girls, he was oddly unaffectionate. Never once did he scoop his girls into his lap the way some fathers did, or bend down to kiss their foreheads. His standoffishness, Frances later guessed, was a reaction to Bertha's abuse at the hands of her own father — as if the old man's lechery had somehow poisoned the whole institution for Bud. Yet there was no doubt in her mind that her father loved them.

For Frances, her favorite times were Saturday mornings. While her brothers and sister went outside to play, she waited patiently for her father to get out of bed, then listened for the rattle of coins in his pants pockets. She would hear him say, "Murl Dean" — he called her by her middle names — "go up there to Wash's store and get us a paper, and bring me back some Prince Albert."

Frances raced out the door with a handful of nickels, then turned up the street to C. V. Wash's general store. Inside, she slapped the coins onto the wooden counter and waited for Mrs. Wash, who wore her hair in a giant beehive, to pull the can of tobacco off the shelf and ring her order. Back home, her father rolled a cigarette, spread out the *Herald* on the kitchen table, and together they read the funnies.

Sunday evenings at eight o'clock, the two of them sat like statues in front of the radio, waiting for the shrill pipe-organ intro of the *Lum and Abner* show, then spent the entire hour laughing. After it ended, Bud turned the dial until he found station XERF out of Acuña, Mexico, which played country and western music. The train songs of Jimmie Rodgers seemed to plunge him into a trance, perhaps recalling his own days on the rails. He looked out the front window and twisted his hair around his finger, the way he did whenever deep in thought. Then, looking up at Frances, he flashed her a big grin, revealing a mouthful of crooked teeth.

Frances liked knowing that of all the kids, only her eyes were brown like her father's whereas Flossie and the boys had Bertha's blue eyes. As Frances got older, she also developed a faint beauty mark on the left side of her nose, just like Bud. Whenever strangers said she looked just like her daddy, she beamed with pride.

She turned ten years old in March 1935. In the weeks before her birthday, the dust arrived in curtains. Cold northern winds from the beleaguered High Plains brought ten major dust storms to Big Spring in February and March, causing residents to joke how they missed the sand they'd so often cursed. A typical West Texas sandstorm blew in at sunrise and was gone by dusk, leaving the atmosphere clean. But these storms were different, coating everything in a stubborn, silty film. Out on the ranches, it suffocated the vegetation and the cattle refused to eat.

A storm in late February had lasted four days. It arrived from the Dakotas in front of seventy-mile-an-hour winds, blocking the sun. On the third day, it mixed with a blizzard in a freak circus of nature that few residents had ever seen. On March 22, the dust was so blinding that cowboys on the Guitar Ranch lost six hundred head of cattle they were driving through the shinnery.

The following week brought the worst duster on record. Around dawn, a red curtain reaching twelve thousand feet rode in on a brisk north wind. At the same time the storm was rolling in, the aviatrix Amelia Earhart was attempting to fly from Dallas to the West Coast when her plane was nearly forced down at Big Spring. She managed to make it to El Paso, where she told reporters, "I was forced to fly blind nearly the whole way. Dust and clouds mingled in a thick yellow haze."

By April, officials estimated that over three and a half million pounds of new soil from the Texas Panhandle and High Plains had landed in Big Spring. "We're breathing payback from 1917," a farmer told the *Herald,* joking that he actually recognized some of the dirt he'd lost in that previous drought.

Out in the oil fields, the drillers and roughnecks slung goggles over their eyes and wet bandannas over their faces, cleaned the dusty sludge from their engines, and defied the weather. By the spring of 1935, production was thrumming at near pre-Depression levels. In the Permian, companies were drilling their holes deeper than ever — over twelve thousand feet — while dispatching geologists and seismic teams to find new underground horizons. The same was happening down along the Gulf in towns like Angleton, Anahuac, and Refugio, where little booms erupted like illumination rounds against the darkness of the Depression.

As more jobs became available, Big Spring filled with workers and families and the merchants eager for their money. On weekends, everyone shucked off their greasers, donned their finest clothes, and walked the downtown streets, past the shops and hotels that the previous boom had built.

Bud and his oldest son, John Jr., put on silk neckties and suspenders and squeezed their feet into stiff three-dollar shoes. The girls wore dresses, made not from flour sacks but with cloth and sewing patterns ordered from the Sears catalog.

As soon as they reached downtown, Frances asked to walk to the Montgomery Ward. They turned up West Third Street and stopped in front of the tall glass windows. There, between the Sky King bicycles and Radio Flyer wagons, were the Shirley Temple dolls that she'd come to see. They stood in a row smiling winsomely, with perfect curly hair, anchor-print dresses, and those true-to-life moving eyes. How badly Frances wanted one. A sign on the window read: ASK ABOUT OUR LAYAWAY PURCHASE PLAN.

But even in these better times, that kind of luxury was out of reach. The previous Christmas, Frances and Flossie had to settle for a cardboard Shirley Temple doll instead, the kind you cut from a workbook and dressed in little paper clothes. But a stiff paper doll could not be held, not like the real thing that stared back now.

For lunch, the family walked toward the

T&P depot, where a Mexican man sold hot tamales from a steam cart. Bud ordered a dozen and the man wrapped them in a bundle of newspaper, then they walked under the viaduct that stretched across the railroad tracks and ate them in the shade.

In the afternoons, while Bud and Bertha visited friends and family in Big Spring, the kids went to the picture show. They'd seen their first movie shortly after Bud started his new job. Their father had marched them down to the front row of the Queen Theater on Main Street, and after taking their seats, issued a gentle warning, saying, "You might think them horses are gonna run right over you, but they ain't." The picture was *Unknown Valley*, starring Buck Jones, and ever since, Frances had not been the same.

Buck Jones, atop his faithful steed, Silver, and wearing his white hat, was a moral force on the lawless frontier. During the heat of the afternoon, the kids planted themselves in the front row to watch Buck Jones battle cattle thieves and gold robbers in such adventures as *The Fighting Ranger, The Thrill Hunter*, and *Outlawed Guns*. And Buck always wound up with the prettiest girl, often some fragile dove he wrangled from the claws of predators. Back home in Forsan, the kids fought over who got to play

Buck Jones. No one, especially Frances, wanted to be the girl.

On Sundays, John Lewis and Bob drove over from Big Spring and helped Bud around the house, or just sat in the yard smoking. One Sunday they butchered a hog behind the garage, salted down most of the meat, then delivered the rest to Bud's sisters.

They often worried about the girls, whose taste in men was poor and oftentimes calamitous. Velva's husband, Willie Bob Henson, was a cowboy and a rounder, so much that John Lewis had warned his daughter against marrying him. "If you're gonna make your bed hard," he told her, "then you're gonna stay in it."

Fannie's husband, Abe, had his own troubles with whiskey and women, even before their house burned down, and the previous year, Ahta's husband, Elijah — Bertha's brother — skipped out on the family, leaving her with three kids and no money. At the time, Ahta revealed that Elijah had beat her when he was drunk.

Months had passed without a word from their other sister, Allie. The previous year, she'd moved to a town called Beeville in South Texas with her husband, Lee Pruitt.

But now, in a letter she sent to Bud, she informed her brother that she'd filed for divorce. Lee's drinking had gotten worse, she wrote, and he'd also taken to beating her in front of the kids. With nowhere to go, Allie had moved in with Lee's friend Emil Holubec, whom she was planning to marry.

The news had stricken Bud, and now his heart ached for his closest sister.

"You don't know how bad I want to see you," he replied in a letter. "It's like it's been twenty years since I heard your dear old voice."

He asked her to come back to Big Spring to be with the family, then ended the letter by reminding her how pretty she was.

"Allie," he wrote, "I just want you to tell me what makes you look so young."

The spring of 1936 arrived as dry and dusty as the previous one, though the storms no longer came as a surprise. The *Herald* gave each one no more than a few lines.

On a Sunday afternoon in late March, a storm moved in from the Panhandle. It was the color of rust and didn't make a noise, didn't even blow — just drifted in softly like dark magic until it diffused the daylight. Bud and the kids were in the yard when Bertha saw the windows cloud over. She called the children inside and yelled for Bud, who was beneath his Ford changing the oil. He waved her off and finished the job, and by evening, he was coughing.

He sat up in bed, unable to stop. What he hacked up was red and black, the dirt from half a dozen states now caked around his lungs. Some came from that storm, the rest from all the others he'd been breathing while out on the road.

He shook with fever all the next day, wrapped in a blanket on the bed and jerking in and out of sleep. By evening, his breathing was short and labored and Bertha said she was calling for help. She took the younger kids and went out the door, leaving Frances, who was now eleven, behind to tend to her father.

Frances pulled a chair next to the bed and sat there watching him sleep. His hair was mussed and matted down over his face, and the room smelled like stale sweat. At one point Bud came to and must have seen the fear in his daughter's eyes.

"Murl Dean?" he said.

"Yes, Daddy."

His hand came out from the blanket and he looked at his watch. "Go over there and turn on that *Lum and Abner,* wouldya?" he said. "I'd like to hear it right now."

She walked quickly to the radio, delighted, and turned its big knob. The house filled with the show's dirgy pipe organ introduction. Up in Pine Ridge, poor Lum was still besieged by the stockholders of the silver mine. The two of them sat there laughing until Bertha returned home with some men, who lifted her father out of bed and took him away in a car.

For a week Bud lay in Big Spring Hospital, fighting to breathe. Doctors said he had double-dust pneumonia, meaning both lungs were full and infection had set in.

Bertha kept constant vigil by his bedside. She slept in his room at night and returned home in the mornings only to freshen up and check on the kids. A neighbor lady from down the road brought them meals, and besides her, no one came at all. During the day, the children managed to find distraction by playing with a nest of baby chicks that had hatched in the chicken coop. At night, they lay in bed and said prayers for their daddy while the wind blew its lonesome song through the slats.

Then one morning, someone else was there, shaking them awake. It was Bertha's brother Grady, coming to inform them that their father was dead.

■ ■ ■ ■

A succession of blurred, horrific days followed. Someone drove the kids to Aunt Ahta's house in Big Spring, where John Lewis sat at the table saying nothing. When Allie and the others arrived, their wailing could be heard a block away.

The kids busied themselves in the yard, playing with their cousins Ruby and Fudge, anything to stay outside. The adults hadn't said anything about their father, how he'd died or where they'd taken his body. Hadn't talked to them, period.

Later, someone drove the children to the funeral home and led them into a small, airless room. Their father lay in a casket, surrounded by flowers, more flowers than Frances had ever seen. The kids gathered round the open box and looked down at their daddy, who did not look like the man they knew. His hair was freshly oiled and he wore a suit and silk tie. But his soul was gone, that was plain. He might as well have been made from wood.

"Look there," Flossie said. "They took the beauty mark off his nose, the one like yours, Murl Dean." And it was true. Frances felt the wind go out of her. She reached out and touched her daddy's face, then instantly regretted it. Bud was thirty-one years old.

After the funeral, Bertha broke down. She refused to get up one morning, then lay in bed for three days, eyes open, catatonic. Allie and the sisters took turns bathing her and pressing cups of water to her dry, cracked lips. "They were so in love," one of them remarked. Then one night, John Lewis was in the room and said, "My god, look at her." Bertha's face had drained pale and her eyes were gathering clouds.

He ordered Allie to bring him a mirror, which he placed beneath Bertha's nose and it didn't fog over. "I think she's dead!" he cried.

But Bertha didn't die, no matter how hard she tried. After a few days she was up on her feet, and not just on her feet, but alive with more energy than usual. A kind of wild, full-moon energy. As the family puzzled over this miraculous recovery, my grandfather Bob appeared and announced that he was getting married.

It happened like this: during the days when Bud was laid up in the hospital, there was a shortage of preachers. Or one preacher, in particular — Brother Deavers from the

Baptist church in Forsan, who'd baptized Bud in Moss Creek and delivered the Sunday sermons the family attended regularly. But when told that Bud was critical and required prayer and counsel, the preacher balked.

"I shouldn't risk my preaching voice," he'd said, assuming Bud's condition was contagious. So another preacher was summoned in his place — Brother Homer Sheats from the Church of God in Big Spring. Sheats's ministry to the sick included a singing duo, in the event a person wished to leave this life in the guiding comfort of a hymn.

The pastor's wife, Velma, comprised half the duo. The day they visited Bud, they also brought a girl from their congregation, whom they introduced as Opal Wilkerson. She was sixteen, plump, with wavy hair and a raw, plains-swept beauty. Like Sister Sheats, she adhered to the rigid codes of the Pentecostal church. She wore no makeup or jewelry and kept her matronly dress below the knee. But her voice, a muscular clear soprano, bespoke the many splendors of the Kingdom.

As they began to sing, her voice lifted the old, familiar hymns — "Leaning on the Everlasting Arms," "What a Friend We Have in Jesus" — and infused them with power. The voice filled the room where Bud lay dying and traveled down the long hospital corridor where it beckoned all those who could hear and lured them to the doorway. Something besides her voice gave the songs new resonance. They seemed to emanate from a place of untroubled joy, a bright force that invigorated those around her. No one in the room was as bewitched as Bob, who'd been sitting in the corner the whole time.

He'd seen her before, working at the laundromat where he sometimes dropped his clothes, and he knew her twin brothers from the oil rigs. He'd filed the encounter in the hospital room away in his mind, and after Bud died, he tracked her down. He stood outside her house and waited for one of the brothers to come home, then made his position known.

Opal fell hard and fast. Bob was older and had seen the country and told many stories about his travels. He owned a car and had money in his pocket, and he was a first-rate cutup. During the early days of courtship, all they seemed to do was laugh. And then, there was his irresistible darkness — his quick and profane temper, the occasional smell of liquor on his breath, and the sad-

131

ness over a dead mother and brother that would bubble up and send him into a mood. With as few words as possible, he'd revealed to her some of that pain and she absorbed it without injury. Behind her hazel eyes was a fortitude the Depression or wind couldn't shake. And it was in those eyes that Bob Mealer saw his future and salvation — his talisman against the family's hellbound luck.

Two weeks after they started dating, they drove down to Allie's house in Beeville, where some of the family had gathered, and Bob broke the news of the engagement.

"What are you waiting for?" someone asked. "Why not get married now?" Bob waved it off until Bertha, who'd finally left her bed, volunteered to pay for the whole thing. As restitution for Bud's untimely death, Shell Pipeline Company had quickly awarded Bertha a modest settlement. Everyone knew her to be generous; even in the hardest of times, Bertha and Bud never hesitated to loan a few dollars, or bring a meal to a family in need. But now, in her manic state, Bertha was spending like tomorrow would never come.

The next day, Bob strutted into the Karnes County courthouse wearing a new three-piece suit and cowboy hat, and Opal wore a long white gown with a pretty blue star stitched on the chest. Afterward, Bertha sprang for the honeymoon. She and John Lewis drove the newlyweds down to Corpus Christi in Bud's car, along with Frances, who managed to talk her way into the back seat. Bertha paid for motels and T-bone steaks, even a boat ride around the bay. Every time Frances asked for an ice-cream cone, someone stopped the car and got her one.

When the wedding party ended, Bob managed to keep a boot in the door of good fortune. He landed a job with Continental Oil and moved his teenage bride to Wink, just shy of the New Mexico line. They rented a shotgun shack in a poor-boy camp and began to raise a family.

But for Bud's children, the Great Depression had only begun.

6

Bertha's transgressions . . . Little Jimmie
arrives . . . a patriarch breathes his last . . .
Frances grows up . . .

After the funeral and wedding party, Bertha
took the kids back to Forsan and told them
to start packing. "We're moving to Big
Spring with your aunt Ahta. I can't stand it
here another minute."

She remained delirious with grief, alter-
nating between laughter and tears. She even
cursed God. "Why couldn't he take
Grandpa instead?" she asked Frances. "Tell
me why?"

Bertha rented their house to another oil
field family, who were happy to find a home
in such a boom. Frances and the kids
packed what possessions they owned, and
before long, a truck appeared outside to
take them to another place.

Bud's sister Ahta and her three children

— Fudge, Ruby, and Jearl Dean — lived along West Highway 80 in a part of Big Spring called Jones Valley, known for its flophouses, tourist courts, and prostitutes. They were staying in the same house she and her husband Elijah had bought before their divorce, which still left Ahta heartsick. And although Bertha was Elijah's sister, she and Ahta found kinship in their respective anguish. Now fortified with Bertha's bankroll, they sought therapy in the bars and honky-tonks around Big Spring.

The kids, meanwhile, were left to do as they pleased. And since no one made them go to school, they filled their days with make-believe. They pretended to be Buck Jones and the Rough Riders pursuing cattle rustlers up the dirt hill, or a pirate gang on the high seas. Ahta's roof doubled as their crow's nest. Then John got the idea of digging a hole into the attic with a claw hammer where they could hide and search for plunder.

The game that Frances loved to play the most was called Going to California. California was the farthest place anyone she knew had ever gone; it was where her daddy had traveled on his adventures to pick cotton and fruit, and where Buck Jones and the Rough Riders patrolled the lawless hills.

Besides West Texas, California was the only place a true cowboy would ever be caught dead, and Frances fancied herself a true cowboy.

Going to California meant loading the red wagon with a jar of water and simply walking west, down the highway and across the pastures until her feet got tired, or the sun began to sink and the fear of coyotes sent her home. Sometimes Frances designed her route to pass by Grandpa's house, which was nothing more than a shack across the highway, where John Lewis had moved after Fannie and Abe's house burned down. Often Grandpa was off working someplace, so Frances climbed through a window and helped herself to whatever was in the cupboards, which was never much. In fact, in the weeks since her father's funeral, Frances had hardly seen Grandpa at all. And when she had, he appeared much older than his sixty-six years. For one thing, he'd let his mustache grow long and bushy. His health had also declined. Carbuncles festered along his neck and back — a condition that had plagued him for years, only more so now. And instead of his long, purposeful stride, he appeared to shuffle, his long legs swelled up with dropsy. It was clear that losing Bud had drained the last bit of vinegar right out of him.

To make matters worse, a few months later, he came home from work to find his own house had burned down. No one knew how the fire started. Everything he owned was gone, along with whatever mementos or records he'd kept of Julia, who'd been dead now nearly two decades. This, along with earlier tragedies and disappointments, propelled my grandfather Bob to finally give name to the Mealer Luck.

Bertha's living arrangement with Ahta proved temporary, and within a few weeks she and the kids were moving again — this time to a tourist court down the highway. Like a lot of tourist courts, which predated motels and catered to the auto traveler, this one featured a dozen tiny cabins clustered around a common area, with outhouses located in back. But the buildings were so poorly constructed, that whenever it rained, the manager issued buckets to catch the water that poured through the roof.

The dirt courtyard and parking lot were busy with families coming and going, many from New Mexico, Colorado, and Oklahoma, people fleeing the dust and heading back to family farms in the east. Others pointed west to California, their cars piled with furniture and pee-stained mattresses strapped to the roofs. Truckers stayed there, too, stealing a few hours of shut-eye before returning to the road. In the dark hours of the morning, their engines stirred Frances from her sleep and she dreamed it was time to send her daddy off to work.

In the weeks since they'd left Forsan, Bertha hadn't once mentioned Bud, even though his absence possessed her like a spirit and sent her fleeing from the house. One evening she returned while the kids were seated around the radio. All of them looked up, astonished. A strange man was standing there holding their mother's hand. He was tall and dark-haired, and shuffled nervously in the doorway.

"This here's Virgil," Bertha said. "We just got married."

Virgil Patton was a roughneck Bertha had met at a bar called the Bucket of Blood. He wasn't the man of any woman's dreams, but he seemed nice and did his best to ignore the kids as much as possible. But just as soon as he came into the picture, he was out again. Frances never knew why, but living with a grieving, depressive woman and her four kids in a one-room cabin probably wasn't Virgil's idea of romance. One night he just disappeared and never came back.

When the divorce papers arrived, Bertha sank even deeper. One day, in an act of desperation, she loaded the kids into Bud's Ford and went looking for Virgil, driving a hundred miles west across the thorn-choked plains to his parents' house in Monahans. Virgil wasn't there, but his folks invited the family in. They gave the kids something to eat in the kitchen while they tried to console Bertha in the next room. At one point Frances heard her mother begin to cry, then say, "I might as well kill us all, just drive onto the tracks and wait for a train."

Frances hardly had time to react before Bertha was coming out, telling the kids it was time to go. Frances wanted to plead for help, grab hold of something and not let go. In a minute, they were back in the car and Bertha was driving down the highway. Frances stared ahead at her mother, searching her eyes in the rearview mirror. Just west of Big Spring, they approached the T&P tracks. The front wheels crossed the rails and slowed. Frances reached for the door to jump. But before she could pull the handle, the car rolled toward home.

They moved again after that — to another tourist court down the road called the

Buckhorn. Shortly after getting settled, Bertha found a new boyfriend, a man named Joe Alvis, who went by the nickname Red on account of his ginger hair. He was older and had a job on the highway. When he went out at night, he liked to dress neatly in a crisp shirt and slacks.

Red hardly came near the kids, other than to pick up Bertha to go dancing. They often stayed out all night and Bertha returned the next morning in a buoyant mood. A few months into their relationship, Frances noticed that her mother was actually starting to act normal. There was, however, a small episode that threatened to derail this progress. One night Bertha fainted and had to be hospitalized. The doctor, considering her history, said she was suffering from hysteria and sent her home to rest. Within a week, though, she was feeling better. Her cheeks filled with color and Frances noticed she was even adding weight.

Several months later, on June 18, 1937, an electrical storm swept over Big Spring. Long streaks of lightning split the purple sky and charged the dry summer air. Frances lay in bed with Flossie and Leamon trying to sleep (John was staying with Fannie and Abe) when, from the other end of the room, came a shrill cry. Frances shot up and saw Bertha sitting on the floor clutching her stomach. She was breathing heavily and soaked with sweat.

"Ooooooh," her mother wailed, then flopped onto her back, seized with pain.

"Run get help," Bertha said, breathless. "Run get Ruby, tell her to bring the doctor."

Frances threw on her shoes and ran out the door still wearing her gown. Ruby was Bertha's fifteen-year-old sister, who'd arrived recently from Alabama. She lived with another family about a mile down the highway, then up another road. Frances took a shortcut through the field to get there, using the electrical storm to see the prickly pear. The wind blew sand into her face and for a brief moment she staggered, just as a lightning bolt flashed fifty feet from where she stood. By the time Frances got to Ruby's, she was so rattled she barely got out the words.

That night, the doctor got to Bertha just in time to deliver a healthy baby boy, whom Bertha had been keeping a secret. With thunder rumbling outside, the child filled his little lungs and howled against the storm. Bertha named him James Lamar Patton. "The funniest case of hysterics I ever seen," she said.

■ ■ ■ ■

Of course, Jimmie was Red's baby, despite having Virgil's last name. He even had bright blue eyes, like Red, and a mop of blond hair, once it started to grow. And for a time, Red showed interest in helping raise the boy. He moved in with the family and they found a new place to live. It was a small, yellow, two-bedroom house on Highway 80, with a bedroom for the kids and one for Bertha and Red, plus a small kitchen with a woodstove. The bathroom had once been modern, but scavengers had ripped out the toilet, tub, and sink, and torn up the floor to get to the pipes. The family used an outhouse instead, while water came in buckets from a tourist court next door. The rent was five dollars per month.

Now with Jimmie in the picture, Bertha went downtown and registered for relief. The government gave her three dollars per week, which went toward milk and basic food items. And on occasion, she helped at the laundromat across the highway, which brought in a few dollars more, but the hours were never steady. She wasn't qualified for much else, and even if she had been, there were few jobs available to women in those days.

Several months after having Jimmie, Bertha fell back into her spells. She stayed out all night, not telling Red where she'd been, then played solitaire all day while the hungry baby screamed. The care of Jimmie quickly fell to Frances. She prepared his bottles, put him down for his naps, and changed all of his diapers.

By now it was fall, nearly eighteen months since her father's death, and Frances hoped to re-enroll in school. She was meant to start the sixth grade. But the more her mother drifted into psychosis, the more it became clear that Frances was stuck with Jimmie.

Once every few weeks, she sent Leamon down to the house of an old classmate named Mary Ellen to borrow magazines. Mary Ellen loved to read as much as Frances, and knowing that Frances couldn't attend school, she was sympathetic. She sent copies of *Silver Screen* and *Picture Play,* with Jean Harlow and Bette Davis on the covers, plus *Amazing Stories,* which featured the space adventures of Buck Rogers.

Frances's reading annoyed Red. Whenever he came home, he expected things, such as water heated for his bath, a plate of supper, and hot coffee. Frances could never move fast enough, could never pay close enough attention to Red's needs. "Always with your fool head in a book," he complained.

One afternoon Frances put a pot of beans on the stove for supper, then went back to her reading. Just that morning, Mary Ellen had loaned her a most incredible book — about stowaways on a rocket ship to Mars — and all day Frances couldn't take her face out of it. She was still engrossed in the story when Red walked in the door.

"Damnit, girl," he shouted. "Can't you smell that? Them beans are burning up." Red was right. Frances had forgotten all about the beans, and sure enough, there wasn't a drop of liquid left in the pot. A whole pound of beans, ruined. Red stomped over to Frances and yanked the book out of her hand. Later, he burned it in the stove, and Frances never learned the fate of those stowaways.

It only took a year for Red to get fed up and leave. One day Frances came home to find him in the doorway, holding a cardboard suitcase.

He told Bertha, "Don't you blame Murl Dean and these kids. It's you I'm leaving." Then Red was gone forever.

Red's exit from the family presented a

brand-new nightmare. For several days Bertha haunted the house like her own ghost, staring into walls, saying hardly a word. She left at night and returned in the early hours, sleeping in strange tormented fits.

One morning the kids awoke and discovered a big breakfast on the table. As they dug in to eat, their mother told them about a dream she'd had the previous night. "You were lined up in a row," she said, "and your heads were cut off."

Frances put down her fork. The minute Bertha left the house, she sprang into action.

"Everybody in the bedroom," she said, and scooped up little Jimmie. Once they were safely inside, she and Flossie pushed the dresser against the door to create a barricade.

Bertha returned home a few hours later and called for the kids. They sat huddled on the bed and didn't make a sound. When Bertha realized what was happening, she banged on the door and tried to open it, but the dresser wouldn't budge.

"What on earth . . . Murl Dean, are y'all in there? Open this door."

"No!" Frances shouted.

"I said open this door right now."

Frances was sobbing. "So you can cut off our heads? So you can take us to the train tracks? I can't let you in, Mama."

She heard her mother's hand let go of the knob. Then a voice — weary and frail — saying, "I would never hurt you kids."

Frances heard footsteps, then the front door opened and closed. Through the wall outside, they heard their mother crying. They pulled back the dresser and Frances went outside. She found Bertha sitting on the front step with her face buried in her hands, her body heaving.

"I'm so sorry," her mother said over and over.

Frances sat beside her and smoothed her dark hair, stroked her back.

"It's okay, Mama," she said. "Everything's gonna be okay."

But things were not okay. The dark visions made Bertha afraid of her own failing mind, so she began staying away from her children, each time a little longer. The kids were basically on their own. Most days, John stayed at Fannie and Abe's. Then Flossie disappeared for a week, only to show up one morning hunting for food.

"Where have you been?" Frances asked.

Flossie shrugged.

Even Leamon left during the day, playing with cousins or friends in town. But he returned each evening before dark because he didn't want his sister to sleep alone.

But Frances was never alone. She had little Jimmie, who was nearly a year old. She had him on her hip every time she hauled those sloshing buckets from the tourist court to boil his dirty diapers. Had him underfoot as she prepared their breakfast and supper. And she brought him outside in the afternoons, to sit on the front steps and watch the world turn without her.

On the evening of June 11, 1938, a powerful storm struck. A tornado ripped through the western edge of town, killing fourteen people, while ten inches of rain fell on other parts of the city in less than two hours. Creeks jumped their banks and devoured homes and farms. Streets turned into violent rivers that flooded downtown businesses. Lightning flashed and thunder pounded. And at some point amid this powerful display, John Lewis Mealer took his final breath.

In room 1303 of the Dixie Camp — a tourist court in south Big Spring where he'd recently moved — John Lewis suffered a fatal heart attack. Neighbors found him the next morning in his thinly furnished room. He was sixty-eight years old.

The news of his death brought family from all around. Bob and Opal drove over from Wink to join his sisters, who were living in Big Spring, Beeville, and Odessa. And together they grieved once again. The next afternoon they laid John Lewis to rest at Mount Olive Cemetery: devoted father and grandpa, lover of jokes and stories, a man who'd witnessed historic advancement, seen an oil empire rise and fall and rise again, yet had reaped little of its reward.

In death, at least, he was given a last recompense. Bertha relinquished her burial plot next to Bud and gave it to his father, deeming herself unworthy to occupy the same ground. "I couldn't face my husband in heaven," Frances heard her say. "Not the way I've been."

Once a week now, Frances left Jimmie with Flossie and went to collect their three dollars from the government. The woman who ran the relief office was named Miss Cronck, and by the fall of 1938, she'd grown accustomed to seeing Frances without her mother. Although the government wasn't supposed to issue money to unaccompanied minors, Miss Cronck overlooked the rules.

Frances collected her check and walked back to Lakeview Grocery, where she cashed it for food.

A good portion usually went toward Carnation milk for Jimmie — four cents per can. And with the rest, she'd learned from her mother how to feed the family for pennies, buying only essentials such as lard, sugar, salt, flour, and baking powder, along with beans, potatoes, and yams, which were cheap and kept for a long time without spoiling. Some weeks, if there was extra, she splurged on luxuries. Nothing tasted better than fired eggs and bologna.

But whenever the pantry emptied, which it did that autumn, Frances had to get creative. Several times, the clerk at the grocery store let her take some bananas that were turning brown. Then she discovered that the bakery next to the Buckhorn camp would barter for day-old goods.

Back when they lived there, she'd overheard the owner complain about not having enough boxes for deliveries. So Frances and her siblings started digging through the alleys in town collecting stacks of flattened cardboard, which they presented to the man in exchange for fried pies. Although a bit stale, they were filled with delicious strawberry, lemon, and coconut cream, which the kids lived on happily until the next week's check.

Then winter arrived and the north wind flew down off the Rockies and sliced across the Caprock, bringing short daylight and more dust. It whistled through the slats of the little yellow house and created a draft that chilled the marrow. The kids wore their Salvation Army coats most of the day, even with a fire in the stove. At night, when the coals turned to powder, Frances buried Jimmie in his crib under a mound of old cloth scraps they'd acquired in their many wanderings. The three kids then huddled under a single blanket, using their body heat to stay warm.

But as the days grew shorter and colder, the stove gobbled up the wood that Frances fed it — mostly knobby mesquite. And when there was no more wood to consume, it grew cold.

Already, it was hard to find mesquite and other deadwood in the nearby fields, which were mostly sand and scrub. And as the weeks wore on and the temperature dropped, travelers in the tourist courts, too poor for coal oil or gas heat, worked to scavenge whatever wood was left. One morning, the kids awoke to a house so cold they could see their breath. Out in the kitchen, Frances started a fire with the last bits of kindling, but it lasted only minutes.

Outside, she combed the pastures for any wood but found none. Then finally she stumbled upon a blown-out truck tire lying in the grass, brittle from the sun. Frances hoisted it upright and rolled it home. In the kitchen, she used her butcher knife to carve the tire into strips, lengthwise along the rim. She sang while she worked:

> If you go to Jones Valley, keep your money
>   in your shoes
> 'Cause the women in Jones Valley got
>   them Jones Valley blues

She stuffed the firebox with rubber and doused it with coal oil from the lamp. Within minutes the stove was kicking off heat, so much that Flossie gasped and said, "Look!" and pointed to the ceiling, where the stove pipe glowed red. Luckily, blown-out truck tires were a common fixture along the oil field highway, and for the rest of the winter, the children never lacked for fuel.

But at night when the stove turned cold, the children froze. There was only one real blanket in the house, and whenever Bertha slept at home she took it and put it on her bed. The children were left with their cloth scraps, which were thin and misshapen and always found their way to the floor. One night as ice frosted over the windows, Frances awoke and heard little Jimmie whimpering in his crib. He'd wet his flimsy covers and now shivered in the dark room. Frances changed his diaper and took the dry scraps from the bed where she and Flossie and Leamon slept, then wrapped them around Jimmie until he fell back asleep. Then she placed his wet covers over her siblings and herself. They pressed their bodies together and waited for daylight.

In the new year, Frances was determined to return to school with the other students. The morning classes began, she left the house early without telling her mother and made her way down to the junior high. Scores of kids were already filing through the double doors and congregating along the sidewalk. Frances got as far as the front steps, then froze.

It suddenly dawned on her that she didn't know what to do once she entered the building. How would she find her class? How would she tell them who she was, and what if they inquired about her mother, or asked where she lived or why she hadn't been in school?

She became aware of the slouchy coat and the dress she wore, then looked down at her shoes, which were lined with cardboard. Worse, her toes poked through the ends where she'd cut them with a knife so they would fit. She sat down on the steps and stayed there until she was alone and the morning bell had rung. Then she stood up and hurried home, praying that no one had seen her chicken out.

After that morning, she began to feel herself getting smaller and smaller each day, as if the wind were slowly wearing her down like a coat of bright paint on a pasture house. Each week more of her seemed to vanish, and when she wondered, *Why won't anyone help us? Why won't they come?* she knew the answer. Because they could not see her. She was not there.

One afternoon she needed some water to boil Jimmie's diapers, so she took her bucket down to the tourist court to fill it. As she lugged home the sloshing pail, the handle dug into her hands, so she set it down to rest. When she looked up, there stood her cousin Earl, plus another cousin — one of Clarence's boys — whom they called Junior.

They must have hitchhiked into town.

"Hiya," she said, surprised.

The boys waved back, then began to giggle. Frances wondered if they'd been drinking. She was about to crack a joke when Earl leaned in and said, "Hey Murl Dean, Junior here says he'll pay you a quarter for a poke."

"For a what?" she said, her lips slightly upturned.

"He'll give you a quarter if you'll . . . you know."

She saw they weren't laughing anymore. In fact, Junior was staring at her in a way that frightened her. She quickly looked down, and when she did, her eyes landed on her bare feet, which were filthy and cracked from the road. She became aware of the welfare dress and saw the dirt beneath her nails. She could feel their eyes moving over her breasts through the cheap fabric. A quarter for a poke.

Frances picked up her bucket and, without saying a word, walked back to the house. Once she was safely inside, she sat on the floor and began to cry. Smaller and smaller, as small as a single grain of sand. She felt, at last, like the silent dust itself.

But alas, she was not invisible. They knew about her — the school and city officials — and they knew about her brothers and

154

sister. They knew about her mother and how she'd left them like animals to survive on their own. Miss Cronck at the relief office had probably told them that much. The officials of the city then informed the officials of the state. In March 1939, when Frances was thirteen, nearly three years after her father's death, they finally came after Bertha. They found her coming out of the post office downtown, and there on the sidewalk, informed her they were taking away her kids.

Just sign these papers, they said. Bertha took the pen and did it.

Back at home, she gathered the children and told them they were going away to Abilene. "To a nice place, where I can visit you whenever I want." But neither of those things were true, since Bertha had waived her rights to ever see them again. The children — except for John, who was signed over to Fannie and Abe — were now wards of the state, handed over to an adoption agency called the West Texas Children's Aid and Welfare Association.

The very next morning, a thin, grandfatherly man appeared at the door dressed in a white suit and instructed them to pack their things. At eighty-four years old, Reverend W. A. Nicholas had been running the agency for nearly thirty years and had

155

placed hundreds of unwanted and orphaned children. In his pocket were calling cards printed with the agency's mission: "To seek homeless, neglected, and destitute children and become their friend and protector. To find homes for them in intelligent Christian families."

Frances cannot recall her mother's response that morning. The only indication that Bertha was even present is a photo that survived among her things, taken in the moments before the children were whisked away. Bertha must have been the photographer, Frances guessed. But in the photo, the children appear to be smiling. Bertha had told them to smile, and they obeyed.

The children were put into the car and taken to the station, where a train sat waiting. The old man herded them into a compartment and sat in the opposite seat.

"You can call me Father Nicholas," he said, but Frances refused to look at him.

"Our father is dead," she replied. As the locomotive lurched forward, the kids clung to each other like castaways flung into orbit, then cried themselves all the way to Abilene.

Three separate vehicles awaited them at the station, each going to a different place. "We've found good homes for you," Reverend Nicholas assured them, and before

Frances realized what was happening, she was sitting in a backseat alone. She cried out for Flossie and Leamon, but they were gone, and that sudden, unfamiliar silence burrowed a hole in her brain and never stopped ringing. Two years would pass before they found each other.

As for little Jimmie, no one heard from him again.

Two years. In that time, she lived with two different families in Abilene, first the Jacksons, and then, when they had to leave town, the Allens. In her case, Reverend Nicholas had been right. Both homes were good, and the families decent Christian people. For the first time in years, Frances had nice clothes to wear and regular meals. Sunday dinners at the Jacksons were formal, white-tablecloth affairs. Her first morning there, she awoke to birds singing through an open window. The bedroom smelled of rosebushes.

And she was finally able to return to school, even though the first day back was dreadful. The teacher called her to the front to solve an equation. As she stared dumbly at the chalkboard, completely frozen, she heard a girl say aloud, "If you don't know the answer, just sit down." And so she sat back down, mortified.

But school got better, becoming the only tangible thing to which she could cling, the only place where she had some answers to the given questions. The other questions, the Big Ones — why her family hadn't come to their rescue, why she and her siblings had been separated to be raised by strangers, what had happened to her brothers and sister — those remained unanswered.

During her second summer in Abilene, 1940, a tiny life raft arrived in the form of a telegram. It was sent in care of Reverend Nicholas from Bertha's little sister Ruby.

"Come home quick," she wrote, "your mother is very sick."

Reverend Nicholas bought Frances a round-trip bus ticket to Big Spring, on the condition she return at once. But when Frances arrived, she discovered Ruby's telegram had been a lie. Bertha's sister had written not because Bertha was sick, but because she was packing to leave for San Francisco without telling her kids.

Frances's surprise appearance startled her mother. Sensing her freedom was suddenly in jeopardy, Bertha lashed out at her daughter. "Nobody wants you here, Murl Dean. Nobody wants another mouth to feed, and certainly not yours."

Bertha then made Frances sit down and write to Reverend Nicholas saying she was coming back. Frances wrote the letter, her whole body quivering, and then bid her mother good-bye. When she got outside she tore it to pieces. She never went back to Abilene.

She bounced around Big Spring for another year. Ruby had married a guy named Joe and given birth to a little girl. Together they lived in a big army tent near One Mile Lake in Jones Valley. They gave Frances a cot in the corner and the four of them got along like regular Indians, bathing and washing their clothes in the lake, and sitting up at night playing music. Joe kept a fiddle in the tent and Ruby a guitar, and together they sang the old Irish reels that Frances remembered hearing her grandfather John Lewis sing, in addition to old standards such as "Banks of the Ohio," "Barbara Allen," and "Little Rosewood Casket." Those sad and woeful ballads washed through Frances like medicine. Joe wasn't even eighteen, but he was good — a kind and thoughtful man who reminded Frances of her father.

On Christmas Eve, Joe said, "What do you think, Murl Dean, will Santa Claus come to a tent?" And even though Frances knew better, Joe made her go to bed early, then hung her a stocking stuffed with nuts and fresh fruit.

But the arrangement was temporary, and soon Frances found herself staying with her grandparents — Bertha's mother and father. The McCormicks had moved to Big Spring, where her grandpa worked with the WPA building roads. In a crowd, he told jokes and funny stories and could play the guitar as good as anyone.

But alone, he remained the predator who'd ruined his own daughters. Alone with Frances, his eyes would move up and down her developing body. Finally one morning, he grabbed her.

"Dance with me," he demanded, pulling her in.

He shouted to his wife in the next room, "Old lady! Go out and get us some food!" then spun back to Frances, his eyes cloudy with lust.

"Dance with me," he whispered, then moved his hips up and down.

Frances pleaded with her grandpa to let her go, but he held her tight. When her grandmother walked into the room, she broke free and ran for the door.

She ran to the only friend she had. Howard

Dodd was a former classmate from Forsan whom Frances had started dating — dating in that way kids do at fourteen or fifteen. Howard was three years older, and like Frances he'd recently lost his father. He now lived in Big Spring with his mother and sister, whom he supported. Howard's mother took Frances in and gave her a room. Not long after, she received wonderful news. Her aunt Velva, Bud's sister, happened to be in downtown Eastland one afternoon when she saw two children she recognized. It was Flossie and Leamon. They told her they were living on a peanut farm not far away, in a town called Rising Star. Both appeared healthy and happy but were desperate for information about their siblings, who'd remained lost to them.

The next week, Frances got a ride out to the farm and reunited with her brother and sister. They sat for hours in a hayloft shelling peanuts and catching up. Flossie and Leamon went to school in Rising Star, and like Frances, had started from the bottom. Flossie was twelve years old and only in the third grade, while Leamon, at age eight, was just starting his second year. The couple who'd adopted them, the Atkinsons, worked them like mules every minute they were home.

They told Frances that their brother John had also visited. Much like that day, they'd sat in the hayloft and talked for hours. When it came time for John to leave, both Flossie and Leamon were devastated. It was like losing him all over again. But comfort arrived the following day when they discovered something John had left behind. On the floor of the barn was a single footprint, which they quickly covered with an old soup pot. Now whenever they missed their brother, all they had to do was return to that place.

"It's always here," Leamon said, lifting the pot to reveal the faint impression. "Just like a picture."

Within six months of Frances's moving in with the Dodds, she and Howard were dating in a more serious way. Howard was dependable, solid, and funny. When Howard asked Frances to marry him, it felt like the right thing to do. His mother arranged the license and booked the preacher down at the Church of God. In mid-September of 1941, at age sixteen, Frances stood at the altar in a blue dress and bobby socks and became Mrs. Howard Dodd.

That December, Japanese bombers attacked

Pearl Harbor and the United States entered the war. With California under constant threat of attack, the enemy poised to swarm the beach at any moment, Frances began to fear for her mother, despite everything that had happened.

So early the next year, Frances and Howard hitchhiked to San Francisco. They thumbed rides through New Mexico and Arizona, traveling mostly at night. Sitting in the backseats of strangers' cars, Howard held Frances's hand and tapped her finger three times, which was their code for "I love you." One night out in the desert, one of their rides blew a tire and overturned into a ditch. The two of them, along with the driver, emerged unscathed, then laughed at their fortune as they hitched another ride west.

In San Francisco, they found Bertha working at a bar on the Embarcadero called This Is It. She was lonesome and distressed by the air-raid drills and blackouts, and in this vulnerable state she was happy to see her daughter. Frances and Howard took a tiny room in a nearby hotel and Howard found a job at the National Biscuit Company. The following year, he enlisted in the navy and went to sea.

Frances took Howard's job stacking ice-cream cones and crackers, then ran a big oven that made vanilla wafers. Howard's mother and sister also moved to the city, only a few blocks away. But Frances couldn't help feeling alone. She missed Howard, but for some reason she missed her father more than ever. Most days she could sense his presence. One night while asleep, she felt him sitting on the foot of the bed. "Don't be afraid," she heard her daddy say, then he stood up and disappeared.

After that, Howard began to drift further and further from her heart. Frances didn't know what was happening, why she seemed to be slowly turning him off. Was it the new city? Why was she behaving like a spoiled little girl? When Howard returned home on furlough, Frances could hardly look him in the eye.

"Why won't you tell me what's wrong?" he demanded, hurt and confused, and Frances couldn't answer. Finally, she took an apartment on Pine Street and they separated.

Frances found girlfriends at work and began going to movies and concerts. At the Golden Gate Theater on Taylor Street, she saw the big orchestras that toured the West Coast — Artie Shaw, Harry James, and Glenn Miller, before Miller's plane disappeared over the English Channel. But Frances was most drawn to a band from back home called Bob Wills and His Texas Playboys. Wills had everything that Glenn Miller and Harry James did: a large horn and string section, drums, plus a piano and vocalists. But he'd taken their same influences — Dixieland jazz, barrelhouse blues, and New Orleans stomp — and blended it with steel guitars and fiddles. The result was a new kind of music called western swing, and it was borne of the East Texas cotton fields and lonesome Panhandle plains where Wills grew up.

The band's slick cowboy attire reminded Frances of Big Spring. The fiddles and guitars took her back to those warm nights in Ruby and Joe's tent. Songs like "Dusty Skies" and "New San Antonio Rose" hit her like waves of memory. And Wills himself was an oddity to behold. He pranced around stage on tippy-toes, twirling his fiddle bow above his head and crying out "Ah-HA!" in a high falsetto that was both unsettling and magnetic. Whereas most people danced to the swing orchestras, fans at Wills's shows tended to cluster against the stage and watch in silence, mesmerized by the tight synchronization of parts.

Wills had recently moved the band from

Tulsa to California, where they played nightly up and down the coast, his audiences packed with Texans and Oklahomans who'd come west during the Depression and to find jobs in the war industries. That success had also carried over into Hollywood, where Wills starred in western films.

Frances saw the Texas Playboys at the Oakland Auditorium, then again at the Golden Gate, each time emerging after the four-hour show feeling exhausted and alive. Already she had a thing for steel guitars, the way they seemed to moan and cry like the wind off the Caprock. She even bought herself an Epiphone Electar and practiced in her apartment after work. For a time, she and her girlfriends were planning an all-gal country band.

Around this time, Howard asked for a divorce and Frances didn't fight him. She signed the papers and for a while slunk into a stupor. One night she read in the paper that Wills and His Texas Playboys were playing in Oakland again, so she went. Wills had a new steel player named Les Anderson, known as Carrot Top, on account of his red hair. He had a smooth and effortless playing style and that night, whenever he looked up from his instrument, his eyes seemed to lock on Frances like big blue magnets. When he sang lead on the Irving Berlin tune "Always," with lyrics about "loving you always," she knew it was silly, but she swore he was singing right to her. After the show, Frances was standing on the dance floor, waiting for one of her girlfriends, when Anderson appeared behind her. They ended up talking for an hour and he asked her out. Those blue eyes were beckoning her — to romance, to new adventure on the road, to a chance to live inside that music.

Frances told him yes, then let herself go.

PART 2

1

Bob and Opal go to Wink . . . Clem and the Salvation Army Band . . . an unbearable tragedy . . . New London explodes. . . .

After my grandfather Bob and Opal's wedding, Bob found work as a roustabout in the newly discovered Keystone field. The town was called Wink, a hundred miles west of Big Spring, where the sand blew wild across the flats. Civilization had ventured there for one reason, and that was oil.

Nearly every waking hour found Bob on a rotary rig assembling pipe, cleaning boilers, elbow deep in wet concrete and drilling fluid. Back at home, he found comfort in his bride.

Times were lean, nothing that Bob wasn't used to, but with Opal it wasn't so bad. The couple moved into a poor-boy camp on the outskirts of town, where the shotgun houses sat close together in a row. Since paint wouldn't last long in that wind, tin sheets protected the houses from the blast coming off the Rockies. The sand and dust were so thick that mothers covered their children's faces with diapers when they went outside to play. Oil field wives learned to conserve water, which was scarce, and to burn trash in pits fired by flare-off gas.

The oil patch families were close and looked out for one another. During the worst of the Depression, a gunman held up the grocery store in Wink and took only milk and bread, saying, "Sorry to do this, but my family's hungry, and I'm going to take what they need." After the man left, the other customers asked the grocer what he intended to do, to which he replied, "If I knew where he was, I'd hunt him up and give him some more groceries."

During their first three years in Wink, Bob and Opal had two daughters, whom they named Zelda and Norma Lou. And when Bertha began to lose her mind out in Big Spring, before she left for California, they took in John whenever Fannie couldn't afford to feed him. Living out in Wink, far removed from the family, it's hard to tell if Bob and Opal knew about Frances and the others. Perhaps they did and figured there was little they could do, what with three kids already and hunger planted on their doorstep. A roustabout's salary didn't go very far. Most meals consisted of side meat and cornbread, maybe some eggs and dried beans. The blowing sand and dust chewed a garden to the nubs and made laundry next to impossible. Whenever Opal hung the diapers to dry, she often returned to find them coated in a red crust.

And yet she was full of joy. She sang as she hunched over the washtub and hung the clothes again. She sang in the little Assembly of God church, where the pastor put her in charge of the choir, her voice part of the great message of hope that delivered a tired and worn-out people from one Sunday to the next.

Opal hadn't come to faith on her own, nor was it instilled through Sunday school or Bible stories, the way many children encounter God. Rather, Opal's father Clem had embraced salvation from one of man's lowest stations — as a drunk in the gutter. And his radical transformation lifted the family from poverty and shame.

Clem Wilkerson hailed from southern Tennessee, near the town of Winchester, where he'd worked as a logger around Sewanee Mountain. He suffered from asthma, and sometimes the air in the forests was so heavy and wet that he slumped against the crosscut saw, choking and turning gray.

"What you need is the desert," a doctor advised. So after marrying Cora McCann and having a baby girl, who they named Agnes, they'd packed the wagon in 1914 and made their slow way toward Arizona. They got as far as Mississippi before another child was born, then settled on a tenant farm in Willow, Arkansas. Another baby was born there and stopped breathing five days later. Pneumonia, they were told. With Little Pauline's grave still fresh, a nearby creek rose up during a storm and carried off everything they owned.

They stopped next in Paris, Texas, where Clem found work as a driver. Five more kids were born in a span of six years, including my grandmother Opal and a set of twin boys, Ed and Fred. After oil was discovered in West Texas, Clem loaded the family into a Model T Ford and headed for Big Spring.

At first he sold fruit in Jones Valley, then found work driving a truck. Prohibition was on, but the boomtown was too wild for a harness. Honky-tonks and pool halls sat on every corner, selling beer, whiskey, and local jake. The jobs Clem held never lasted long, and neither did his pay, but he was fortunate in that the openings always out-numbered the available men.

It didn't take long for him to gain a reputation as the town drunk. People called him "Sleepy," for the way his eyes half closed whenever he was on a stink. Even in later years, when he was sober and upstanding, the name lingered.

By 1932, two of Clem and Cora's children had grown and married. Agnes fled to the East Texas oil patch with a roughneck from Lamesa. Herman, the oldest boy, married the daughter of Arkansas boomers who were living in tents in Jones Valley, where the tar-paper shacks flapped in the wind. Herman's new wife was also named Opal, just like his sister. But she was a bit shorter and rail skinny, so folks called her Little Opal to avoid confusion.

Little Opal was headstrong and zealous and bent on reforming her drunken father-in-law. She even followed him into the beer joints to shame him into going home.

"I won't stand to see you hurt this family any longer," she wailed.

Clem never needed a daughter-in-law or anyone else to provide him shame; regret and self-loathing came all on its own. But occasionally, he managed to elude her. And one sweltering night in that droughty summer of 1932, he claimed such a victory and spent his last dollar celebrating.

Midnight found him downtown on Third Street, sitting on the curb. The night's revelry had left him feeling ill and dizzy, and as he sat wheezing from the dust he heard what sounded like horns. Sure enough, from down the block he saw them come into view — the Salvation Army band. The ragged brass section marched two steps behind a tall preacher who was dressed in shirtsleeves and carried a thick, black Bible. The procession stopped directly in front of Clem and paused their music.

For fishers of men working the night beat, the bites were few, especially from drunks. But Clem must have revealed something only the preacher could detect, some radiant spiritual hunger. The preacher introduced himself as a sinner and a man of flesh and blood. He explained how we are all children of God, and even the most wicked of men can be redeemed through his grace. All it took was a confession of faith, from the same mouth that had drawn so much wine, for a man to be saved from himself. He asked Clem something like "Are you willing to surrender the old burdens of flesh and be set free?"

With the preacher's knobby hand clutched around his forehead, Clem repeated the simple prayer of salvation. Right away, he felt something tear free from his body, like wind blowing through every pore. He saw in his mind his old spirit, grooved from the Devil's dark touch. In its place now were only lightness and an almost nauseating sensation of love. All was forgiven, all was made clean. Not a trace of booze remained, for he'd drunk of what Jesus called "the living water" and would never be thirsty again. He left the preacher's side and ran all the way home, blubbering like a newborn babe.

The change was instant and dramatic, and never did Clem Wilkerson drink another drop. He joined a small Assembly of God church in Big Spring and brought Cora and the kids. And there in the dark bowels of the Depression, he and his family discovered something that John Lewis had failed to find for his own — a community.

They found people who looked like them, who'd traveled from the same Georgia hollows and Tennessee forests chasing the same broken dreams, people who'd been busted and bankrupt, who buried children and other loved ones in places long ago abandoned, and who'd carried that trauma knotted up inside them. They'd arrived with little education and few possessions, harboring past crimes and shameful, hidden desires. But like Clem, they learned how to surrender these troubles to the Lord and regain what Jesus promised in the Gospels: wholeness, humanity, and self-respect. They also learned how to pray, as a way to both communicate with God and give voice to their own worries and regrets, and by doing so come clean with themselves and begin to change, to heal.

The church became their refuge and lighthouse. It provided a safety net for those who needed it. The church nurtured them when they were sick, fed them when they were hungry. When a loved one died, their fellow church members wept with them and gave them comfort and encouraged them to live. And the church provided strength when they were feeling susceptible to the world's destructive ways.

Clem became fired up for Jesus, and soon, so did his children — especially Opal, who at twelve years old discovered her gift of song. Whenever the church doors were open the family was inside, either for worship, choir practice, or to feed a funeral party. After the preacher finished his sermons, Clem stood before the congregation and reminded them of the man he used to be and how the Lord had molded him, like a lump of clay, into the man he was today. He always ended this testimony by quoting his favorite psalm: "Oh that men would praise the Lord for his goodness, and for his wonderful works to the children of men!"

Clem became a deacon in the church and a leader among the congregation. He was not a learned man, but after his transformation, his desire to know Scripture pressed him to improve. Each night after supper, he charged his two daughters still living at home with teaching him how to read, using the Bible as his primer. He became a devoted student of Scripture and opened his home each week for Bible study and prayer. Those meetings often built to a peak, with people receiving the Holy Spirit and speaking in tongues, just like they'd done on the day of Pentecost.

But all the racket embarrassed Herman's wife, Little Opal, who was raised hard-shell Baptist and didn't tolerate such antics. "You can hear them from down the block," she complained to Herman. "They sound like a bunch of *fanatics.* Folks are gonna run them outta town."

But even Little Opal couldn't resist. Soon she was kneeling at the altar of that little church and receiving salvation, her voice joining the jubilee.

Around that time, Little Opal and Herman had a girl named Mary Lou, who they raised in that tiny church. By the time Mary Lou turned five she was a missionary for Jesus just like her grandpa, reciting scriptures she learned in Sunday school or asking friends to come to church. And Mary Lou was a regular prayer warrior, whether it involved laying hands on her sick baby dolls or blessing the family supper.

But in November 1937, a terrible accident occurred. Herman and Mary Lou were driving along the highway to meet Little Opal at a family gathering when Mary Lou somehow opened the car door and tumbled to her death.

They buried her at Mount Olive Cemetery, just a few rows down from Bob's brother Bud. Brother Sheats led the service. Afterward, Little Opal stood over her daughter's grave and asked those gathered round, "If I'm a Christian, then how can I lose my child?" And no one could give her an answer that took away the pain.

After the funeral, she fell into a panicked state of grieving. For an entire year, she still made a plate for Mary Lou at supper. Each morning she walked two miles down Highway 80 to the cemetery, where she lingered much of the day. After a year, she looked up and realized Mary Lou's name was misspelled on her tombstone, having gone unnoticed all that time, and this sudden stroke of cognition seemed to snap her from her spell. She called the monument company to correct the mistake, then returned to the living world with a new kind of faith, one fired and tempered like iron.

Not long after, tragedy tested the family again — this time with Agnes, the oldest of Clem and Cora's kids. Agnes and her husband, Arthur Hahn, were raising four children in the boomtown of New London, in the heart of the East Texas field. The previous year, on March 18, 1937, their oldest boy, Granville, had awoken with mumps, so Agnes kept him home. Granville attended first grade at the nearby New London School, which the town had built for a million dollars as a totem to its oil boom, a colossal middle finger extended toward the Great Depression.

While the school boasted the first football stadium in Texas with electric lights, the administration chose to skimp on its heating system. Instead of using fuel oil, they piped raw natural gas straight from the field.

On that day in March, the pipe sprang a leak and slowly filled a crawl space beneath the school. When it finally exploded, shortly after 3 p.m., the blast lifted the steel-framed building into the air. Nearly three hundred students and teachers perished, most of them crushed and burned when the roof came crashing down on them.

Over a mile away, still in bed, Granville felt his own house buckle, then heard the caterwaul of anguished cries as the town registered the carnage. The New London explosion still stands as the worst school disaster in American history.

But at home, the real tragedy was Granville's daddy. Arthur Hahn was a mean and calloused man, with glaring blue eyes that dared anyone to try him. Whiskey made him into a monster, and he took out his wrath on the two oldest boys, Granville and Wendell. During supper, Arthur would spring across the table and cuff one of them for the slightest transgression, knocking him to the floor. Once, when Wendell tried to escape a beating by crawling under the house, Arthur spotted him easily under the floorboards, trembling like a rabbit. Without saying a word, he boiled a kettle of water on the stove, then stood atop the child and emptied it over his body.

Naturally, Agnes lived in terror of her husband, who'd once fantasized to her about setting the house on fire while their children slept. So he probably did them all a favor the day Agnes gave birth to their fifth child.

It was a girl, whom Arthur insisted they name Betty Kansas.

When Agnes asked, "Why Kansas?" her husband replied, "Because that's where I'm going to get rich," then walked out the door and never came back.

For six months after Arthur left, no one heard from Agnes or the kids. Back in Big Spring, her father, Clem, became so worried that he sent her sister Dorothy on a bus to New London to check on them. When Dorothy walked in the house she found Agnes and the children practically starved. Agnes was nearing a nervous breakdown, trying to nurse Betty Kansas and care for the others. The older kids were eating wild blackberries and poke salad they foraged in the woods. Most days, a neighbor's milk cow provided their only nourishment.

Dorothy called their father, who dispatched Herman to East Texas to get them. Herman had started a cattle-hauling business and owned a truck with a forty-foot trailer. Two days later he arrived in New

London, placed Agnes and the baby in the cab, then drove four hundred miles back to Big Spring with kids and luggage loaded in the carriage.

As for Arthur, he was gone. A few years later, his parents and brother came looking for him, with the sole purpose of telling him that he was a millionaire. It turned out some land that he'd bought years before in the Permian Basin was full of oil, and companies were clamoring to drill. But Arthur never appeared, not even to claim his fortune (his brother, through some crafty legalese, stole the land and squeezed Agnes out of the royalties).

In the months after Arthur disappeared, some recalled reading about a big explosion on a Kansas oil rig that killed many men. And since nothing ever surfaced about his whereabouts — no census or death records, no mention of him in newspapers or city directories — that's where I'll choose to leave Arthur Hahn: in Kansas, consumed in a tower of flames.

Bob and Opal return home . . . Clem becomes a prayer warrior and cousins heed the call . . . Gloria Jean, terrified of worms . . .

Out in Wink, Opal's heart ached for her family. At least once a month, when the homesickness became unbearable, she and the girls caught a bus to Big Spring and stayed with her parents, attended church and fussed over new babies. In 1939, two years after Mary Lou's death, Herman and Little Opal had welcomed a second child, a boy called Homer, in honor of their pastor Homer Sheats. And just eighteen months later, a third child arrived, whom they named Evelyn.

After a few days with family, it crushed Opal to have to leave. Each time the bus left Big Spring and crossed the empty flatlands it felt like dismounting the earth altogether. There was nothing there to see. Nothing out the window but dead grass and mesquite and pump jacks that seesawed in the fields like alien mules. Occasionally, a working rig appeared close to the road and she glimpsed human life, but mostly they were just roughnecks of some kind, their khakis and coveralls slicked with oil. Men like her husband. The heat waves off the blacktop made their hard hats appear as white bobbers on a rust-colored sea. She hummed softly to Zelda and Norma Lou, both asleep in her lap, and knew when she was close to Wink by the grit in her teeth.

She was terribly lonely, but what good would it do to complain, especially to Bob? In these years, a man was blessed to have a job and a home he could afford to keep. So she resolved to wait it out, through the long nights when Bob worked turnovers, and through the dirt that greeted her each morning on her clean kitchen floor in that tin-covered shack that whistled like a cheap horn.

Then, in December 1941, the war solved all her problems.

The government began hoarding steel pipe and other raw materials needed to fight Hitler and the Japanese, which cut exploration by nearly half. Rigs were shut down,

186

men enlisted en masse, and the land emptied as if by magic. The same was happening around Big Spring, and due to the shortage of help, Bob easily found a job there. That summer, the family loaded their belongings into Herman's cattle trailer and finally headed home.

The war overseas brought the extended family closer together, but the price they paid was their young men. Three of Clem's boys went off to serve, including the twins Ed and Fred, and on Bob's side, there was his sister Allie's son, Orville Pruitt. Ed and Fred fought across Czechoslovakia in the Sixteenth Armored Division and came home decorated soldiers, while Orville died in Lorraine, France, in November 1944, his body never recovered. Allie, who still grieved the sudden loss of her brother Bud, became so distraught when told about Orville that she drove her car onto the T&P tracks and waited for the afternoon train. Luckily, someone found her and brought her home, but that evening she came down with a headache that never went away. With no corpse to bury, she erected a shrine to her only son in his quiet sunlit bedroom, one that she maintained for the next forty years.

Bob and Opal found a cheap duplex on the

west side of Big Spring, and while Bob disappeared on the rigs, Opal spent her days at her parents' house. Clem and Cora lived just across the train tracks on the north side of town, alongside the city's black and Hispanic residents, who in those days seldom crossed the T&P line.

Her parents' house was big for the neighborhood, three bedrooms on two stories, but every inch of space was accounted for. Three of Opal's sisters lived there, Veda and Dorothy, plus Agnes and her five children. And to make extra money, Clem and Cora took in boarders. A man named Smith, who'd been shot in the back during the war, rented a small bedroom, while a Mrs. White, whose husband was still overseas, occupied a second room with her daughter, Gloria Jean. One of Herman's drivers, an alcoholic named Jug, rented a bed on the back porch.

The family's nerve center was the white-stucco Assembly of God church at the corner of Fourth and Lancaster, which Clem had helped build after being saved. Opal joined a singing trio that included the pastor's wife, Velma Sheats. They performed on the radio station KBST as part of a Sunday afternoon gospel hour. And later that day, the three of them rode around in Brother Sheats's open-air Willys car, which was equipped with a bullhorn speaker, the women singing harmony while the pastor pitched the evening meeting.

A typical service kicked off with two hours of singing, which laid the red carpet for the Holy Spirit to work its wonders. As a denomination, the Pentecostal church prided itself as a bottom-feeder, its message of salvation tailored toward last-chancers. The moment these people accepted Christ, they tended to believe passionately what the scriptures promised about how "old things are passed away . . . and all things are become new," for in their own hearts and minds they were transformed. *This* they could feel.

And if God could deliver them from hopelessness, they believed, there must be no limit to his works. Their faith was absolute, like a child's, and so they didn't hesitate to ask for wonders otherwise unthinkable, and to believe in their outcome. When faith was this concentrated and aligned, it set the stage for miracles.

Opal liked to tell the story of Frank Mack, a traveling preacher who'd married her friend Inez while they were living in Wink. As a boy, Mack was stricken with polio and for two years lay paralyzed in bed, hardly able to eat. His mother prayed constantly, begging for a miracle. With the little money she had she ordered a prayer cloth anointed by a faith healer she'd heard on the radio. The day it arrived, she laid it across Frank's body. "Lord, he's just an invalid," she prayed. "But he belongs to you, and we're trusting you to heal him."

"As I slept, God touched me," the preacher later wrote, "and the next morning I felt new strength. When my parents came to set me in the chair, pain shot up my legs like piercing needles. Glorious pain where there had been no feeling at all!"

Mack became a preacher at seventeen, advertising himself as "Once a paralytic and now a flaming evangelist!" One night in a motel room in California, he felt as if he were suddenly "covered in liquid fire." It was God calling him to lay hands on the sick, which he began doing in crusades across America.

My family told stories about witnessing their own miracles, about people receiving the Holy Spirit and performing extraordinary feats — like the man who ran to the piano, on which he'd never laid hands, and started playing as if he'd done it his entire life. About deaf people who could hear, about cancer and diabetes vanishing under prayer. "With his stripes we are healed,"

Scripture says, and this my family believed.

After his transformation, Clem became a prayer warrior who laid hands on everything. He prayed over scraped knees and stomach aches, a wart that wouldn't go away. At his county job, where he worked as a foreman on road crews, he prayed each morning over the men and their trucks, and the men were always grateful. He even prayed for the hoboes who came off the trains and knocked at his door for a meal, handing each one a prayer tract with every plate of beans. He prayed because there was no shortage of God's grace. It was inexhaustible, and often it came when you needed it most.

When Clem was helping build the new church in 1938, the congregation ran short of money and couldn't finish the roof, so he climbed up and prayed over it. While sitting there, a stranger rolled up in a rickety pickup.

He shouted, "Hey! Somethin's tellin' me y'all could use some help."

"Boy, we sure could," Clem said.

The man reached into his shirt and pulled out a check for five thousand dollars. No one ever saw him again.

But the greatest miracle Clem ever helped facilitate happened right under his own roof. Mr. Smith, the wounded war vet who rented a room, was confined to a wheelchair when he first arrived at Clem and Cora's house, and right away Clem saw him as a challenge.

One Sunday before church he laid his hands over the jagged scar across Smith's back and began to pray, then loaded his chair into the cattle truck and took him to the evening service. With Brother Sheats and the deacons gathered round, they prayed long and loud, with the congregation backing them up with fire. And sure enough, Smith stood to his feet and pushed away his chair. The crowd parted and he began to walk — with a slight limp, but upright all the same. That night when they got home, Smith refused to go to bed but limped through the house until dawn, stepping over sleeping children and mumbling a stream of hallelujahs.

Just as the disciples had spread the news of Jesus's wondrous works, this burden also fell upon the church. So following the singing portion of the service, members of the congregation stood and shared the ways — both large and small — that the Lord had moved in their own lives.

A woman they called Sister Taylor, who was quite large, gave thanks one morning after being struck by a car. "Bounced up on the hood and hit the windshield," she proclaimed. "Tore that car to pieces and it didn't even hurt me. Praise Jesus!"

For Zelda, Homer, and the rest of the cousins, testimony time was rich theater. They sat together and tried not to giggle, but each child knew their time was coming. The signal was usually a tap on the shoulder from someone sitting behind them: "Don't you have anything to say about the Lord's wonders?" they asked, and the children had to be ready.

One such morning, Zelda stood and said, "I love the Lord this morning because He first loved me and saved me from the miry clay." The congregation nodded with approval.

This was not her own testimony, of course, but one of the many she and her cousins had memorized. Each time they heard a new testimony they liked, they wrote it down in a notebook and gave it a number for easy recall.

"I love the Lord because He picked me up when I was worthless and set me upon the rock," said Iris Hahn, quoting number 4, which naturally followed number 3 since it derived from the same scripture.

"Praise Jesus!" someone shouted.

Next came Norma Lou, who rolled out number 7: "I love the Lord because He gave me victory. I was down and out in the mully-grubs and He lifted me with His love. Now I've been smiling all week."

"Amen!"

Then Homer stood and started on number 1. "I'm so happy because God so loved the world that He gave His only begotten son . . ."

It was that old ace in the hole from the Book of John. But halfway through, Homer seemed to lose his train of thought, and to his cousins' horror, began barreling through the whole system, snatching pieces of numbers 2, 5, 8, and 10 and jumbling them into one baffling ramble.

"And what shall a man give in exchange for his soul . . . uhh . . . Jesus the same yesterday, today, and forever . . . uhh . . . Verily, verily I say unto you. Amen."

Stranded and out of options, Doris was forced to tell about a half-bottle of perfume she'd found in the alley, which was actually true and wonderful but didn't fetch any hallelujahs.

As usual, Brother Sheats preached on the wages of sin, how heaven was good and hell was hot, and how there was no in-between.

"Don't be on the fence," he warned. "God

doesn't take fence walkers in His Kingdom. You're either in or you're out! Don't come in here and talk about being in when you've been out all week long, fighting with your husband, letting your kids run wild, harboring wickedness. God will root out a liar."

He paced the floor. "If I'm talking about you, then you need to hit the altar. Grab hold of the *horns of the altar* and get right with Jesus."

The Baptists down the road believed that once a person accepted Christ and was cleansed by his blood, their ticket to heaven was guaranteed. But not the members of the Assemblies of God, who in those days believed that even the slightest blemish of sin carried the whole weight of damnation. Each tiny transgression had to be reckoned with and expunged.

The cousins all agreed that it was Homer who usually led them into sin. After all, it was Homer who'd thrown the neighbor's cat down the outhouse hole, forcing the fire department to come and save its life. And just the previous week, he'd sent Betty Kansas next door to the neighbors under the auspices of cleaning their house and forced her to steal cigarettes. But there were other sins, too, sins that were all their own.

They told lies, of course — whoppers.

Lately, they'd been walking to the Hotel Settles to ride the big brass elevators, telling everyone they were from New York City.

"Is that so?" the bellhop asked. "Tell me then, what's it like in New York City?"

"Oh, you know," said Homer. "Tall buildings. Lots of people. Fancy people, mister." Iris even wore a hairnet to look extra convincing.

Aside from telling bald-faced lies, there was envy, the silent destroyer. Envy that swelled up like fever whenever friends from school talked about seeing picture shows or going roller-skating, both of which were forbidden. There was resentment, too — toward the visiting evangelists whom Clem invited over for supper, fat men with hearty appetites who filled their plates with fried chicken and potatoes, beans and biscuits, and left the kids to fight over the crumbs. Resentment for their boarder Jug, who did the same thing, even after the kids had seen him in Shine's Drug Store eating hamburgers and french fries. Jug even had his own bed, while they slept on the floor.

Each child sat filthy before the Lord. So when Brother Sheats ended with the altar call for people to get salvation, they marched down the aisle and received their weekly dose.

While there, they prayed for the souls of others — whoever in the family happened to be drifting away from God, according to Grandma Cora. That list usually included my grandfather Bob, who never took to church the way Opal and her family did. And the children prayed for their uncles Ed and Fred, Opal's twin brothers, who were off fighting in the war. They asked the Lord to watch over them in battle, but also prayed for their souls lest they be killed and face judgment. Or worse, the great Rapture took place and left them in a foreign land as fodder for the Beast. For it was well known that Ed had been drifting into deep water. In the months before shipping out, Grandma Cora discovered he was going down to the Settles to jitterbug with the lonely wives of servicemen. He'd even won a contest, which shamed her deeper.

"Pray for Eddie," she would say, her face stricken. "He's down there a-dancin' again."

In addition to the two Sunday services, Brother Sheats preached the midweek sermon on Wednesday night. An evening service on Friday provided distraction from weekend temptations such as football games and honky-tonks. Saturdays were reserved for a revival or camp meeting, which occurred monthly and kept the spirit sharp.

"Got to stay on the firing line," the pastor reminded his flock.

Church was so prominent in the children's lives that it naturally influenced their play. The iron bedstead in Clem and Cora's bedroom doubled as Zelda's organ, while Norma Lou and Evelyn made accordions out of old funny papers. Opal's little sister Veda, who was nine, insisted on preaching, and her sermons always blazed with hellfire.

Every service needed an altar call, and no altar call was complete without a sinner. This role usually fell to Gloria Jean, the little girl who lived in the rented room with her mother. The two of them didn't go to church, which made Gloria Jean the obvious candidate. By now, Veda had sent the poor girl to hell and back a hundred times.

Since Zelda took piano lessons, she got to lead the music. Her fingers danced atop the iron rails through standards such as "I'll Fly Away" and "On the Jericho Road," with Norma Lou and Evelyn squeezing the paper accordion beside her. After singing, Homer gave a short testimony on how God so loved the world, and then Veda took her place behind the bed, which doubled as the pulpit.

"I'll start with Scripture," she said, then cleared her throat. " 'Verily, verily I say unto you, unless your righteousness exceeds that of the scribes and Pharisees, you shall not enter the kingdom of heaven.' " She looked up from her Bible. "What that means is you go straight to hell."

"Excuse me," Doris interrupted, raising her hand.

"What is it?" Veda said.

"I'm the visiting missionary from Africa, and I'd like to tell about the marvelous wonders the Lord is doing with the lost and native tribes."

Well, everyone knew that visiting missionaries held status in the church and were always permitted to speak about their work. And with Veda always hogging the pulpit, being the missionary was the only way to get a word in edgewise.

"Fine," Veda said. "Sister Doris will tell us about the marvelous wonders the Lord is doing in Africa."

Doris stepped to the pulpit and told how Africa was a land of lost souls and false gods. "But the Lord is saving them through his love and mercy because he doesn't want us going to hell." She smiled, satisfied.

But Veda had no patience for mercy. "The Scripture tells us, '*Do not be deceived; God is not mocked: for whatsoever a man soweth, that shall he reap.*' He who sows to his own flesh reaps corruption. *Think about it.*"

Sensing her message hadn't sunk in, she pounded the mattress. "Give up the flesh! Give up your pride!" She then cast her eyes on Gloria Jean, and the girl's body went rigid. This was her cue. "There are *some* of us out here who've been *seared* with the hot iron of sin," Veda continued, "whose heart has grown calloused toward the Lord. You've drifted *far* from the shore, become a slave to the nightlife, bare-legged and painted like a Jezebel. So let me ask you, if Jesus came back right now, where would you wind up?"

The girl stammered, but Veda cut her short.

"In the fiery furnace!"

The children shrieked. "And what does the Bible tell us about hell? That it's a place of eternal darkness and flaming winds. A place where worms eat your flesh. *Forever!*"

The part about the worms always made Gloria Jean cry. She stood up with tears in her eyes and walked to the bedstead, where, once again, she was saved. The other cousins soon followed, for no one is truly free of sin. And seeing them bowed and broken, Veda raised her hands in victory.

"Revival has come!" she proclaimed.

"Glory to God! Preach the word and revival will come!"

Of course, Gloria Jean wasn't the only hell-bound soul in need of saving. The neighborhood outside was full of people who didn't go to church — or at least, not to *their* church — and Veda and the cousins saw it as their duty to rid them of the stench of sin. So when the weather was nice, they held open-air revivals in the alley behind the house, near the little grocery. For the unwitting boys and girls whose mothers sent them to the market, the narrow pathway became a trap of righteousness.

In one afternoon, Homer must have dunked twenty kids in a fifty-five-gallon barrel filled with water, proclaiming, "I baptize you in the name of the Father, the Son, and the Holy Ghost. Now go and tell everybody how you've changed your life."

As the children wrestled free and fled down the alley, soaking wet, the cousins applauded them on their good fortune. However, they soon suspected Homer's zeal for baptizing was less than godly, because once he ran out of sinners, he started dunking cats.

3

The final fall of Josh Cosden . . . the arrival of Raymond Tollett . . .

After a hard decade of dust and Depression, the Second World War ushered in an era of phenomenal growth in Big Spring. It began with the federal government opening a small air base for the training of bombardiers, which grew over the years to employ thousands of people who settled down to raise their families. On the other side of town, Joshua S. Cosden's refinery — mothballed during the Depression — found new life as it began churning out fuel in support of the war. Like the air base, it continued to grow, pioneering products that would change America. The only unfortunate thing was that Josh Cosden himself wasn't around to see it.

After losing his company back in 1930, around the time John Lewis and his family traveled back to Georgia, Cosden and his wife Nellie retreated to New York and Palm Beach, where his health began to deteriorate. In August, the *New York Times* reported the oilman recovering in a private hospital on Park Avenue where doctors were treating him for a serious lung infection. Later reports described Cosden in critical condition down in Florida.

But in 1933, like a ghost, he'd appeared on the veranda of the Big Spring refinery, his blond hair now the color of steel, and took back his company in a receiver's sale. Once again, his wealthy friends gambled on his trademark resilience, this time loaning him half a million dollars to purchase the lion's share of stock to take ownership. Cosden then set out to refurbish and expand the plant, but within two years, he'd lost control again. After a nasty legal battle with his stockholders, the court barred him from the company.

By the fall of 1940, Cosden lay fighting for his life at Cook Memorial Hospital in Fort Worth, his lungs giving out. At one point, the tycoon looked up at his doctor and tried brokering one last deal: "Make me a well man again and I'll make a million," he said. "You'll get your cut."

Doctors released him into the care of his

longtime butler, William Hudson, who joined him on a Southern Pacific train to Palm Springs, California, where the air was clean and dry.

Some years earlier, after losing his company and perhaps seeing for the first time the limits of his ability, a reflective Cosden sat down with writer C. B. Glasscock for a rare interview. "When a man plays for high stakes every day of his life for years," he said, "when he races horses, fights economic and political hazards, and lives constantly up to the limits of his energies and his physical, intellectual, and material resources, it burns him up. I have lived that way and I have enjoyed it. But I have learned that I want something else. I want peace and tranquility."

In Palm Springs, Cosden would try to find that comfort. A quiet desert oasis would allow him to regain his strength, to rally the Wall Street brigades for another western campaign, and perhaps seize the throne once again. At some point during the night, the train entered Big Spring and rumbled past the dim lights of the refinery, where Cosden's name still hung over the gates. By the time it reached Wilcox, Arizona, shortly after lunch on November 17, Cosden had suffered a fatal heart attack and was dead.

He was fifty-nine years old.

Authorities placed Cosden's body on a train to El Paso, then notified Nellie in Westchester, New York. Just days earlier, the last of her husband's holdings had been auctioned off to pay his remaining debts, and from every indication the family was broke. In Westchester, Nellie was working as an interior decorator to stay afloat.

A second call was placed to the refinery headquarters, now located in Big Spring. In a modest office beside the cracking units sat its new president, a thirty-two-year-old former FBI agent named Raymond Lee Tollett.

Only five months into the job, Tollett boarded the first train to El Paso and claimed Cosden's body. He left enough money with Hudson to have the remains cremated in San Antonio whenever Nellie arrived, then returned to resurrect what he called "the pile of junk and rust" now in his stewardship.

When Tollett had come aboard that summer, having briefly served as secretary-treasurer, he inherited a company battered by years of Depression, insolvency, and mismanagement. Liabilities exceeded credits and its common stock was worthless. At its lowest point, one of the refinery's suppliers refused to leave an order of $7.80 without being paid in cash.

Tollett's first line of business had been to move the headquarters from Fort Worth to Big Spring so he could monitor its everyday affairs. Using money from the sale of an oil property, he built a one-story office building in the shadow of its columns. He then persuaded Hiram J. Halle of Universal Oil, to whom he was already heavily in debt, to extend more credit.

"Give me fifteen months," he said, then orchestrated a comeback more spectacular than Josh Cosden could have ever dreamed.

Raymond Tollett stood over six feet tall, with a head of slicked brown hair that by his late twenties had already begun to recede. The hairline made him appear older, but so did something in his eyes, which were deep set and dark and with no distinguishable color. He was a gangly man, with long arms and big hands, Lincolnesque. But this awkwardness was offset by his penchant for expensive tailored suits. In fact, the word most people used to describe Tollett was "dignified," which probably meant more in Big Spring than in New York or Chicago. Yet even in those cities, alongside eastern bankers and businessmen, he stood out as distinguished and cosmopolitan, especially for a Texas oilman.

Tollett's presidency of the Big Spring refinery came on the heels of an already dizzying life and career, one entirely self-made and scratched from the Red River bottomland where he was raised. As it did for Josh Cosden, Big Spring would provide Tollett a redemptive second act, and over the coming decades, he would be hailed as both a genius who thrust the little town and its "country oil company" into global prominence and a benevolent leader to his rank and file.

In exchange, the town pledged its loyalty and, later, its protection during Tollett's long black spells with depression and drink. Unlike any resident before or since, "Mr. T.," as everyone called him, was the undisputed king of our patch of oil and sand, and his influence would have profound impact on my family's future.

Tollett was born in Oklahoma in December 1907, around the time John Lewis and his family were fleeing the boll weevil. His parents lived in Temple, just across the Red River from Texas, where his father, Franklin, worked as a tenant farmer. When he was two, the family moved to the other side of the river, near the town of Burkburnett, on a bluff overlooking the water. During heavy rains when the river flooded, teams were forced to cross at the low point just below the family's four-room shack. Later, Tollett described the thrill of "watching wagons slipping and sliding down the north clay bank, and teams pulling hard and fast across the wet sand, coming to Texas."

By the time he was ten, two younger brothers had arrived, and his parents often struggled. The family now lived in town, and to help his father, Tollett went to work. Each night after supper, he walked to the rail depot to meet the southbound train, purchased sixty newspapers, and sold them within the hour, "pocketing a net profit of $1.20."

Burkburnett's oil boom began late the following year, 1918, just as the one in Ranger ran its course. Twenty thousand people streamed into town, where enormous derricks sprouted from every open space. Whenever it rained, oil and sewage ran through the city's streets near the Tollett home, mixing with the bottomland mud churned by the wagon teams. Behind the house, Raymond had planted a small orchard of fruit trees. One day he arrived home from school to find rig builders hacking it down with axes. He also discovered his father had sold the family's cow, calf, and two ponies — all to allow room for a drilling rig that rose fifty feet from his bedroom window. Water pumped from a well on their back porch soon became undrinkable.

The family moved to nearby Wichita Falls, where the madness was just as intense. When Tollett turned eleven, his father asked him to quit school and find a full-time job to help the family. He was big for his age and easily found work delivering telegraphs for Western Union.

His telegraph runs acquainted him with wildcatters and promoters who'd become overnight millionaires, and the fleeting nature of the enterprise meant he also watched a few of them lose it just as fast.

One bitterly cold evening in the winter of 1919, he saw a man downtown whom he recognized as G. Clint Wood, who'd made his fortune in oil fields from Beaumont to Electra. But the word around town was that Wood had suffered a sour turn of luck and was broke. As Tollett watched, a beggar approached Wood and asked him for money, a World War I vet whose thin clothes left him shivering in the weather. Wood reached into his pocket and gave him a few coins. As the man turned to walk away, Wood called to him again, pulled off his expensive overcoat, and draped it over the man's shoulders. Although G. Clint Wood would go on to reclaim his millions and then some, his selfless gesture that night made a deep and lasting impression on Tollett.

The year he turned fifteen, Tollett took his first job in the oil patch as a roustabout, and at some point struck upon a lucrative side game. There in Wichita Falls, he gathered up cabaret girls and drove them out to the derricks, where crowds of whooping men paid money to watch them dance.

A Sunday school teacher urged him to return to school. In 1925, he graduated as the valedictorian of his class but couldn't afford a coat. Likewise, after winning an appointment to the U.S. Naval Academy, Tollett hadn't the money to buy his mandatory uniforms, and was forced to turn down the opportunity. He took a job as a mule skinner instead. One day on a delivery, he looked up and saw the building for Cline's Commercial College. Tollett walked inside, applied for a scholarship, and got it. By the next year, he was clerking for local oil companies.

In the early 1930s, Tollett married a local nurse named Leta Marie and the two welcomed a daughter, Kay. Now with a family to support, Tollett began studying accounting through correspondence courses, until one of his bosses urged him to pursue law. With the Depression slowing the oil business, he began borrowing law books from attorneys in town and reading them at home at the kitchen table.

Tollett was blessed with remarkable intelligence and also possessed a photographic memory. His only formal law training was three months of twice-a-week night classes in Dallas, two hours away. When he sat for the Texas bar exam in February 1932, during an era when only 5 percent of candidates passed, he aced it on the first attempt. The following year, he resumed his mail-order accounting lessons and passed the CPA exam.

To earn extra money, Tollett began tutoring people for civil service exams. When one of his students expressed interest in joining the FBI, Tollett took the entrance test on a lark. Several weeks later, he was shocked to learn that he'd passed.

The early 1930s were the bloody heyday of the Depression-era gangster, when the Bureau's "flying squads" famously pursued its list of public enemies. In 1934 alone, agents would take down Bonnie and Clyde,

John Dillinger, Baby Face Nelson, and the Mealers' erstwhile neighbor in Georgia, Pretty Boy Floyd — all in hails of gunfire.

After a recommendation from a congressman from northern Texas, the Bureau offered him a job as a junior agent in August 1934. After doing accounting work in Jacksonville, New York, and New Orleans, Tollett was given special agent duties just in time to help crack one of the biggest cases in the country.

Alvin "Creepy" Karpis ran with the Barker gang, whose list of crimes included killing a sheriff in Missouri in 1931 before orchestrating two high-profile kidnappings in St. Paul, Minnesota. Police and FBI had managed to capture or kill every member of the gang except Karpis, who remained the last person on the Bureau's list of public enemies.

In April 1936, agents in New Orleans discovered Karpis hiding in an apartment of Jefferson Davis Parkway, and for two days, Tollett worked surveillance across the street while colleagues prepared the raid. On May 1, around 5 p.m., agents nabbed him as he walked out to his car.

Tollett's work on the Karpis case, while not glamorous, earned him the top job in New Orleans. But after three months he was demoted after female stenographers in the office claimed he was verbally abusive. The Bureau transferred him to Los Angeles, where he worked briefly on a missing persons case. Disillusioned and unhappy, Tollett soon resigned and took his family back to Texas.

An old friend had arranged an accounting position at the Wrightsman Oil Company in Fort Worth, where he was working when he received a call from Henry Zweifel, who was Josh Cosden's lawyer. Zweifel had taken over as president of the Big Spring refinery in its last receivership. After two years at the helm of the company, he wanted to return to private practice and was looking for a replacement. Tollett joined as secretary-treasurer in July 1939 and began vetting potential candidates, but within a year, he accepted the job instead.

After traveling to El Paso to meet Cosden's body in the symbolic passage of power, Tollett returned to Big Spring to resume his resurrection of the failing refinery. The company's stock was valued at less than a dollar per share on the market. No suppliers would extend any credit. Worst of all, Tollett thought, was the morale among the workers. Upon his arrival, his first task had been to assure his employees.

All throughout the Depression, workers had kept the plant operating despite having no proper tools or supplies. The equipment was outdated and unsafe, and offices and control rooms had no heat in winter. When Tollett gathered his employees outside a warehouse, many thought they were about to be fired, but instead the tall man in the sharp suit assured them the company would survive. Not only that, it would grow. Go home and tell your wives that your job is safe, he told them. The men walked away amazed, not just by Tollett's confidence, but by the fact that he'd known every one of their names without ever having met them.

After moving the headquarters to Big Spring, Tollett narrowed its market to Texas and bordering states, rather than the entire Midwest. The war in Europe soon led to increased fuel demand, and the refinery became the first in the nation to establish a "pipeline on wheels," sending whole train-loads of gasoline and crude to the East Coast for shipment overseas. A deal with the T&P to pay pipeline rates rather than rail rates ensured that many thousands of trainloads followed. By 1942, the company earned its first profit, and two years later, it paid its first dividend.

But in order to truly succeed, Tollett had to reimagine the whole enterprise. He set about expanding Cosden Petroleum Corporation into a "custom refinery." He added a modern catalytic cracking unit that manufactured every grade of gasoline and diesel while tailoring formulas for individual companies. Soon all the majors, such as Shell and Phillips, were mixing their fuels at Cosden for their southwestern markets, rather than trucking it from refineries along the Gulf. The company then opened its own filling stations in towns across Texas and New Mexico, its billboards featuring a friendly traffic cop mascot. And once Tollett discovered that the oil under Big Spring was fabulous for distilling asphalt, he installed a new vacuum still that churned out forty-five premium grades, which soon covered a third of all Texas roads.

By the end of the 1940s, Tollett had primed the company and the town, which had weathered the ups and downs of two oil booms, for its most productive decade. But even as Big Spring and the refinery thrived, both would have to contend with his demons.

4

The skies dry up, a seven-year plague . . .
Bob trades in oil for sand . . . Homer
battles the Lord and Little Opal, finds
deliverance in the road . . .

The 1950s arrived with a drumbeat of war
in Korea. At the air base on the west side of
town, the federal government began training pilots. Thousands of new recruits poured
into Big Spring and soon the rumble of T-33
jets filled the skies. To keep the warplanes
and vehicles moving overseas, the Cosden
refinery dispatched an endless stream of fuel
and chemicals into its pipeline and rail cars,
while drillers harnessed two of the biggest
oil discoveries to date. East of town was the
Kelly-Snyder field, part of a colossal limestone reef called the Horseshoe Atoll that
had once lined the ancient Permian sea.
Geologists were saying it could hold over a
billion barrels of crude. To the west, toward

Midland, crews were spudding into the tongue-shaped pools of the Spraberry Trend, believed to contain ten million more.

My grandfather Bob found steady work with small, independent outfits in the Permian such as Rowan Oil and Norwood Drilling Company. No longer a lowly roustabout, he now worked as a floor hand threading the lengths of drill pipe that pushed thousands of feet into the earth. Some of Norwood's wells tunneled more than two miles down.

Between the oil wells, refinery, and air base, tens of thousands of people were moving to Big Spring, whose population by the end of the fifties swelled to an all-time high of thirty thousand. As in many small towns in America before the arrival of interstates and shopping malls, its downtown remained a place where people could open businesses and make a living. Weekends still brought a swarm of shoppers, people watchers, and traffic so thick that drivers circled the blocks in search of a place to park. The atmosphere was often so festive it carried the feel of a circus. In fact, one afternoon while Homer was selling newspapers along Third Street, a crowd gathered outside the Hotel Settles. Looking up, he saw Benny and Betty Fox, professional acrobats, dancing the Charleston on a platform suspended from the fourteenth floor, nary a net to save them. Their act was called "The Dance of Death."

As the town's borders expanded, both physically and figuratively, the ones my family occupied remained stubbornly narrow. They believed in what Jesus said in John 17 about the world hating those who obey God's word, and therefore they were *not of it*. To them, this earthly life was a mere vapor before the promised solidity of paradise.

As Zelda, Homer, and the cousins grew older, the laws of the church, designed to shield them from the outside world, only grew tighter around their lives. They stopped playing church. Puberty dragged them out of innocence and made them fair game in the tabernacle. The sermons condemning backsliders and painted Jezebels began to feel more personal, as if the eyes of God and the preacher were in every passing headlight. And guilt, which had once pressed only lightly on their souls, now rooted itself like a tree, rattling its branches at the slightest sign of pleasure.

In high school, the rules of the church ostracized them. Since most school functions were forbidden — dances, concerts, and football games where drunks and gamblers preyed — the cousins socialized only with one another and other members of their denomination.

Attempts to assimilate were struck down, like the time Zelda and Doris signed up for a talent show at the Municipal Auditorium. Without telling their parents, they dressed like hoboes, blacked their teeth, and sang the popular song "Side by Side" with the new preacher's daughter, Beverly Eldrige, playing piano. They easily took first prize, which was a trip to Houston to appear on television. Opal and Agnes, although surprised by the news, saw their daughters' talents as gifts from the Lord and gave their consent. But when Pastor Eldrige caught wind of this contest, he forced the girls to forfeit.

Not everyone felt oppressed by the church's restrictions. Despite its many rules, it remained the center of life for so many people, including Zelda and Doris, who would grow up cherishing the memories from those years, like the out-of-town camp meetings with close friends from their congregation, or the impromptu donut-eating contests they held down at the coffee shop. Zelda was never ashamed of her faith and its demands — especially not at gym class, where she and Norma Lou had to wear frumpy culottes instead of shorts. And the girls who giggled in the hallways and whispered "Holy Rollers" never shook her convictions. "I treasured God's love in my heart and knew I always wanted to serve Him," she later wrote. "I have never wanted to turn from following Him."

But Homer, like my father in later years, felt the church's laws like a pair of hands clasped around his throat. He resented not being allowed to date outside the Assembly of God. He stewed every Friday night when school friends drove to Midland and Odessa, while he went to church for the fourth time in a week.

"Mother, can't I stay home? I have to study," he said one Wednesday night, but Little Opal was firm.

"You have to go."

"But I have a test tomorrow, and I don't know the material."

"The Lord will show you," she said.

Once at church, the preacher's clothesline sermons skewering short hair and war paint seemed pedestrian and unrelated to Scripture. And the scare tactics trotted out each night seemed like canned fiction — like the story about the man who felt compelled to hit the altar but didn't, only to die in a car wreck on his way home — "He could've entered heaven, friends, but he chose hell instead." Each week, Homer vowed he'd never fall for it again.

Yet every time the preacher issued the call for salvation, Homer's protest began to crack. Always, at that critical moment, his resistance turned to fluff and he was consumed with shame. He stood up with the rest of his cousins and made his way to the front.

But the doubt and anger always crept back in, usually by church on Wednesday.

He told his mother, "I've confessed my sins so many times, I imagine the Lord is sick of hearing from me."

"Then go on to hell," she said.

His stubbornness toward the Lord enraged her.

"I wish the Mississippi River'd rise up between me and that church house," he shouted. "With no bridges!"

"You'll be lucky if God don't strike you dead."

The battles with Little Opal over the state of his soul grew epic. By the time Homer was fifteen, he'd had enough. One Sunday morning he made a final stand and refused to go to church, and before Little Opal could grab her belt, he was out the door.

He jumped into her Buick and sped downtown, where he rented a cheap room at the Duncan Hotel.

"Gimme the whole week," he told the clerk, and plunked down seven dollars. For the next hour he sat on the bed listening to a ball game on the radio and stared out the window, until Herman, driving the cattle truck, spotted the Buick outside and persuaded him to come home.

As they pulled out of downtown, his father assured him, "You don't have to go to church anymore. I'll talk to your mother."

But Homer knew better. "She won't have it," he said, and he was right. Little Opal let him backslide a few more weeks, then tightened the clamps again. Homer was planning a more permanent escape when, most ironically, he was saved by the Lord's own wrath.

In 1950, the rain just quit.

A drought arrived, and like the one in 1917, it came during a time of record petroleum production, as if the Lord forbade the land to give all its bounty at once. But this drought was epic and unprecedented. It first took root in Texas, then spread west into New Mexico. Within a year it had made its way to California before

222

gripping Oklahoma, Kansas, and the rest of the Central Plains — a swath of destruction that covered ten states and persisted for much of the decade. But Texas suffered the greatest, and the longest — seven years without rain. The drought surpassed 1886, 1916, 1917, 1919, and all the Dust Bowl years combined, described by officials as the most devastating in six hundred years. It bored its grooves into the land and left it, along with those who depended on the land, forever changed. For the families who survived its indignity, the 1950s do not bring about wistful nostalgia. "The time it never rained," as writer Elmer Kelton so eloquently described, was a time when "many a boy would become a man before the land was green again."

In Big Spring, cotton was the first to go. The crop of 1949 had been phenomenal, a bumper yield. The following year, despite the rains drying up, the harvest remained fair, thanks to moisture still left in the ground. But that was all.

Old men in Big Spring still talk about how it happened, how the farmers seeded their fields in May after a short planting rain — the last — and how they watched the gentle shoots break ground, only to awake in the

223

middle of the night to the sound of their fields going airborne. They'll explain how the farmers would chart a lone gray cloud across the sky all day as if it were filled with money. "That one smells like rain," they'd say to their wives, only to watch it vanish like a soap bubble against the sun. Or how farmers gathered by the dozen under brush arbors, put their hands together, and prayed for some relief; how they listened to the men from up north talk about seeding the clouds, which never worked; and how most years the gentle shoots were left defenseless against the hundred-degree winds that chewed them to stumps.

And say you got a rain shower after the seedlings sprouted, as happened in 1953, which allowed the cotton to actually turn out buds. Well, that just brought the aphids and grasshoppers, the false wireworms and thrips, and especially the godforsaken weevils. Sometimes, you didn't even know you had weevils until you walked out after breakfast and found half your field beheaded on the dirt.

The government brought in soil experts and the men listened to them. Grow cover crops to keep the soil down, they said, to keep it from "migrating" the way it had in the past. Some farmers were already planting milo for cover, because milo was feed crop and akin to gold in the drought, and with milo, you might actually break even. The experts were full of other ideas — sunflowers, black-eyed peas, hybrid sorghum, weeping love grass, and guar, just to name a few. Several men planted whole fields in guar only to go bankrupt when the the buyers didn't show up come harvest. But most were stubborn and stuck with cotton, and they went down with it, too.

They traded their two-row tractors for four-row machines that had the power to dig deeper, to bust the hardpan and allow in some moisture — if only it would rain. But it didn't rain, and now the fields were dug deeper and gave the winds more soil to suck up. Hadn't they learned anything from the Dirty Thirties? Well, here were the Filthy Fifties, where at noon the streetlights were coming on again. The sand was so thick it seeped through the sealed windows of the hospital operating room and coated the instruments, even penetrated the vaults at State National Bank, despite a foot of concrete and steel.

"Does the wind always blow thisaway?" a newcomer asked when he got to town.

"No," the answer came, "sometimes it blows thataway."

But the farmers kept hitting their fields with their four-row tractors, "deep breaking," they called it. By 1955, the blades were cutting nearly two feet down, but all they turned up was sand. The topsoil was gone — *two feet* of precious soil, now replaced by a desert. By 1957, the damage was complete. Over 70 percent of Big Spring's farms were gone. As one man told the *Herald,* "I don't know how to farm this land anymore. Seems no matter what we do, these fields blow a little more every year."

For the cattlemen, the drought was its own nightmare. First the range grass began to die, and once the grass was gone there was nothing. The established ranches, and especially ones with oil money, leased grazing land up in Kansas and northern Colorado to preserve their herd. Others paid out the nose for hay and alfalfa, which they bought on credit from the bank, and you'd hear them moaning in the coffee shop how they'd pitched their net worth out the back of a pickup.

The ones with plentiful water got cocky and kept their animals too long, until they stripped the grass roots and left the land bare, ready to blow. It could take five years for good range grass to return, and in the meantime, mesquite crept in and sucked up the groundwater. Weeds also found a foothold and filled the empty spaces: turpentine weed and broomweed, along with locoweed, which contained a chemical called swainsonine that would poison a cow's brain and send it thrashing in the fields. The corymbs of snakeweed, which thrived in drought, caused miscarriages and stillborn calves.

As a last measure, some ranchers poured cheap molasses on dead grass and safer weeds to stimulate the cattle's appetite. They took butane torches to prickly pears to burn off thorns so the cows could eat. But a cow that takes to prickly pear will often become addicted, eating only cacti even when standing in grass up to its knees. Eventually the thorns ripped holes in their mouths and invited screwworms, which, as one rancher observed, would eat their heads clean off if left unchecked. Cattle that grew weak became exposed to disease and opportunists. A rancher in Tom Green County once watched a pack of javelinas descend on a heifer that had lain down, eating her alive as she bawled in the dust.

Most cattlemen sold out before things got that bad. Ranches unloaded their livestock by the hundreds of thousands, trying to salvage at least something. By 1955, you could drive fifteen miles through neighboring Bordon County, where range made up nine-tenths of the land, and not see a single cow. In Martin County, on the other side, dairy farmers teetered on bankruptcy. Sniffing blood, competitors from Minnesota sent down convoys of insulated milk trucks to steal their business, using two drivers so they wouldn't have to stop. In three years, half the dairy outfits had vanished.

Cowhands who'd worked on ranches thirty and forty years had to be let go. Facing the reality of paying rent for the first time, many joined the legion of busted farmers working for wages in the oil patch. Each morning at dawn, a group of them gathered along the Garden City highway to lay pipe south of Midland — fifty men squeezed into a windowless doghouse, winched to the back of a truck and hauled to the gas fields. For $1.20 an hour, they coated pipe with creosote and dropped it into a trench, but the dirt was so dry that the holes filled right back up. Gang pushers paced behind them like prison guards on horseback, and whoever complained didn't get asked back, for thirty other men were waiting for each job.

On Saturday nights, the out-of-work cowboys gathered at the Stampede out on the Syder Highway, where Hoyle Nix and His West Texas Cowboys played western swing. They danced the Paul Jones and waltzed with pretty women, drank their whiskey in the parking lot and picked fights with the Yankee airmen, who didn't seem to know or care that a way of life was vanishing before their eyes.

By the end of the drought, barely a third of the county's original herd remained. Ranchers moved north to less cursed ground, or else sold out and moved to town. All told, by the end of the decade, Texas lost nearly a hundred thousand farms and ranches, and 30 percent of its rural population.

Where vast herds of cattle had once grazed on the West Texas range, the pastures now yawned empty, and where cotton had blanketed the fields like snow in autumn, the wayward soil now piled against fence rows, so high you could walk clean over the tops.

As much of a tragedy as this was for everyone, it presented great opportunity for my family.

When the soil began to blow, my grandfather Bob was wrenching pipe on a derrick floor along the lucrative Spraberry Trend. But by 1952, the Spraberry's promised bonanza was proving elusive. The geology was tricky; the underground pressure began to taper off, and wells that had started off strong slowed to a trickle. When that happened, rigs were laid down and men lost work. By that time, the bounty from the Korean War had also run its course. Now the war needed pipe and other supplies, which slowed production even more. Oil prices froze.

Then, in April 1953, the Texas Railroad Commission, concerned over the excessive flaring of natural gas that the oil wells produced, ordered the Spraberry shut down completely. The small independents Bob worked for lost millions. In fact, one of them — Rowan Oil Company — sued the Railroad Commission, and the Texas Supreme Court sided in the company's favor. The commission reopened the field several months later but limited production to ten days a month. The smaller outfits couldn't survive.

The major companies responded by simply going overseas. At the time, the oil fields of Saudi Arabia, which began producing during World War II, had a virtually endless supply of light crude that was both refinery-friendly and cheap — costing only twenty cents a barrel to produce, compared with seventy cents in Texas, and without all the government interference. Pretty soon, cheap foreign imports flooded the American market, and when that happened, Bob threw up his hands and said to hell with it.

"From now on," he told Opal, "I work for my own self."

He'd been thinking a lot about Davey Jones, Abe's brother, the gambler who'd taught him and Bud how to play dice. Davey had quit farming and now owned a fleet of dump trucks, which Bob helped drive on occasion. In fact, the whole Jones gang drove dump trucks. When the twins, Eldon and Weldon, grew ornery and wouldn't stay in school, Davey bought them trucks. Same for his wife Maudie, his son Wayne, even his daughters, Katie and Maxine. If the girls had a date, they went in a dump truck, and it was always a Ford.

Meanwhile, out in Odessa, Abe and Fannie's boys — Earl and Raymond and Troy — had started their own company with a front-end loader and a used bobtail truck. In just a few years, Jones Brothers Dirt and Paving Contractors would build many of the new interstates across Texas.

Whenever there was no work in the oil patch, Bob hooked up with Davey's crew for jobs hauling caliche, mainly to build drill pads and lease roads. They answered a work call like a gang of Huns, the whole convoy growling down the highway in a swirling white cloud. Seeing all those trucks idling at the gate, Maudie and Maxine behind the wheel, shirtsleeves rolled, the foreman always gave them the job.

But Davey remained a compulsive gambler. He'd still disappear for ten days at a time, losing work and his family's savings in a string of poker games. Landlords evicted them, and the bank seized their things. When the bank came for the family car, Davey one-upped the repo man by tossing a Sears catalog inside and setting it ablaze, leaving nothing but the steel wheels. Once, years earlier, when Davey had gambled away their cotton money and the cupboards went bare, he took a calf — which the kids had trained as a pet, teaching it to jump on a chair and sit like a dog — and had it slaughtered. When the children returned from school they were horrified and refused to eat it.

Yet the kids loved their father, despite his disease. For one, he knew how to have fun: when the army drafted the twins, they dressed Davey in fatigues and snuck him onto the base, where he won hundreds of dollars shooting dice. And his big heart was boundless: no week passed without Davey bringing home a stray kid he found on the streets and asking Maudie to feed him, despite their having little to eat themselves.

His behavior drove his wife to fits of rage. She pulled a gun on him one afternoon and Davey responded by drawing a pistol of his own. It took the sheriff locking them both up overnight just to keep the family together.

Opal couldn't stand the mention of Davey Jones's name in her house. But Bob liked Davey, and better, he liked his trucks and the freedom they allowed.

"The Joneses answer to nobody," he'd say.

Bob decided to get a truck of his own. He'd noticed a better opportunity than building highways or hauling caliche — but it wasn't glamorous or even vaguely romantic, like the oil patch. In fact, it might even be harder, if that was possible.

It was dirt, but not just any kind.

The millions of tons of topsoil that blew off the farmers' fields now settled in great mounds across the county: along the shoulders of the highway, in ditches and creekbeds, anywhere it could drift. The worst was on ranches, where it gathered against the tumbleweeds that stuck to fence lines, piling so high it buried the barbed wire. Cattle were walking right over it, onto highways and neighboring property. Ranchers couldn't clear it away fast enough.

And what would Bob do with that dirt? Big Spring was growing by the day, thanks to the air base and refinery and oil fields that needed labor. In order to house all the new arrivals, homes were being built with tidy yards and landscaping that required topsoil. Not only that, but all over town sprinklers churned in vain against the drought to keep the grass alive. At the golf course and country club and in neighborhoods far and wide, the turf needed cover to hold down the moisture, and Bob had the best possible kind, already chock-full of nutrients — soil that had borne a million bales of cotton, in fact. It just happened to be momentarily displaced.

He took his idea down to State National Bank and secured a loan for a brand-new truck, a Ford F-500, powder blue, with a manual-lift bed that held five cubic yards of the stuff. Then he phoned the *Herald* and placed an ad, using the colloquial term for the sandy West Texas loam:

YARD DIRT
Red Cat-Claw
or

Sure enough, the ad found its audience, and people began calling. Bob wasted no time getting to work, for there was money to be made in this drought.

While Bob took advantage of thc blowing soil, Opal's brother Herman cashed in on the dying range grass. Since the late thirties, he'd established himself as a trusted cattle hauler and was well liked by the local ranchers. So when the drought turned their pastures to dust and forced them to sell their herds, Herman had eight trucks and trailers waiting, each capable of holding up to thirty cows.

In 1951, he and his brother Tooter hauled forty-two loads off the Tom Good ranch, north of town, and took them to Oklahoma, where the bluestem was still green. Homer was twelve at the time, and his father brought him along to keep him out of trouble. Homer worked as a swamper, sweeping manure from the trailers and helping offload the animals. His father never spent money on motels, so at night they pulled the trucks over and spread cots beneath the trailers to sleep. After two weeks, one of the biggest ranches in West Texas was empty.

After clearing Tom Good's pastures, they moved to Lorin McDowell's place. Back when McDowell's father had settled their sixty sections in 1883, the grass was belly high on his horses. Like most early stockmen, he ran as many as two hundred head over a single section, relishing the bounty. But decades of overgrazing had whittled down the range. And now with the drought, the land was growing patchy, the grass bitten down to the roots. McDowell set aside twenty-five purebred Brahman cows, and the rest — over a thousand head — he loaded onto Herman's trucks for the feedlot in Fort Worth. From there, they traveled by train to the Sandhills of Nebraska, the Flint Hills of Kansas, and the Black Hills of South Dakota, where McDowell had taken his father's stock during the drought of 1917, the experience that had left the furrow on his brain.

Clint Murchison, one of the richest oilmen in Texas, built a feedlot in Lubbock that held ten thousand head, which he fattened on cottonseed cake and alfalfa that he grew up north. For three years Herman ran Murchison's cattle from Lubbock to the slaughterhouse in Fort Worth. If they had to drive through the night, Homer took the wheel on the long, straight stretches so his father could sleep.

"Just wake me when it's time to shift," he told him.

But in 1954 Murchison opened his own slaughterhouse in Lubbock and the work vanished overnight. By then ranchers and speculators from out of state were capitalizing on the drought. They bought cheap cattle at the local auctions and shipped them to where food and water were still plentiful and prices were higher. One of the biggest markets to open up was in Southern California, where everything moved through the Union Stockyards in Los Angeles.

When Homer turned sixteen, in June 1955, he drove his mother's Buick downtown and got his commercial license, which allowed him to start driving — legally — for his father. Only a short time had passed since he'd run away from home and sat in the Duncan Hotel, listening to the radio. Summer break had just begun, and the drought was stretching into its fifth scorching year, with the worst yet to come. The morning after Homer's birthday, he awoke to find a truck idling out front, its trailer loaded with twenty-five whiteface Hereford bulls. His father stood at the door.

"You're taking these to L.A.," he told his son, and handed him a map. And for the next two years, whenever school wasn't in session, it wasn't the Mississippi River that rose up between Homer and the church house, but eleven hundred miles of wild American road, a blacktop Chisholm Trail that carried him toward salvation and delivered him from himself.

The job was straightforward: get to the stockyard before any animals died from the heat, see that they were fed and watered, and collect payment from the commission house — which amounted to about five hundred dollars. On the way back, he was to pick up a load of hay in Yuma to sell in Big Spring, and he was to stop someplace to rest.

There was only one way to get to Los Angeles, which was straight west on Highway 80, so Homer pointed the wheels in that direction and was gone. The truck was a 1952 Ford F-8, its cab and trailer painted fire-engine red. The forty-foot trailer was made with oak sideboards bolted to metal posts, with two dividers to separate the livestock and keep them from clustering on the hills.

With a full load, the whole enterprise

labored at forty miles per hour, just on the straightaways, plus his father had insisted that Homer stop every two hours to check the tires and oil, and to make sure that none of the livestock had laid down. An animal that lay down could be trampled to death in a matter of seconds, and to get one back on its feet often required offloading the entire trailer.

Although he'd just received his commercial license, it wasn't the first time he'd driven alone. With so much stock to move, and ranchers wanting it done quickly to mitigate loss, his father had already allowed him to take several loads. Mostly they were short hauls to sale barns in Fort Worth and Amarillo, which he traveled to by back roads to avoid the police. And once there, no one bothered to ask his age.

At sixteen, Homer weighed 120 pounds drenched in saltwater, but he carried himself like someone older and bigger, and it helped that he knew his job. Only once had his livestock ever arrived dead, and that was the previous winter when he'd hit a blizzard on the way to Amarillo. Heavy snow had blocked the highway and paralyzed traffic, forcing Homer to pull over and wait for a grader to clear the road. It took half the night to get moving, and when he finally reached the stockyard, he found a calf frozen to the trailer floor.

But when it came to handling his vehicle, Homer was one of his father's best drivers. He'd hung around the shop long enough to learn how to repair most mechanical problems, as long as it didn't require pulling apart the engine. Plus he could back the forty-foot trailer into a chute this wide and hit it flush the first time. When you can park a truck like that, no one will tell you anything.

He pulled over at a truck stop just west of Odessa and peered through the slats, looking for lay-downs. In his hand was a homemade Hot-Shot, built with a lantern battery, doorbell, and two beer openers wired to the end of a walking cane. Any bull touched with one usually bawled and got up. But from what Homer could tell, they were all still standing.

In the other hand he carried a ball-peen hammer, which he bounced off the tires and listened for an echo — a sign they needed air. As he walked the perimeter of the truck, one of the bulls let forth a steaming piss that splattered through the slats and soaked his shirt — a smell that would linger on him the whole way west.

He reached Van Horn by late afternoon, the horizon a jagged line of mesas as he ran the lip of the Chihuahua Desert. Outside it was blazing, but the Ford had no air conditioning, not even a firewall to protect his feet from the engine — an old V-8 that threw off heat like a potbellied stove. Even with the windows down, the temperature inside the cab reached 115 degrees. The water in his thermos tasted like copper, yet he couldn't seem to drink enough. The only relief came from a small floorboard vent, which, if positioned right, blew a stream of hot air up his pant leg and dried the sweat.

He approached El Paso near dusk, the whole sky shot through with orange and red. After that, it was desert all the way to Banning, California, and he intended to drive all night while the weather was cool. By now the air had turned crisp, and with the change in temperature, he became aware of the tingling in his skin and the way his jaw was clinched tight.

Before leaving that morning, his father had given him a handful of Benzedrine pills from a bottle he kept in the medicine cabinet. Homer had swallowed five back in Sierra Blanca to get him through the desert, and now he could feel them kicking in. They were less potent than the pills used by the other drivers, who bought them by the fruit jar across the border in Juarez. Even still, five bennies were enough to turn on your jets.

He reached into his front pocket and pulled out a rumpled pack of John Ruskin cigars and lit one. He then clicked on the radio and turned the knob until a familiar voice rang through the speakers.

"This is your old neighbor and friend Bill Garrett doing the talking, down by the silvery Rio Grande."

It was station XERF, located across the border in Acuña, the same one that Bud and Frances used to listen to back in Forsan. It used a 250,000-watt transmitter to blast everything from "Honky Tonk Blues" to Charles Jessup's hellfire to the far corners of the earth — as far as Korea, some said. By a stroke of luck, Homer had caught Garrett's country and western hour, and the cab soon filled with the sound of Ernest Tubb singing "Walking the Floor Over You." He puffed his cigar and watched a jackrabbit cross his beams, then vanish behind the black veil of desert. He placed a steady foot on the gas.

It was dawn when he crested a hill outside

Benson, Arizona, and pulled to a halt. On the side of the highway sat a small, makeshift truck stop where his father had told him to rest. The owner, a retired pharmacist, motioned the truck toward a series of corrals where a tall Indian offloaded the bulls and gave them hay and water. Nearby sat a single-room bunkhouse that had once been a chicken coop. The Indian had covered its walls with thick adobe, which kept the room remarkably cool, despite the surrounding desert.

"How long you plan on being here?" the pharmacist asked.

"I reckon till dark," Homer said.

Including stops to check the cattle and refuel, he'd been driving for twenty-four hours. The pills had worn off and his vision had taken on a shimmer, like trying to navigate through an aquarium. He pulled off his boots and lay down without even drawing the sheets. When he awoke, the sun was sinking behind the bony hills.

He entered Los Angeles late the following day, roughly twenty hours after leaving Benson. The second leg of desert had largely been mountains, which the truck had labored to climb. It took half an hour in low gear just to get up Telegraph Pass, east of Yuma, where the narrow highway scaled a craggy divide. The temperature was over one hundred degrees, and halfway up the engine overheated, forcing him to pull over and prop an oil can between the hood and fender to let in some air. At the top, he made sure that none of the cattle had collapsed on the hill. Then he swallowed two more pills, kicked the Ford out of gear, and let it fly down the mountain.

By the time he pulled into the L.A. stockyards, it was nearly dark. Two young cowhands waiting near the chutes helped unload the cattle and arranged their boarding. In the next few days, after the bulls had put on weight, an agent from the feedlot or packinghouse would walk over to the gate, eyeball the herd, and throw out a number for the whole lot. That's how they did it in California. With the drought hammering prices back in Texas, it meant the buyer who'd purchased them in Big Spring for thirteen dollars a head would wind up making nineteen or twenty, even after subtracting the freight.

But the buyers weren't the only opportunists in the drought. Our family had certainly benefited from it, although you'd never catch Herman or Little Opal describe it that way. In the past four years, Herman had added four drivers and six new trucks.

They'd moved into a bigger house out by the auction barn, and Little Opal drove around in a brand-new Buick. Even Bob was earning good money selling blow soil, enough to pay off his truck and shop for another. Between cattle hauling and dirt dobbing, they'd found their coin in the wreckage.

The miseries of the older generation were receding into lore, becoming like the words in the country song, "the good old days when times were bad." Homer's cousin Bobby — Bob and Opal's youngest boy, my father, now two years old — would grow up never knowing what his parents had endured, and he'd be spoiled rotten as a result. Homer suspected as much because people said the same about him, although they'd never seen the way his mother tore after him with a belt.

Homer couldn't collect his payment until the office opened the next morning. Knowing how his father felt about spending money on motels, he chose to sleep with the truck. Herman was such a miser that when Homer called collect to say he'd made it to L.A., his father refused the charges and hung up the phone. The operator had told him all he needed to hear.

Homer unbolted the nose of the trailer

and flipped it back so it made a shelf, pulled an old army blanket from behind the seat, and climbed up to his bed.

The sound of the stockyard coming to life awoke him at first light. He sat up stiffly, having hardly slept, and listened for a moment to the chorus of diesel engines and lowing. He walked through the catacomb of sheep and cattle pens until he found the office, where inside sat an old man of about eighty.

"Say, was that you I saw sleeping on your trailer?" the man asked.

"Yessir, it was."

"Son, do you realize how dangerous it is around here? Someone's liable to come and do mischief to you. From now on, have your daddy draw a bank draft on me and we'll pay it once it gets here. That way you won't have to wait."

Homer thanked the man and stuffed the freight check into his shirt pocket, then snapped it closed.

Back at the truck, he left the trailer sitting in the parking lot, then drove to the Pacific Electric train station. He bought a trolley ticket to Long Beach, where he spent the morning at the amusement park.

The Pike, as it was called, had the nation's biggest roller coaster, the Cyclone Racer, which whipped out over the beach with such force that Homer swore he'd end up in the salt. He rode the Double Wheel and then the Rotor, a spinning thing that felt like being rolled down a hill in a barrel. And when his stomach couldn't handle the dips and twirls anymore, he walked to the midway and took in the crowds.

The Pike boardwalk was a barking, strutting panorama of every kind of fun that Little Opal and the church railed against. A fat man chewing on a cigar tried to guess his weight. A freak tent advertised a bearded lady and a girl born with a monkey's face. Sailors emerged from smoke-filled beer halls and crowded the doors of tattoo parlors, displaying fresh ink. In every direction, it seemed, were packs of tanned barefoot girls, their swimsuits covered in sand from swimming in the surf. The smell of fried onions drew him to a burger stand, where he ate lunch, then bought a plank of saltwater taffy to take back on the road.

He'd neglected to sweep out the trailer after offloading the cattle, and once back on the city streets, the truck spewed a cloud of pulverized manure that left a wake of ill will. With an empty load he made Yuma in eight hours. He ate a chicken-fried steak at a truck stop and rented a bunk for a dollar, then slept so long the owner shook him awake, fearing he was dead.

Outside of town, he bought from a farmer a load of prairie hay, which his father would sell at a markup. Fourteen hours later, after stopping only to use the restroom, he pulled into the company lot in Big Spring and phoned his father to pick him up. When he awoke the next morning, he found the truck parked out front with another full trailer, its cargo bawling and ready to go.

That summer, as Homer helped empty the West Texas range, something happened: all that anger knotted up inside him, toward his mother and the church, began to slowly unwind. The country was less crowded back then, and all that emptiness out the windshield had the power to pry open the mind. Jack Kerouac once described his book *On the Road* as a story of "roaming the country in search of God." And while Homer didn't know it at the time, that's what he was doing, too, searching for his own vision of God, which could only be found outside of Big Spring and the church where he was raised.

Each trip west became its own adventure, bringing new experiences to file away. Once, outside of Tuscon, a motorcycle gang harassed him for twenty miles and refused to let him pass. Homer finally backed off the truck, then put the pedal to the floor. By the time the bikers saw him coming, he'd swung the trailer straight into their pack and pushed them against a fence.

He picked up hitchhikers to pass the time, mostly sailors headed home. And once, in the Imperial Valley, he watched a calf kick a man square in the mouth and knock out most of his teeth. As business increased, the desert crossings grew more frequent and required more pills to fuel them. Sometimes he brought a friend from church, Charles Miller, to share cigars and keep him alert, especially in the dark hours when the medicine twisted his brain. During a feverish push through Arizona, Charles made the mistake of dozing off, awaking abruptly to Homer, his eyes beaming like silver dollars, smacking his head.

"Leon, wake up! Wake up, Leon!"

"I'm not Leon. Leon's your cousin. I'm Charles!"

"Sure you are, Leon. *Sure you are!*"

Out in the western wilderness, the road was a splendor of temptation and treasure. Once, after leaving Big Spring, the two decided they would stop in Juarez on the way home and get a prostitute. The plan intoxicated them across four states, as they imagined aloud every possible scenario.

They hardly ate the entire trip in order to save their money. Coming back, they parked the truck in El Paso and walked across the international bridge. And once in Juarez, they entered the first beer hall they saw and sat down like a pair of sheep. "Two Cokes, please," they told the waitress.

Before the drinks even arrived, a pack of prostitutes swooped onto their laps and ordered themselves round after round of beers. When money ran low, Homer and Charles tried to flee, but a *federale* met them at the door and shook them down for the rest.

On the second attempt, weeks later, it was seven in the morning when they crossed the bridge. To their dismay, the only two women still up drinking were a lesbian couple who paid them no mind.

The third attempt was a success. The girls were around eighteen and spoke little English. The drive back to Big Spring was silent, until Charles spoke up and said, "I just don't get it. We go in there to have sex with these girls, then you spend the rest of the time tellin' 'em about the Lord."

"I just asked her why she lived that way, that's all," Homer said.

"A whore in Juarez and you're worried about her soul."

But it wasn't the girl's soul that Homer anguished over.

There was plenty of time to get right with God, because as soon as summer ended, he was back in church four times a week, paranoid the preacher was sniffing him out like a calf that took a wrong turn.

For another year he watched classmates couple up for homecoming and prom and listened to friends brag of conquests at the drive-in. For Homer, the car keys were off limits if his date was not a member of the Assemblies of God, and most of the girls in his church were kin. By the time Homer graduated in May 1957, he'd been allowed to attend only one football game, and that was to sell peanuts for Future Farmers of America.

"You can go," Little Opal said, "as long as you don't watch the game."

Directly after graduation, as classmates shucked their caps and gowns and drove to the pastures to drink beer, Homer climbed into his truck and delivered a load of bulls to California. At least he had the road.

5

Big Spring on the global stage . . . a corporation for the people . . . Iris and Raymond . . .

While the drought buried the fields and emptied the pastures, out on the east side of town, Raymond Tollett and the Cosden Petroleum Corporation rose to their zenith.

For the refinery, the Korean War was a boon. Government contracts for jet fuel, diesel, and other products padded the bottom line even as oil prices vacillated and hurt smaller companies. It was during this time that Tollett made his greatest gamble.

A decade earlier, during World War II, raw materials such as rubber had been in short supply and cheap alternatives were sought using the glut of American crude. Companies began deriving benzene and styrene from petroleum to make synthetic rubber and nylon. Eventually, they developed plastics, and the age of petrochemicals was born.

By the time the Korean War began in 1950, the boom was on. Europe, still recovering from the war, had only two production plants in the entire continent. In fact, 85 percent of the world's supply was being made along the Gulf of Mexico near Houston and Beaumont, where both petroleum and water were plentiful. Big Spring was isolated in West Texas and staring down the worst drought in centuries, but it made no difference to Tollett. Despite being an oilman at heart, he believed Cosden Petroleum could be a player in the petrochemical age, and he seized the opportunity.

It happened almost by accident. In 1952, Cosden opened its signature BTX unit, which was the first in the world to produce benzene, toluene, and xylene — the chief ingredients in war-grade jet fuels and explosives — simultaneously in one integrated plant.

It was during this project that engineer Dan Krausse discovered a fluke: the oil found below them in Howard County, in addition to being great for asphalt, had a unique ethyl benzene and xylene content — ideal for making plastics. That process was generally done by shipping the raw materials to a separate plant, which cost time and money. So Krausse and his team designed a facility that fed directly into Cosden's BTX stream, thus pioneering the world's first petrochemical assembly line. Dozens of products, ranging from gasoline to clear polystyrene pellets, could now be produced in the same location and from the same barrel of oil. The technology was later licensed around the globe and made the company a fortune.

At the same time, Tollett was buying small oil companies and regional refineries, making Cosden one of the largest inland fuel refiners in the country. The company sank hundreds of miles of pipe, drilled on a vast patchwork of leases, and invested in burgeoning oil exploration in the Gulf.

By 1958, despite a sluggish oil market, the plastics game had catapulted Josh Cosden's bankrupt refinery into the lower ranks of the Fortune 500. Tollett, with his FBI pedigree and dapper demeanor, became a dean of the industry. He was head of the influential American Petroleum Institute, whose headquarters were in Manhattan, in addition to chairing top trade organizations around the country. He represented Big Spring in the biggest cities of the world — New York, London, Istanbul, Beijing — and whenever he returned home, the people regarded him as their king.

For the head of a publicly traded corporation, Tollett ruled with rare egalitarian candor. He gave engineers free rein to experiment and explore, creating a culture that more resembled a modern tech startup than a mid-fifties oil company. Early in his tenure, he helped employees start a labor union so he "could have someone to negotiate with." And he routinely awarded bonuses to any carpenter, truck driver, or janitor who submitted a persuasive suggestion on how to improve efficiency, then published their photos in the company newsletter.

"I believe in respect for the dignity of the individual," he wrote in a monthly column that appeared in the *Cosden Copper.* "I believe that a workman on the lowest paid job may be just as honorable in every respect as I could be."

The company plane took sick employees to doctors and specialists across the country, and to family funerals out of state. The boss made daily visits to jobbers laid up in the Big Spring Hospital, and whenever a baby was born, he sent them a monogrammed Bible. To ensure a stream of future talent, he hired teenage children of employees for summer internships. And in 1956, he began providing older workers college scholarships to study chemical engineering while keeping them on the payroll.

For men who reached a milestone twenty-five years at the company, Tollett took them to Juarez for steaks, drinks, and girls (if they wanted), then handed them a gold ring and a month's paid vacation. And whereas Josh Cosden had given his workers a glazed ham for Christmas, Tollett cut them a bonus check instead, which he signed himself — hundreds of them — along with a personal note of gratitude, a task that often took him all night.

He sat on the Big Spring school board and fought to have its district line moved to include the refinery just so Cosden could pay taxes. He served as president of the hospital board, presided over the Boy Scouts and Chamber of Commerce, raised money to build the YMCA, and championed local charities. He urged his employees to follow his lead, to run for local office, to sit on boards, to fulfill his vision for a "citizen corporation." On any given day, his office was a constant parade of people seeking loans, advice, or referrals, and few walked away empty-handed.

"Ray loves the underdog," said one of his executives. "He'd rather loan money to a skid row drunk than to an engineer."

My father's cousin Granville Hahn was a shining example of this philosophy. Granville, who'd survived the New London School explosion by coming down with the mumps, and whose own father had abandoned his family, was driving a truck for the Cosden tire shop in 1950 when he grew curious about the refinery's laboratory. He approached one of the chemists, a German named Mr. Franks, and asked to spend his lunch hour observing him. Granville's education had ended with a GED, but within weeks he was blending high-octane fuels and no longer driving a truck. When Tollett opened the polystyrene unit, he made Granville its lab foreman and engineer. By the time Granville retired he held twenty-five patents, including the three-layer sheet of plastic used in every Solo party cup. But these kinds of success were not unique during Tollett's rein.

Even as the company grew, Tollett made himself available to the rank and file. He toured the units on Sunday afternoons, dressed down in khakis, and called each man by his name. He asked about their families, about recent fishing trips or vacations, then asked if they were happy. And for all of those things, the people gave him their energy, creativity, and the very best years of their lives. In this way, everybody won.

And with help from the air base, Tollett and the refinery helped make Big Spring a cosmopolitan oasis. In heavily conservative West Texas, where anti-intellectualism was the norm and anti-communism was taking root, Big Spring fostered an upper middle class that was cultured and well educated. Engineers and executives vacationed in Europe and Asia, their wives hosted book clubs, and the civic groups they chaired brought in scholars, artists, and dignitaries. In 1955, Tollett paid for the Harry James Orchestra to play the Cosden Club. A decade later he brought Duke Ellington. Prince Farman-Farmaian, cousin of the Iranian shah, was a close friend and frequent guest of the refinery.

By this time Tollett was already divorced and remarried. In 1945, his first wife, Leta, accused him of infidelity and left him, taking their daughter Kay. A short while later, he met a stewardess on an American Airlines flight to Chicago, and they'd fallen in love. Iris Goodbrake was ten years his junior, a stunning beauty with dark curls and deep green eyes that gave off a glimmer of derring-do. She'd grown up in Effingham, Illinois, near St. Louis, where her father worked for the railroad. After studying at the Art Institute of Chicago in the years before the war, she became a nurse — a requirement for being a stewardess back then — and started to fly.

Iris and Raymond married in November 1945 at Chicago's Stevens Hotel. After a brief honeymoon in Mexico, they returned to Big Spring and settled down in Tollett's home on Hillside Drive. By the early fifties, Iris had given him three children: Raymond Jr., Jason Blake, and little Ann.

With Iris running the household, one that included a full-time cook and housekeeper, the Tolletts became Big Spring's royal family, and for the next decade, nearly every ball gown that Iris wore to fund-raisers, every European jaunt or summer camp away, would be carefully chronicled in the *Herald* and discussed in beauty shops across town.

The children grew up under the ever-expanding shadow of their father's company. The two boys joined him on his Sunday strolls through the units and came to know the workers. Weekends were spent in the ultramodern Cosden Country Club, which opened in 1955 on the lake in Pioneer

Park. Enclosed in glass with views of the water, it featured a three-hundred-person ballroom and dining area, swimming pool and tennis courts, and a lounge where the walls and ceilings were deep pink. Vacations were taken aboard one of the two company planes, a de Havilland Heron or a Dove, and usually accompanied by their father's secretary, Helen Green, always dressed immaculately in white gloves and a hat. The Beaux Arts Hotel in New York, the Ambassador in Chicago — these were places they knew well.

Each Sunday, the children sat in the front row at St. Mary's Episcopal Church, where their father served as lay reader. At home, he changed into a silk smoking jacket and retired to his study to read or write his monthly column. His busy schedule required him to budget his reading, allowing himself six works of nonfiction each year, usually the latest business or management books, and six works of fiction and poetry. John Steinbeck and Carl Sandburg were his favorites, and his shelves contained first editions of *East of Eden* and Sandburg's collected poems, both signed by the authors. Each night before bed, he spent exactly ten minutes with the Old Testament, wherein, he wrote, "lie the keys to all wisdom."

To his children, he was like an old lion in a fairy tale. Mornings they pressed into his bathroom to watch him dress for work. Their father was a clotheshorse. As partial owner of a men's store downtown, each year he selected eight suits and one tuxedo out of stock. His pocket handkerchiefs were imported Indian linen, his silk smoking jackets custom-tailored in New York, where he also ordered his hats. Each week, the housekeeper pressed his underwear.

Fresh from the shower, he dipped his fingers into a jar of white cream and rubbed it into his hair, which by then was turning thin and gray. "This is mother's milk from an elephant," he told them.

"And what is that?" Jason Blake asked, pointing to a small scar on his back. "That," his father replied, "was where I was shot by a jealous prince."

Once or twice a year, their father's old FBI buddies appeared at the house. The men dressed like him, in dark suits and dark hats and shoes that reflected back the sun. And they all seemed to drive the same large black sedans.

"Come have a look," one of the men said to Jason Blake, then popped open the big trunk. Inside was a tommy gun, just sitting there.

The moments the children spent alone with their father were rare, and therefore cherished. Often he was traveling, building his company, giving his time to the community. And whenever he was in town, evenings were usually spent at the club, hosting guests and providing entertainment previously unthinkable for a small town such as Big Spring.

Always hosting, always with a glass of scotch in his hand. Drinking came with the job, and Raymond was famous for it. Colleagues marveled at the amount of whiskey he could put down and still function at full capacity. And when hangovers threatened to slow his pace, Tollett went to see his best friend, Dr. Marion Bennett, who administered a quick IV of fluids. On a flight home from New York one morning after a long night of partying, one of his executives watched him pull from his briefcase a syringe of B12 and spike it into his leg.

"Didn't even bother to lift his pants," the man said. "Shot it straight through that thousand-dollar suit."

Tollett's first wife was a teetotaler, but Iris could match her husband round for round. In photos the children still keep, the two of them are partying in Louisville for the 1953

Kentucky Derby, as guests of the Aetna Oil Company: at the Seelbach Hotel for cocktails and the Old House for champagne; Iris at the President's Ball at the end of the evening, in a double strand of pearls, heels kicked off and dancing alone.

Tollett possessed a driven man's sexual appetite, only heightened by whiskey, and during the early years of marriage his wife was game to satisfy.

Back in the early fifties, they'd caused a huge scandal at the Big Spring Country Club, where they were members. Leaving one night, the two of them stumbled out to their car, which the valet had pulled to the front door. But before Iris could get in, Raymond threw her against the hood and hiked up her dress. As people filed out to leave, they encountered quite a scene. The club's directors were mortified, going so far as to hold a hearing and revoke the Tolletts' membership. Raymond was furious.

"Have it your way," he told them. "I'll build my own place." And that was how the Cosden Country Club came to be. The year it opened, it was so popular that it nearly sank the Big Spring club.

But like slow poison, the drinking crippled their marriage. Raymond was always gone. He wore no wedding band, and there were rumors of women on the road. Iris, alone most of the day and hounded by suspicions, drank to calm her nerves. In the fall of 1957, she was driving the kids home from the club when a patrolman saw her swerving and pulled her over. While Iris went downtown, charged with a DWI, the children cut through a pasture and found their way home.

Her case went to trial. On the stand, her doctor testified that he prescribed sedatives for her nerves. And on that particular day, she was distraught after her husband's plane briefly disappeared from radar during a trip out west, which explained her erratic behavior. It was a false alarm, of course, but all the same, it took the jury less than twenty minutes to acquit.

But the drinking continued. Iris said that she felt inadequate around Raymond, that her husband didn't see her as an intellectual equal. Iris could also be a flirt, especially after she had a few rounds and was feeling vindictive. Tensions came to a peak in the summer of 1958, when Raymond filed for divorce. The announcement appeared in the public-notice section of the *Herald* — just one line and nothing more. But for whatever reason, the couple remained married.

That fall, Tollett carried Cosden Petroleum into the Fortune 500. But earlier that year, he marked another personal milestone when the family moved into the dream house that he'd painstakingly designed, located just down the block on Hillside Drive.

He'd chosen the site for its eastern views, where he could step out at night and see the refinery — the "Jewel of the West," he called it — sparkle under the stars. The *Herald* published a big spread about the home. Photos showed Ray Jr. posing with his model airplanes, Jason Blake demonstrating his puppetry in the birch-paneled den, and little Ann dressed in bobby socks at the breakfast bar, along with their housekeeper, Annie Mae Huey.

In the pictures, Iris strained to smile. She hated the house, found it cold and fussy, but played along with the pageantry. "The new R. L. Tollett home was designed with the family in mind," the reporter concluded, making only vague reference to the separate master bedrooms.

6

Frances rides with the Texas Playboys, weathers a broken heart . . . home beckons . . . Tommy enters with his steel guitar . . .

For Frances, who came to California in search of her mother Bertha and had already married and divorced, the fifties began with a desire to return home. The wide-open skies of West Texas beckoned her, as full of sand and familiar misery as they were. Most likely it was a broken heart that pushed her toward familiar ground. For that's what she was — heartbroken. When the love had gone away, the old roots had found purchase again and were trying to fill the empty space. But if she went back to Texas, where would she go? What would she do? What was even left for her now in Big Spring?

So much had happened in California. She'd come west like everyone else, looking for adventure and a new beginning — and look where she landed. Sometimes she couldn't believe it herself.

She'd fallen deeply in love with the guitarist Les Anderson after that night at the Oakland Auditorium, and a dizzying romance ensued. By then, in the mid-1940s, Bob Wills and His Texas Playboys were one of the biggest acts in the country, especially along the West Coast, where they outdrew Tommy Dorsey, Benny Goodman, Harry James, and all of the big bands combined. With the war still on, people had money to spend. Soldiers coming home were seeking diversion in girls, cold beer, and dancing, and Wills's western swing carried a reckless energy they craved. The crowds were growing so big that auditoriums turned people away, even on Monday and Tuesday nights. It was in this whirlwind that Frances found herself wide-eyed and in love.

She practically lived with the band, riding up and down the coast on their bus, the Playboy Limited. During the week she danced till two, then rode all night back to San Francisco to get to the Biscuit Company by nine, often without any sleep.

Her girlfriends Mary, Dorothy, and Tiny also dated guys in the band, and the four of them were inseparable. They kept each other company on the road and during the long days in the studio. In spring 1946, the girls were in the basement of the Mark Hopkins Hotel in San Francisco when the band started recording their most famous sessions, known as the Tiffany Transcriptions, using a stripped-down lineup and a single microphone. Frances had no idea she was witnessing country music history. All she remembered was that Noel Boggs, one of the steel players, pranced around with castanets under his shirt trying to make her laugh, and the prank upset Wills — who insisted on total silence in the studio.

When the band embarked on national tours, which carried them away for weeks at a time, the girls endured the lonely nights by cooking for one another. On Christmas Eve the previous year, they'd came up with the idea of serving dinner to the entire orchestra in Frances's tiny apartment on Pine Street. She probably had three chairs in the whole place, but nonetheless, there stood the greatest western swing band to ever take the stage, eating glazed ham and listening to her Jimmie Rodgers records.

Les had worn a wedding ring on his finger the night they met, but told Frances it belonged to his mother. She'd believed him, but a month later, one of her girlfriends said he had a wife and kid out in Fresno. Frances told Les she never wanted to see him again, but something always drew her back. She'd be sitting in her little apartment feeling lonesome when he'd call wanting to come over. She'd let him in, even though she knew it was wrong, even as she thought, *There are names for women like me.*

She supported Les when he quit the Texas Playboys in 1946 and started his own band. The Melody Wranglers appeared regularly on radio shows like *Spade Cooley Time* in Los Angeles. Frances traveled beside him, rode for hours on buses to meet him, even started the Les Anderson Fan Club because he asked her to. His nickname for her was Poncho, and whenever he called her that and looked down with those big blue eyes — well, who could blame her?

But by 1949, it was clear that Les wasn't going to leave his family, not that Frances would ask him to anyway. And when he did go back to his wife, she didn't protest. Heartbroken, she moved to Richmond to live with her friend Dorothy so she wouldn't be alone, then fell deeper into the music, which was the only thing that felt good.

Back then, the country and western circuit was small and informal, and through Bob Wills, Frances knew just about everyone who toured California. She knew many of the Grand Ole Opry stars, such as Ernest Tubb, Red Foley, Little Jimmy Dickens, and Roy Acuff. For a while she hung out with Hank Garland, who must have been eighteen or nineteen but could play the guitar better than anyone she'd ever heard. It was around the time he released "Sugarfoot Rag," which sold over a million records. In the summer of 1950, she saw him play an Opry package show in San Jose with Tubb, Minnie Pearl, and others. Afterward, Billy Byrd, Tubb's guitar player, passed along a note from another Hank who'd played that night — Hank Williams, who'd seen Frances backstage and wanted her in his room. They all shared a laugh over that, but what Frances didn't say was that her heart still longed for Les.

By then, the old gang had broken up and scattered. Bob Wills was no longer the juggernaut he'd once been during the war. By the late forties, his drinking had spiraled out of control and practically wrecked his band. In September 1948, he committed sabotage when he fired his lead singer, Tommy Duncan, after overhearing him complain about Wills's drinking. Many of the core band members also left, forcing Wills to get along with a smaller outfit that struggled to find the old chemistry. In the summer of 1949, worn down from touring and no longer earning the same money, Wills moved the band and his family to Oklahoma and eventually back to Texas, seeking a sense of home he'd never be able to find.

Meanwhile, Tommy Duncan started his own band, the Western All Stars, that featured Millard Kelso, Joe Holley, and Ocie Stockard from the old Playboys lineup. And for a while it featured Ernie Ball on steel guitar, the same Ernie Ball who'd later make a fortune selling his own brand of guitar strings called "Slinkys." Frances became good friends with Ernie and his wife, Gail, and the three of them stayed close even after Ernie was drafted for Korea and shipped to Arizona.

By 1950, Frances had been in California for nearly ten years. During that time her mother, Bertha, had married a man named Tex, who was a former boxer and merchant marine. Tex had also served in World War II and suffered an accident that paralyzed half his face, though Frances never asked for the details. But she knew that drinking turned him violent, and more than once he'd wheeled around from a barstool and challenged three men to a brawl, licking every one. When nobody was around for Tex to fight, he beat up on Bertha. During one of these episodes he punched out her front teeth.

Bertha didn't believe in divorce, and despite her children's pleas she stayed with Tex and endured his abuse. Together they lived in a crummy residence hotel on Sixth Street near the Tenderloin, where each Christmas Bertha cooked plates of food for the lonely drunks and old people living in the building. Once, after getting some insurance money, she shocked Frances by giving most of it to a soup kitchen.

Bertha could be so generous, yet she remained this enigma to her children. Never once had she discussed those years in Big Spring after Bud's death, her relationships with Virgil and Red, or the trauma she caused them all. Decades would pass before she offered up an apology, but in the meantime, the love that Frances had for her mother was left to drift in a gulf between them. That love had initially brought her to California, but it could no longer hold her there.

Frances was closer to her siblings. Leamon and Flossie had eventually moved to the Bay Area after leaving their foster home in Texas, and John soon followed. For ten years, the four of them were hardly out of step. She and John attended dances together — Bob Wills and the Maddox Brothers and Rose, among others. John had even married one of her girlfriends, Geneva.

But still there was no word of Jimmie, and there never would be. Bertha never mentioned the boy, as if he'd never been born, but for the rest of Frances's life she'd look for him. At any moment she expected him to run up and grab hold of her dress, just like he did in those years when all they had were each other. Jimmie would be going on twelve now. Her heart ached to think of him alone and missing them, wondering why his family had chosen to leave him behind.

Not long before, Frances had met the singer Hank Thompson and become friends with his wife, Dorothy. The couple were based out of Dallas, where Thompson and his Brazos Valley Boys played dance halls when they weren't touring the southwestern circuit. In 1950, Wills had also moved the Texas Playboys to Dallas, where he opened a giant club called Bob Wills's Ranch House. If Dallas was good enough for Hank Thompson and Bob Wills, then it was good enough for Frances. Texas was calling her home. So, without a job or place to stay, she packed her things in a box and boarded a Greyhound bus before she could change her mind.

Along the way, she stopped in Phoenix to visit Ernie and Gail Ball. Ernie was stationed at Williams AFB, where he played in the air force band. That night, as the three of them sat around reminiscing, Frances suddenly began to second-guess herself. Was she crazy for leaving California? Did her friends think as much?

At some point, Ernie mentioned Bob Wills's brother, Billy Jack, and what a killer band he had out in Sacramento. Frances said that she'd dated Billy Jack's steel player, Tommy Varner, after meeting him at one of their shows. They'd gone out a few times and she really liked him. He was shy and a bit austere, a real musician's musician. But for some reason she'd never called him back. By that time she'd made up her mind to move to Texas, and after that, Texas was all she saw.

She hadn't thought about Tommy in weeks, but suddenly, the three of them were talking about him like he was someone important. Then Frances said, without even thinking, "That's the kind of man I'd marry, even if he wasn't a musician."

"Then get on the phone and call Tommy!" Ernie shouted, triumphant.

"Oh," gushed Gail, "you must!"

"Go back to Tommy!"

The next afternoon, the box of Frances's clothes arrived at the bus station in Dallas, but Frances was not there to claim it. She had boarded a bus in the opposite direction. She was going to find Tommy.

Out in Sacramento, she found him in a smoky ballroom behind a pedal steel guitar, creating the music that offered the only real sense of home. Everything familiar lay tied up in that lonesome sound, and right away, she began falling deeper in love with the man who could roll it off his fingers.

It was a different scene now, with different players. Bob Wills had taken his Texas Playboys to Oklahoma and Texas and left Billy Jack, his youngest brother, to keep the sound alive on the West Coast. Billy Jack Wills and His Western Swing Band took residence in the Wills Point Ballroom on the outskirts of Sacramento, which Bob had opened back in 1947. It featured a large dance hall with a beer garden and restaurant, while outside was an Olympic-sized swimming pool, burger stand, and picnic grounds. There were also six apartments where members of the band and their families lived.

For six years, Billy Jack had paid his dues as his brother's bassist and drummer. He even penned the lyrics to one of Bob Wills's most enduring hits, "Faded Love." But Billy Jack was twenty years younger than Bob and favored more modern rhythm and blues. His band covered the black musicians whose songs were laying the foundation for rock and roll, artists such as Ruth Brown, Wynonie Harris, and Larry Darnell. And the swing numbers Billy Jack composed jumped harder and faster and were fused with elements of R&B, bebop, and early rockabilly. This sound was drawing a new generation of fans the way the Texas Playboys had done before the war.

Frances took an apartment in downtown Sacramento and found a job at the American News Company sorting magazines. She spent Saturday nights at Wills Point watching the band, and the rest of her nights with Tommy. Although Tommy was only twenty-three — three years younger than Frances — he'd been married twice already and had two daughters from another past relationship. The girls lived with their mother in Bakersfield. And from one of her girlfriends, Frances heard that Tommy had been involved with yet another woman before her — the sister of Dean and Evelyn McKinney, who'd sung backup with the Texas Playboys.

Wanting no more surprises, Frances asked Tommy about the woman and was told she had moved back to Alabama.

"You don't see a ring on her finger, do you?" he said, to which Frances replied, "Well, I don't see a ring on *my* finger, either." The next day, June 12, Tommy showed up at her work and drove her to the courthouse. She became pregnant almost immediately.

She was seven months into her pregnancy, big as a house, when the U.S. Army sent Tommy a draft notice for the war in Korea. He reported for basic training, but while home on leave before shipping out, Tommy went AWOL.

Being on the lam meant he had to quit Billy Jack's band and lie low. In the meantime, Billy Jack brought in Vance Terry as a replacement while Tommy took gigs down in Vallejo and wherever else he could find. He was still AWOL in February when Frances went into labor. Their daughter, Mary Ellen, was born at the military hospital but Tommy stayed at home, afraid of being caught. When the baby was four months old, Tommy went on tour with Duncan's band, the Western All Stars, only to find the military police waiting when he returned. Within a matter of days, the army had shipped him to Korea, where he worked on the front lines filing paperwork. Meanwhile, Frances's brother Leamon was fighting with the Third Infantry. In the summer of 1953 he participated in the war's most famous battle, the taking of Pork Chop Hill.

The Red Cross helped Frances find a babysitter so she could keep her job. She dropped the baby off each morning and picked her up in the evening, walking everywhere because she didn't own a car. This period lasted a year, and despite the tired feet and sleepless nights, it was a time that Frances would always cherish, for she'd never experienced a stronger feeling of love than she did for Mary Ellen. At night, the two of them cuddled together in bed and Frances told her everything she knew.

Despite his having gone AWOL, the military released Tommy with an honorable discharge in early 1954. He slotted right back in with Billy Jack, this time playing bass fiddle, since Vance Terry still had his old job on steel guitar. By then the band had gained momentum, thanks to appearances on KFBK in Sacramento, which beamed all the way up the Pacific coast. In rural Washington and Oregon, where television had yet to penetrate, people came by the hundreds to see them play.

Frances and Tommy soon moved into Wills Point, taking a cramped one-bedroom apartment just below the dance floor. It was then that Tommy's old girlfriend, a woman named Frida, started driving up from Bakersfield to drop off his daughters, who were ten and eight. Frida was unstable and prone to hysterics, and her very presence put Frances on edge. At the same time, Frances's sister Flossie also appeared from Texas, pregnant, and announced she was staying. With no room to fit everyone, Frances and Tommy rented a motel room across the road just so everyone had a place to sleep.

Their arrival at Wills Point also coincided with Bob Wills's return to California. Back in Texas, he'd started worrying that the club was losing money and prestige under his baby brother's marquee. So in spring 1954, Wills resumed control of Wills Point and eventually took over as bandleader. In early 1955, he put Tommy back on steel guitar, since Terry had left to attend college, and scheduled a major tour. Before Frances knew it, her husband was a Texas Playboy and constantly on the road.

With Wills at the lead, the band traveled

for most of the year, playing dance halls across six states. Back at home, Frances cared for Mary Ellen, plus Tommy's daughters whenever they visited. After Flossie had her baby, the two sisters settled into the shared chaos of diapers, feedings, and little ones underfoot. The kids ate Creamsicles by the big pool and played in the creek behind the club. Each night at bedtime, the ceiling shook from the touring acts playing the ballroom above.

During this time, Frances was carrying her second child. Despite being surrounded by her sister and a houseful of kids, she grew lonely without Tommy, and found her only comfort in the letters he sent from the road.

One afternoon at the mailbox she saw an envelope addressed to Tommy. The handwriting appeared to be a woman's, so Frances opened it and began to read. Sure enough, it was from a woman named Betty in Oklahoma, who, judging from the contents of the letter, was clearly her husband's girlfriend. Frances stood outside their apartment and read it again and again, then something happened. She felt dizzy, then the next thing she knew she was lying in bed with a crowd of people standing above her. A doctor was checking her pulse, asking about the baby.

The next week she sent a telegram to Betty, pretending to be Tommy, and asked for a photograph. She wanted to see this woman; she had to know. A few days later she received a glamour shot of a pretty, buxom brunette. The gauzy evening dress she wore looked just like the one hanging in Frances's closet.

It just so happened the band was coming home for a showcase at Wills Point. The day they arrived, Frances asked her sister to take the kids so she could confront Tommy. He was dressed in his band uniform, about to go upstairs for a sound check, when Frances pulled out the photo and threw it onto the table.

"Are you cheating on me?" she asked.

But Tommy deflected the question. When he refused to answer a second time, Frances grew so angry that she grabbed the nearest object, a black stiletto heel, and went after him. The blow laid his head open wide and caused him to crumple to the floor. Blood was pouring from the wound when he came back up and smacked her so hard that her feet left the ground. He then cursed as he fumbled in the kitchen for a rag to stop the bleeding. Within minutes he was gone, leaving her alone with a throbbing head and the sound of his steel guitar playing upstairs.

Neither of them wanted a divorce. Besides, the arrival of their daughter Janet worked to realign their commitment. Then tragedy brought them together. In the early morning of June 16, 1956, the family was fast asleep when a teenage kid, for reasons unknown, lobbed a Molotov cocktail through a window at Wills Point. The explosion woke them up, and the smell of smoke and crackling of flames told them the fire was above their heads on the dance floor. They had just enough time to scoop up the kids and dash out into the night. Within minutes, the fire engulfed the entire compound, taking with it everything they owned.

The Red Cross arranged an apartment for them, where they stayed for several months while plotting their next move. Before the fire Bob Wills had returned to Amarillo and left Billy Jack the ballroom again. But for Billy Jack, what he'd created at Wills Point could never be reproduced. Shortly after it was destroyed, the band broke up and went their separate ways, and within four years, the younger Wills brother retired from music altogether. As for Tommy, his career ended, too.

The family moved to Bakersfield, where

Tommy's two daughters lived, along with Frances's brothers, Leamon and John. Leamon was selling Kirby vacuum cleaners and John worked as a roughneck north of town in the Kern River field. Tommy enrolled in college on the GI Bill and began studying music, hoping to become a teacher. They took a one-bedroom house, which they furnished with a five-dollar sofa they found at a flea markct. But they still had no clothes and no money; everything had been lost in the fire. At their lowest point, they couldn't even feed themselves. One afternoon, a Catholic food pantry turned them away because they weren't members of the church.

Her brother Leamon was having success selling Kirbys door-to-door, mainly to suburban housewives. One morning, looking to help the family, he picked up Tommy and took him along on his route, hoping Tommy could land a sale of his own. When her husband walked through the door that evening, he had a big smile on his face.

"Don't tell me," she said. "You sold a Kirby!"

"I sold a Kirby," Tommy said. He'd found himself a new career.

Tommy began selling lots of Kirbys, so many in fact that there was talk of opening

his own distributorship in Bakersfield. Leamon got one, and before long, even John
quit the oil patch and was knocking on
doors, demonstrating the power of the
Suds-O-Gun, Crystalator, and Handi-
Butler attachments. Push-button house-
keeping — now that was the future! *It does
everything but the cooking, ma'am! Don't buy
a mere vacuum cleaner, buy a clean home for
the rest of your life!"*

They were a Kirby family now of all
blessed things.

They moved into a bigger house and enrolled Mary Ellen in school. And now that
they lived near Tommy's two girls, Frida
brought them over for longer stays. The girls
arrived with dirty hair, their clothes wadded
up in cardboard boxes. Frida also dropped
them off unannounced at odd hours of the
day, then took them back without telling
Frances. The family lawyer told them to
stand up to Frida, so one day Frances
refused to let them go. Frida hauled off and
punched her in the face, breaking her nose.

Tommy and Frances went to court and
got custody. Frida moved to Los Angeles to
live with a boyfriend, but the judge still gave
her a month that summer with the girls.
When they came home after their first visit,

Frances noticed something was wrong. The older one, who was twelve, refused to go outside and wouldn't sit near a window.

It was through her younger sister that Frances discovered that while in Los Angeles, Frida had propositioned two older Hispanic men, who took the girls to a motel. "Do whatever they tell you," Frida told her daughter, "just get their billfolds." The younger girl sat outside while the men had their way with her sister. They so traumatized the older girl that she was convinced they were coming to kill her, so she was hiding from view.

Frances and Tommy went to the district attorney and tried to build a case, but nothing ever came of it. The DA finally told them that in order to truly protect the girls, they should leave the state. So that's what they did. A Kirby distributorship opened up in Phoenix, and Tommy jumped on it. Leamon soon followed and opened his own shop, working the opposite side of town, while John found a store out in Tucson.

The family moved into a brand-new house in Scottsdale, right across from an elementary school. And over the next ten years, while Tommy sold vacuum cleaners and parts, Frances stayed home and raised the girls. Each year her old life seemed to recede more and more into memory. Bob Wills and Billy Jack, Gail and Ernie Ball, the stars of the Grand Ole Opry — it was like she'd met them all in a dream.

She began to see her life in distinct chapters — Texas being the first, then California, and now Arizona. The funny thing about this chapter was that it was the hardest one of all, but the one in which she took the most pride. She would be the best mother in the world, she vowed. She would give her children the love and devotion that she never received from Bertha. And this is what she did, but it came with a price.

PART 3

1

The drought ends . . . hard times again . . .
Bobby and Preston . . . Clem finds his
reward . . . enter Grady . . .

In May 1957, the rains finally returned to
Big Spring. After seven long years of
drought, a series of soakers laid a carpet of
lush green grass along the dry and brittle
range. After that, the family's fortunes
began to shift.

By now, the ranches were mostly empty,
and the long hauls to feedlots and packing-
houses were fewer and fewer. Beef production in the United States declined, while at
the same time, Americans were eating more
steak and bologna than ever before —
ninety-five pounds per person per year. To
meet the demand, cheap imported cattle
began flooding the market from Argentina,
Ireland, Australia, and elsewhere, and drove
prices on American cattle into the base-

ment. They arrived to the coasts, where local truckers hauled them inland, leaving Herman without an edge. By the end of 1958, his business was in trouble.

First he failed to make payments on several trucks he'd purchased during the boom. But instead of giving them up and cutting his losses, he chose to refinance using his entire fleet as collateral. He even threw in the family house. The only thing left off the note was Little Opal's Buick, only because Homer talked him out of it.

Soon the bank came for the whole lot. It was Homer who met the repo team at the office, pulled the logos off the truck doors, and handed over eight sets of keys. Back home, his mother and sister packed their house and prepared to move.

His father, meanwhile, climbed into the Buick in a fog of despair and headed to San Angelo. In the glove box sat a .45 that he planned to use on himself once he arrived. But luckily his brother Tooter had noticed the gun missing from the office and talked his way into the passenger seat. By the time they hit town, he'd coaxed him off the edge.

After that Herman looked for other trucking jobs. He heard the Fort Worth stockyards were hiring drivers and drove all the way there, only to come home dejected, telling his children, "They weren't even interested in me. I told them all of my experience and they weren't even interested."

For a few months he sold cantaloupes door-to-door, until his feet swelled so badly he could hardly walk. Then one day he noticed that the owner of the Jumbo Restaurant on Gregg Street had a FOR SALE sign in the window. It was a small lunch counter with just six stools and a couple of tables, but the price was decent. With a small loan from Little Opal's mother, Herman went into the restaurant business.

The property included a tiny shotgun apartment behind the kitchen, where the family lived. It was too small to accommodate their nice furniture, so they stacked it against the walls to create a walking path. And to make back their money, they stayed open twenty-four hours and worked in shifts. Herman hardly slept from all the stress. After two months, he collapsed from a heart attack.

It was Charles Miller, Homer's friend, who discovered him one morning sitting on the edge of his bed, clutching his chest and going pale. Charles called an ambulance, which broke down on the way to the hospital. It took Herman, in the throes of cardiac arrest, to explain to the driver how to fix

291

the engine. By the time they reached the ER, he was in a coma. The doctor told Little Opal, "If he lives, it's because of higher hands than mine."

Little Opal looked down at her husband, covered in tubes and wires, and felt panic and despair. What on earth were they going to do? She drove back to the restaurant, telling her daughter, Evelyn, "I owe my mother two thousand dollars and we have to pay it back. There's no way forward but straight ahead." With Herman lying in a coma, flat broke and buried in debt, they went back to work.

The only good news to come that year was that Homer got engaged. He'd met Stina at a youth conference in Odessa. She was a good Pentecostal girl, tiny as a mouse, but equipped with a sidelong humor that Homer found irresistible.

They exchanged vows at the altar of First Assembly, then stunned everyone by announcing they were leaving the church and becoming Baptist.

To live a simple Christian life, one possessed of love, joy, and compassion, rather than judgment and condemnation — that's what they wanted, and the Baptist church that many of Homer's friends from school attended seemed like a good enough place to start.

But the old flock didn't let him go easy. For months they called and warned him against leaving, as if to illuminate his road to damnation. They stopped him on the street and asked to pray. And in crowds and family functions, they greeted him with cries of "John the Baptist."

Homer told his wife, "This must be what it's like to escape a cult."

No one was more wrecked by Homer's decision than Little Opal. She bore the weight of his breakaway, feeling as if she'd failed to hold him in glory. Four times a week, she threw herself on the altar at First Assembly and pleaded for the return of her wayward boy. "Bring him home, Lord," she cried. "Please bring home my son."

Sitting in the pews watching all of this was my father, Bobby.

Opal had gotten pregnant with Bobby when she was thirty-three, an age considered ancient for childbearing in those days, especially when her eldest child was already old enough to drive. When Zelda heard about the pregnancy, she planted both hands on her hips, and cried, "Oh, *Mother*!" and refused to tell her friends.

Bobby arrived in December 1953, when the sky was red and the sum value of the land — including the oil and the men who pumped it — amounted to nothing. The following year is when Bob quit the oil patch and colluded with the drought, his eyes full of sand dollars.

The minute that powder-blue Ford appeared behind the house, you'd think the stork had dropped two babies on the family instead of one. From that day forward, dump trucks and blow dirt took hold of Bob's heart and never turned loose.

He attacked the sand drifts with vigor, backing his truck against the first one he passed on the highway. With no proper tractor yet, he chipped away at the piles with a pair of No. 2 shovels. Holding the other one was Preston, my father's older brother, who could sling dirt as far as any man.

"Fill her to the top, Big Boy. There's plenty more."

At eight years old, Preston was already tall, broad in the shoulders, and tough as an old tire — no doubt his father's pride and joy. Bob called him Big Boy Blue.

Preston was quiet, never spoke without necessity. When Bob worked the rigs, Preston would roam the pastures with his single-shot .22, hunting rabbits. Sometimes Bob climbed off the derrick floor and found him standing in the shade of the doghouse, a dead jackrabbit between each finger, never sure how long he'd been there. It's possible the only people to ever get ten words out of Preston were the ticket clerks at the Ritz Theater. On Saturday mornings, when a Gandy's milk carton got kids under twelve in for free, he was forced to actually argue his age.

Because of this, Bob liked to tease Preston during the long silent moments while they drove.

"You say somethin'?" he'd ask.

But Preston stared ahead, the dust boiling all around him.

"I — I —" he'd stammer, as if surprised by his own voice. "I didn't say *nothin'*."

Bob sold the loamy "cat-claw" sand for ten dollars a load. Business was brisk, enough to buy a used Ford tractor to make the work easier, followed by a second truck, also a Ford. He got it cheap because it was missing its passenger door.

"I'll get you that door once I get going good," he assured Preston, who didn't seem to mind the extra air-conditioning. In fact, Preston loved working with his father. Each day at noon, Bob pulled into a little grocery and came out with two cans of pork and beans, a pound of bologna and bread, and a white onion. Together they ate their lunch in the cab under a blanket of dust, chasing it down with a jug of warm water, never happier.

That first year, Bob preached the wonders of dump trucks to anyone who'd listen, especially to the men in his family. "Can't ya see? It's the best work going," he told his brother-in-law, Bill White, who'd married Opal's sister Veda. "Everywhere you look: sand! And it's yours for the taking."

Bill was fresh out of the service and needed a job. He'd never considered dirt, but Bob was sure crazy about it, and soon Bill was following behind with a wheel-barrow and spreading it on people's yards, collecting the fee. Then one day Bob spotted a 1952 Ford for sale over in Stanton. It was just a stripped-down chassis, no bed, for five hundred dollars.

"If you don't jump on it, I will," Bob said, and Bill jumped on it.

Luckily for Bill, Davey Jones had a spare bed lying around his house, which they welded onto the frame. It held only two cubic yards of dirt, but after mounting several two-by-sixes on each side, it was able to carry a few more. Now Bill was in business.

Around that time, Zelda graduated high school and married an Alabaman named Charles Odom, whom she'd met at church. Charlie worked in the grocery business and had recently taken a manager's job at the air base commisary. Norma Lou had also married. Earlier that ycar, she fell for an airman out at the base named Bill Glaesman and demanded to wed at sixteen, which Opal — who'd done exactly the same thing — reluctantly allowed. As soon as Charlie Odom and Bill Glaesman joined the family, Bob had them driving dump trucks, too.

Together with the Jones gang, they hauled caliche and hot mix to build the feeder roads to Interstate 20, which would eventually divert traffic around Big Spring and drain the lifeblood right out of it. And during slow weeks, when public interest in catclaw was lacking, Bob and Preston combed the pastures filling tow sacks with cow patties. Preston sold them door-to-door as fertilizer while Bob crawled behind in the truck, encouraging the shy boy to speak.

Bob bragged to anyone about Big Boy Blue, but once at home, it was clear who was the darling. Like his brother Bud, Bob had never showed affection toward his kids. But he was different with my father, who crawled up into his lap like a little kitten to be petted. And seeing this made Preston seethe.

But what upset Preston even more — as if it were a direct attack — was Bobby's gift of smooth talk. The kid was a born bullshitter. Preston grew sick of hearing the adults recall the hilarious things Bobby would say, like the time he rode his tricycle up to Zelda's Charlie, milk bottle tucked into his pocket, and said, "Give me a stick of gum and I'll let ya be my pal." Or after his first day of school when Opal asked, "So, did you learn anything today?" and Bobby shook his head and replied, "Nope. Teacher said she can't teach me nothin'."

Or what about the time Opal turned around in the department store in Midland to find Bobby peeing in the display toilet? All she did was laugh; Preston knew if that had been him, he'd still have trouble walking.

Naturally, Preston took it upon himself to ensure the boy didn't grow up a sissy. Once Bobby got old enough to run, Preston gave him a three-second start down the alley before firing on him with a BB gun.

"Get fast, little brother."

And when Bobby dug tunnels into Bob's dirt piles, a few of which he kept at the house, Preston waited until he was deep inside, then jumped on them in hopes of burying him alive.

At the age of six, Bobby walked down to the altar at First Assembly and gave his heart to the Lord, and when he returned to his seat, he saw that his mother's eyes were raw with tears.

Bobby loved the Lord with all his heart, loved Him the way she had taught him to love, and the Lord revealed Himself to the boy not as a cloud tyrant or a punisher, but as a master worthy of his mother's voice.

He realized early that his family was special in this regard. A feeling of pride washed over him whenever his mother and sisters stood before the church to sing, because the second they opened their mouths, the whole congregation fell under a spell. Even the Baptists and Methodists asked them to sing at their weddings and funerals.

His mother sang all the time, and not just at church. She sang as she cleaned the house and cooked their meals and when she bathed him in the evenings, her voice like a meadowlark in the winter trees. And it was

Opal's voice that carried the family singa-
longs each Saturday night at his grand-
parents' house. After supper, the grown-ups
gathered in Clem and Cora's tiny living
room to sing songs like "On the Jericho
Road" and "Just a Little Talk with Jesus"
that lent themselves to harmony. With no
air-conditioning in the house, they threw
open the windows, and some nights they
looked out and saw their neighbors sitting
on lawn chairs listening.

There was power in the music, and as
Bobby sang along he felt what he'd later
recognize as the Holy Spirit moving restless. Zelda's voice alone was so potent that
when Bobby was a toddler, his sister used
to gather her girlfriends and say, "Watch
this," and sing a slow, woeful ballad, not
even a religious one. Each time it brought
the boy to tears, as if slowly turning a knob.

In the moments down at the altar, or singing with his family, he wondered how
anyone would not choose Jesus. Already he
was aware how the family divided according
to differences in belief or practice. Really, it
came down to who drank and who didn't.
From hearing the grown-ups talk, he knew
that a few of his cousins and uncles liked to
have a beer or two after working in the oil
patch.

And his own father drank — an open secret in their house. Out in the toolshed, Bob kept a six-pack of beer in a cooler, which he sipped whenever alone and working on his trucks. He'd never dream of asking Opal to keep alcohol in her fridge, so when the beers got hot he drank them over ice. Opal knew about the drinking, though she never made a fuss. It was clear that she and Bob had reached an agreement over the fate of their respective souls. "I'm letting God work on him" is what Opal said. Nonetheless, every Sunday morning, she still paused at the front door and said to him, "We're headed to church if you wanna come."

As if on cue, Bob started swatting the air, then offered his standard reply: "Y'all go on. I gotta stay here and kill these flies."

After church the family gathered at Clem and Cora's for Sunday lunch. They stood milling around the dining room table with its ring of mismatched chairs, which, to Bobby, seemed as long as a bowling lane. Soon the parade of food began, the trays piled with fried chicken and potatoes, bowls of gravy, beans, biscuits, sliced cantaloupe, and pots of coffee. Three pies sat cooling under cloth napkins, all of which Cora and her daughters had prepared before church. Opal, like her mother, could feed Coxey's Army on a moment's notice.

Then Bob appeared, smelling of Varsol and degreaser, and endured a good-natured ribbing from the other men.

"Missed you at church this morning."

"Oh, had a flat on that ol' Ford."

"Killing flies again?"

Here they were, the men of his childhood — Clem, Herman, Homer, Ed and Fred, Bob, Charlie, Bill, Tooter — back-slapping, talking trucks and the absence of rain, sharing lewd jokes under the earshot of wives. Then everyone squeezed into the small dining room, kids were summoned and told to hush, hands clasped in a great circle around the table, and Clem began to pray, *Oh Heavenly Father.* Within what seemed like minutes, the men were leaned back and reaching for toothpicks, calling on pie, shuffling the children back outside with a warning to not soil their clothes. And how many times did Bobby return inside to use the bathroom, to apply monkey blood to a skinned elbow, and find them all singing?

It was an idyllic world, but one that came undone on April 6, 1961, the year Bobby turned seven. It was a Thursday night. Sister

Sherrill from church had stopped by Clem and Cora's for supper and fellowship. After coffee, she and Clem sat in the living room while Cora finished up the dishes. Clem was reading aloud from the Bible on his lap, one hand raised to proclaim the Word, when suddenly he fell silent. Sister Sherrill ran into the kitchen.

"I say," she told Cora, "I do believe something's the matter with Brother Wilkerson."

The funeral was massive. Hundreds crowded the tiny tabernacle at First Assembly, so many they clustered near the doors and along the walls. The flowers concealed most of the stage and baptistery and filled the air with heavy perfume. An open casket held center stage. Inside was their beloved deacon, the man the Lord had scraped off a downtown street as a living testimony to them all, the man who built the very roof that sheltered them now from the howling April wind.

"He died proclaiming the Gospel," they said.

"Ain't a better way to go."

None of them doubted where Clem's soul had gone — to that great family reunion, as he liked to say — and so there was joy swirling beneath the sorrow.

But Cora and her family could not be consoled. They filled the first rows of pews and sat in disbelief. Opal and Zelda did not sing. Brother Homer Sheats drove down from Lubbock, where he pastored another church, and delivered the sermon. Already he had ministered to Bud during his dying hours back in 1936, then laid to rest both little Mary Lou and John Lewis.

The family buried Clem at Trinity Memorial Park, three miles south of town. Riding out to the cemetery, Bobby and Preston stared out the back window and marveled at the long trail of cars. It snaked over the hills and seemed to have no end. As Clem's casket was lowered into the ground, the wind shifted suddenly and turned the sky to rust, then ushered in one of the worst sandstorms of the year. People ran for cover and drove home with their headlights on. Clem was seventy-five years old.

Not long after Clem's death, Bob's family moved into a two-bedroom shotgun at the dead end of Owens Street. The property had a large back lot where Bob kept his dirt, and where Preston waited with a BB gun to shoot the cats that left it stinking of ammonia. On weekends, he disappeared with his father beneath one of the Fords like a prairie dog into its hole, while Bobby showed little interest in mechanics. Instead, he walked over to a large pasture at the end of the street where his friend Grady Cunningham waited.

The pasture is where Bobby had first spotted Grady the day they moved into the house. Grady stood amid the mesquite and sticker burrs watching their car as it pulled into the drive. He was pale and fleshy, with freckles covering his face and arms. His hair, as red as a Martian sunrise, stood frizzy in the wind.

Grady was eleven months older than Bobby. The two boys had attended different schools and, until now, had never met. Like Bobby, Grady was the youngest child, an accident baby. His father, Luther Cunningham, worked as an engineer on the T&P railroad, spending his days along the three hundred miles of track between Baird and Toyah. One of nine brothers, Luther was a solemn man who took great care to make himself invisible. He did not vote and never served on a jury. If he ate in a restaurant, it was never the Settles or the bustling Crawford, but the dark railroad cafés where he could vanish into the scenery. On rare days when he was home, Luther liked to sit in his chair and be left alone.

His wife, Louise, was pragmatic and dutiful. She'd guided their two daughters, Nancy and Linda, with a firm but gentle hand and a steady Church of Christ upbringing. But in January 1953, when the girls were twelve and nine, their baby brother arrived and hijacked their mother's affections.

Born into the epic drought, Grady was allergic to water. Baths left festering welts on his skin, which Louise lathered in salves and creams to keep him from screaming. The best doctors in Big Spring and Lubbock were flummoxed. Luckily, the railroad provided Luther with good insurance, so he and Louise took Grady to the Cleveland Clinic, where a specialist diagnosed him with a rare form of psoriasis. Heavy doses of topical steroids soon allowed him to bathe.

As Grady grew older, Louise dragged him to a chiropractor for rounds of colonic hydrotherapy, which was thought to flush allergens from the body. Yet for the rest of his life, things like cows, dust, and certain grasses — the very furniture and fixtures of the plains — caused his skin to blossom with hives.

While Louise coddled her sickly child, Nancy and Linda faded from the foreground. Louise smothered the boy with so much attention that he grew anxious and hyperactive and raged whenever deprived. In order to control him — and also out of spite — his sisters tied him to a clothesline pole whenever they hung the laundry.

Each Sunday Louise drove her kids to the Church of Christ on Fourteenth and Main. The church clung to lean restorationist principles, singing their hymns a cappella because the New Testament never mentioned instruments. There was little hellfire in the Church of Christ — they had no truck with the Holy Spirit or bizarre heavenly tongues — but on visits to First Assembly with Dad, Grady found inspiration.

Like Bobby's older sisters and cousins, Grady liked to play church, and he liked to preach. But the way he did it was like a jester tap-dancing on the altar. Out in the pasture, he cleared a place under a mesquite tree and muscled over a block of sandstone for his pulpit. With a Bible in his hand, he took on the persona of a Pentecostal flamethrower, hurling sharp rebukes at his assembled flock — usually Bobby.

"Repent!" he shouted. "Repent, or be *h-e-l-l-l-bound*!"

In second grade, Grady started wearing his

sister's clothes. He put on one of Linda's dresses and smeared his mouth with bright red lipstick, then walked over to Bobby's to play. Nancy and Linda complained to their mother, worried how the neighborhood might respond, but Louise defended her son, saying, "I see nothing wrong with a little dress-up." Bob found it hilarious. Whenever he answered the door, he turned and yelled to Bobby, "Sister Grady's here to see you."

Bobby never saw Grady as "peculiar," which was how adults sometimes described him. The cross-dressing turned out to be a phase anyway, and the boys continued their games in the pasture, playing baseball with other neighborhood kids, catching horny toads along the grass line, and sneaking Luther's cigarettes — an act that filled Bobby with guilt.

As the summers passed, Bobby sprouted tall and lean while Grady grew pudgy. Once, while horsing around, Bobby tried wrapping Grady in a scissor hold, only to find himself pinned with Grady's knees in his chest. His strength was surprising. During another scuffle, Grady stunned Bobby by ramming a sharp pencil into his stomach and drawing blood.

In school, teachers often punished Grady

for rude behavior. He burst into song during their lessons — "They say don't go . . . on Wolverton Mountain" — daydreamed, and slacked on his homework. And Grady could be sadistic, especially to girls. He shot spit wads into their hair and once trapped a classmate in the restroom by pushing a desk against the door.

Bobby noticed how few kids wanted to play with Grady and how he walked alone in the hallways. Even Opal noticed this, saying, "Bobby, you must be that boy's only friend," in a tone that suggested both pride and pity.

Sensing loyalty, Grady nudged closer to Bobby. He invited him to his regular chiropractor appointments, showing him the table where they inserted the tube that power-washed his insides. After Grady disappeared with the doctor, Bobby sat there with Louise, trying to read a magazine over the humming of the machine.

Sometime the following year, when the boys were ten and eleven, Grady took Bobby to meet a new friend. They rode their bikes over to a small clapboard home that had a tidy yard and potted flowers up the walkway. An old man answered the door and Bobby recognized him as a friend of Grady's father, Luther, who also worked for the railroad. The man took Grady down into the basement but made Bobby wait outside. After half an hour, he saw his own way home.

Bobby was never invited back. "He doesn't like you," Grady told him. The old man had a homely-looking wife and two grown children who apparently never suspected a thing. The same went for Luther and Louise, who allowed the man unlimited access to their son.

At one point, a *Herald* photographer even captured them after a fishing trip. In the photo, Grady stands next to the man and his wife, who is smiling. The man wears a straw hat and a chambray shirt tucked into a pair of high-waisted jeans, his geriatric flesh sagging beneath his clothes. Grady — stout, his hair styled in a buzz cut — clutches a fourteen-pound catfish caught off a trotline. The caption reads, "Luck at Lake Thomas."

Sadly for Bob, orders for red cat-claw steadily declined after the drought ended, forcing him to realign his vision of an empire built on sand. Already he was a man alone. His cohorts in the dirt game, Charlie and Bill, had long sold their trucks, having grown fed up with the hard work and low returns. Bill even started prank-calling Bob to toy with his temper.

"Mr. Mealer, I hear y'all are having some weather up there. Is there any chance of you bringing me a load of snow?"

"Snow?!"

For a while Homer drove one of the Fords, but he quit after he got married and found a better job. So the day Preston turned fourteen, Bob went down to the DMV and got him a hardship license, and they drove together all summer. But once school started back, Bob had a hell of a time hiring help. The best he managed to find was a dimwit rounder named John Lee who abandoned one of the trucks by the highway rather than change a blowout. Bob was furious.

"Why didn't you call me from a station?" Bob demanded.

"I didn't have your number," John Lee said.

"The heck you mean you didn't have my number?"

"Well, not with me, Bob. Got it wrote on the ol' wall at home."

Out in Odessa, the company that Fannie and Abe's boys ran, Jones Brothers Dirt and Paving Contractors, was growing fast — so fast that Raymond and Troy were buying

311

horse ranches and Cadillacs. Thanks to the Spraberry oil boom in the fifties, Odessa added over fifty thousand people, and the Jones boys were building new roads to fit them all. Hearing that Bob was having hard luck, they offered him a job, but Bob was reluctant to accept on account of a recent tragedy.

Just two years earlier, in August 1962, Norma Lou's ex-husband, Bill Glaesman, was hauling equipment down to Sanderson for the Jones boys when he dozed off at the wheel. He managed to wake up just as his semi left the pavement, but the act of correcting the rig caused it to jackknife and flip. Bill was thrown from the cab and his truck caught fire. He died on the side of the road.

Bill and Norma had split up a couple of years earlier, and Bill had left her with a boy, Rodney, to raise on her own. But despite this, Bob was fond of Bill, and news of his death landed pretty hard — it was the only time Bobby can remember ever seeing his father cry. The Jones boys, for their part, had paid a handsome settlement to Bill's new wife, even though the crash was no fault of theirs. For Bob, the tragedy still lingered. But as much as he hated to do it, he hired out his trucks to his nephews and moved the family sixty miles to Odessa.

Compared to Big Spring, Odessa was grubby. While it was bigger (pop. 80,000) and offered more in terms of shopping and restaurants, its landscape was mostly flat and choked with mesquite. Over the years it had become the well-established work camp of the oil patch, while Midland, which sat between it and Big Spring, seemed to attract a higher class of people. Midland was smaller (pop. 62,000) but it had the oilmen, executives, and tall buildings, while Odessa settled for hard-luck roughnecks, truckers, and broken-down cowboys. People like the Mealers.

Bob and Opal took a two-bedroom house on the north side of town, right on West County Road where oil trucks and semis rumbled past night and day. Bobby started fifth grade while Preston enrolled at Odessa High and joined the football team. At seventeen, he stood over six feet tall, and his brawny physique rippled with power. "Much of a man," Bob liked to say, reaching up to slap the boy's shoulders the way he'd kick a set of Michelins. Preston played junior varsity defensive end, and his strength and quickness off the line left his coaches marveling.

"You'll be All-American if you stay with

it," they told him.

His friend Larry Gatlin played varsity quarterback and sang in a gospel trio with his brothers. They sang at the Mealers' church, Bethel Assembly, and performed on local radio and TV. Larry's daddy was a driller, but Larry liked to talk about college and a life somehow absent of sand and sulfured coveralls. Some nights, after hanging out with the Gatlins, Preston would lie in bed and imagine himself a college man, the All-American.

But Bob and Opal never approved of Preston playing football and offered little support. During his senior year, his grades slipped anyway and he quit the team. After graduation, Larry Gatlin accepted a scholarship to the University of Houston to play wide receiver, then moved to Nashville and cut albums with the help of Dottie West and Kris Kristofferson. Preston got married and went to the oil patch, laying pipe for two dollars an hour.

On Saturdays, after hauling gravel all week for the Jones boys, Bob took Bobby and Opal to visit his sisters. Fannie had long divorced Abe Jones and now lived on her own. Her sons had bought her a nice house and provided her with plenty of cash, which she liked to hide in strange places around her home. Once, Bob looked up and saw a bouquet of hundred-dollar bills rolled up in the dining-room chandelier.

Like their father, Abe, the Jones boys courted danger. One night after a dance in Stanton, Earl got into a fight with a man who pulled a knife. It was Fannie who ran between them and took the blade in her arm. Were it not for a set of drapes in her backseat, which Earl fashioned into a tourniquet, his mother would have died in the parking lot.

Bob's sister Allie had remarried, this time to a roughneck named Tom Henson. Unlike her previous husband, Lee Pruitt, Tom was a kind and compassionate man who loved children. Whenever Bobby and his cousins came to visit, Tom would dress up in silly costumes and parade around the house, much to their delight. But Allie still grieved over her son, Orville, who'd been killed in the war. She still kept the door to his bedroom closed, but the kids knew what was inside: Orville's trunk draped with a flag, along with pictures of him on the wall framed with curtains, like in a church. And they knew about the headache his death had given his mother, one that still greeted her each morning.

Bob's other sister, Velva, was still married to Tom Henson's brother, Willie Bob, a cowboy who broke horses on the Jones boys' ranch. For decades he'd blown the family's money on whiskey and honky-tonks. But because he was never vicious, his children loved him. For years Velva had accompanied her husband to the bars and dances, but stopped in the mid-forties after she found the Lord. One night my grandmother Opal had taken her and the kids to a revival in Odessa, featuring Opal's old friend Frank Mack, "once a paralytic and now a flamin' evangelist." At the end of Mack's sermon, Velva and the kids walked to the altar and repeated the sinner's prayer. That night she told Willie Bob, "I don't want no more of this life. You do what you want to, but I'm going to church."

Velva's relationship with the Lord helped steel her for tragedy. A couple of years after being saved, their boy Jack drowned in a caliche pit. He'd just finished his paper route and ridden his bicycle to the swimming hole, where his friends said he jumped in but never surfaced. After Jack's body was recovered, one of the boys rode the bike back to the house, clothes and a pair of tennis shoes still tied neatly around the handlebars, just like Jack had left them. That very morning he'd told his mother, "This is gonna be a great summer if Daddy don't drink."

At Jack's funeral, Opal sang "Does Jesus Care?"

Does Jesus care when my way is dark
With a nameless dread and fear?
As the daylight fades into deep night
    shades,
Does He care enough to be near?

After two years of living in Odessa, Bob grew fed up, and in 1966 decided it was time to come home. As if looking for an excuse, he convinced himself the Jones boys were cheating him. He'd pore over his ledger each night and conclude that the number of paid loads versus money in his pocket did not jibe. He grumbled about it daily, fumed and bristled when he sat down for his supper. Then one afternoon at work he finally exploded, dumped both beds right there at the job site, then announced to his driver, "We're going to the house!" There, he instructed Opal and Bobby to start packing their things. He hauled their belongings back to Big Spring in the dump truck.

But the town had changed by the time they returned, or at least the family had.

Marriage, careers, and the steady clip of time had dislodged roots and scattered them like tumbleweed. Preston and Norma Lou were living in Odessa, and Fred, Doris, and Iris were all gone. Others had settled in far-flung oil towns and returned only for holidays with their children, who had to be reminded of the names of family.

The death of Clem had served as just another blow to the fracturing alignment. They still gathered at Cora's for Sunday lunch, but the house was quieter now. The table seemed more in proportion with the room, and before sitting down to eat, the *Oh Heavenly Fathers* were raised by whomever. After the dishes were cleared, they still gathered in the living room to sing, but their songs now carried a hollowness in the middle that only reminded them of the precious octaves no longer in the choir. Everything felt different, like a strained reenactment of better days.

2

The fall of Raymond Tollett . . . a town
mourns its king . . .

For Raymond Tollett and Cosden Petroleum, the 1960s opened with a surge of
expansion. In 1960, Cosden began licensing
its plastics technology overseas — first to
Sinclair and Koppers, then to Naphtachimie
in France. Within a few years engineers were
flying from Big Spring to Japan, Germany,
Belgium, Poland, Italy, China, and Czechoslovakia to open new refineries. And since
only 3 percent of a barrel of oil could be
used for petrochemicals, the company operated over a thousand filling stations nation-
wide to market its gasoline.

Tollett, now twenty years at the helm, had
taken the $6 million company and increased
its earnings to $83 million. Each year
brought new innovations, dreamed up by
his scrappy team of engineers tinkering in

the middle of nowhere, their zeal for the company and its president resembling that of a sports team. What people at Cosden loved most was their underdog status, which drove bigger companies crazy. And thus, it was only a matter of time before the big boys came for a piece.

Out in New York, J. Peter Grace was fascinated by Cosden. He was president of W. R. Grace and Company, which his grandfather — former New York City mayor William Russell Grace, who accepted the Statue of Liberty from France — had founded in the 1850s as a shipping firm. When J. Peter took over in 1945, he began branching into chemicals, and by the mid-fifties had snatched up companies across the country. He started pursuing Cosden in 1956, smitten by its vast supply of crude, pipeline, and chemical capacity. Grace proposed a merger, but Tollett and his directors declined. Another attempt failed in 1959. Finally, in January of the following year, Grace staged a takeover by purchasing 51.9 percent of the company's common stock, making Cosden a subsidiary.

At first, it proved a convenient arrangement. Tollett remained president and the company kept its name. Better, with oil prices at a momentary low, the security that

Grace provided guaranteed expansion. A new product line was introduced, and Grace even built a fertilizer plant across the street that further bolstered Cosden's portfolio.

But the two men clashed. According to executives, Grace wanted to plunder the employee pension plan, which Tollett had put into a trust that couldn't be touched. The sluggish market eventually put a damper on what Grace originally thought he could make off the refinery. Fed up and occupied by more lucrative ventures, Grace sold its controlling shares in April 1963 — and with it, Tollett's claim on his prized Jewel of the West.

The buyer was American Petrofina, the U.S. operator of the mammoth Belgian oil company, which laid down $90 million for the entire outfit. All existing shares, including Tollett's, were liquidated. The old stockholders, who'd met each year at the Cosden Country Club over T-bones and tumblers of whiskey, gathered one last time on the top floor of the Petroleum Building to await the final sale. Nearly two thousand miles away in Jersey City, a team of lawyers gathered signatures and it was done. "We have just been informed that the closing has been completed," the chairman in Big Spring announced, after hanging up the telephone. "This meeting is adjourned."

And with that, the *Herald* reported, "last rites had been pronounced over the old patriarch."

A new enterprise, the Cosden Oil and Chemical Company, rose in its place. Employees were guaranteed their jobs for at least a year, after which there were no promises. At first, Tollett refused to sign the letter of intent that fixed the sale, then reluctantly agreed. The new owners asked him to stay on as president and director, but with largely diminished powers — as a glorified manager who answered to New York and Brussels.

Instead of giving his decision, he hopped on a plane to Kenya, far off the radar, where he could think in peace. He took a long safari, spent days walking the beach in Mombasa, and contemplated what he'd lost and what there was left to gain.

At fifty-six years old, he was already a rich man. Personal oil dealings, plus his salary, had left him a millionaire, and the sale of his Cosden shares had netted him even greater wealth.

But years of stress had worsened his drinking. His benders now lasted weeks instead of days, and within the past year, he'd hospitalized himself on several occasions to get sober. His executives were exasperated, worried that his habit was dulling his famously tactical mind. He forgot things easily and moved a bit slower. Worse, he was showing up drunk at the refinery on Sundays and palling around with the men. Come Monday, the union boss was pulling out his hair from all the promises Tollett had made to everyone. The drinking had become a corporate liability.

Tollett knew his days as a magic man were limited, if not already gone, just as they'd run out on Joshua Cosden himself. If there was any legacy left to build on, he concluded, it was repaying the town for all they'd given to him.

He returned to Big Spring and resumed his job, but without a contract. "I'll serve from term to term at the pleasure of the directors of American Petrofina Incorporated," he wrote in his monthly column. As for his reasons for staying in Big Spring, he said, "It just so happens that I like the people."

He bolstered his Siblings Foundation, named in honor of his children, which gave grants to "people of all creeds and colors" to start small businesses and pay college tuition. And he aggressively raised money for other Big Spring charities, such as the

United Fund, which funneled money to over a dozen local groups. For this particular campaign, Tollett expected each of his clients, from truck contractors to caterers, to donate their share.

This was something Homer had discovered personally several years earlier while working as dispatcher at Steere Tank Lines, whose trucks hauled gasoline out of Cosden. One afternoon he picked up the telephone to find Tollett on the other end.

"If I don't have a check for the United Fund on my desk by five o'clock this afternoon, no Steere trucks are entering my refinery," he said, then hung up the phone. Homer scrambled to get the check and arrived with minutes to spare.

Waiting for him at the gate was Tollett, who'd invited the *Herald.*

"This," he told the reporter, slapping Homer on the back, "is what I mean about a company doing its part."

With Tollett, everybody won.

In 1964, Tollett was pushing hard for a bond election to expand the school district, which he'd championed for over twenty years. Shortly before the vote, he appeared on the local television station KWAB (which he had helped bring to town) to drum up support for the cause. But by the time the cameras rolled, Tollett was so drunk he could barely get through the segment.

As his drinking became more public, so did other sins. He was cavalier with his mistresses, bringing them into the Petroleum Building, where he pampered them in his top-floor suite. He brought them to the airport, where he ordered his pilot to fly them around town so they could carouse. Many people witnessed this behavior: employees, members of his church, and friends of his wife, and it both incited their disapproval and saddened them. Around this time Tollett also began leaving hundred-dollar tips under his plate in restaurants.

Meanwhile, up on Hillside Drive, Iris drank in bitter isolation, save for a few lovers herself. The infidelity cut both ways but always seemed to cut her the deepest. "Everyone in town talks about me," she told a friend, and she wasn't wrong.

People concluded the only reason Iris and Raymond stayed married was to destroy one another. Explosive fights at the home sometimes brought the police. But the officers remembered the policeman's ball that Mr. T. threw down at the club for them and their wives, with all the food and dancing they could stand. They issued a gentle warning and the boss apologized for making them come out so late. "Just a little misunderstanding," he told them. "Just a little too much to drink." The door would close and a tired voice on the other side would tell little Ann to go back to bed.

A terrible fight erupted one evening over dinner. Iris's niece Judi was seated at the table, and so were the kids. At one point Iris said she was leaving, jumped up, and stormed out the door, then Raymond did the same. But he reached his car first, and when Iris made the mistake of crossing behind him on the way to hers, he backed up.

The car knocked her down and rolled over her leg, snapping her femur in half. Her niece heard the commotion and ran outside with the kids. They found Iris pinned beneath the tire, screaming, while Raymond stumbled into the house to call an ambulance.

Doctors installed a steel rod in Iris's leg, but not before spending an hour picking the gravel out of her skin. The paper reported the incident as "a mishap at the family home" and gave the gossip mill a delectable feast. Raymond told the police, "I must've hit the accelerator instead of the brake."

■ ■ ■ ■

In order to shield the children from their drinking, the Tolletts sent them to boarding school. Ray Jr. began attending New Mexico Military Institute in Roswell in 1963, with Jason Blake following two years later. After junior high, Ann went away to a Baptist academy in San Marcos, Texas. Each month, their father would fly out for a visit, or else send the Heron to fetch them.

Summers weren't left to chance. Ann left for Western Life Camp in northern New Mexico, while the boys attended Camp Silver Spruce, in the mountains outside Durango. Once they got older, Tollett insisted the boys travel abroad, but always on their own. Each of their trips was carefully detailed in the morning *Herald.*

Kept out of the society pages, but known throughout town, were the trips Tollett was taking to dry out — spates of voluntary detox that were becoming more and more frequent. For years, Raymond had relied on his friend Dr. Bennett to help recover. But after Bennett's sudden death in 1958, Tollett's benders became uglier, and to get sober, he turned himself over to Bennett's old partner, Dr. Clyde Thomas, the refinery's chief physician and surgeon.

Dr. Thomas practiced out of Hall-Bennett Memorial Hospital, where my aunt Zelda worked as a secretary. As soon as Tollett arrived, Dr. Thomas sent for two bottles of scotch and began the slow withdrawal, then remained at the hospital around the clock. To ease the shakes and prevent seizures, he administered heavy doses of benzodiazepine, usually Librium or Valium. For Tollett, the treatments could be anguishing, marked by fevers and violent hallucinations. But after about a week, he always resurfaced on the other end, donned one of his custom suits, then drove back to the refinery and made it rain.

"Such a brilliant mind," Dr. Thomas once said. "And yet we'll never know what it's truly capable of doing."

On May 22, 1968, the evening *Herald* delivered shocking news. After nearly thirty years at the helm, Tollett was stepping down at the refinery, effective immediately. The paper was working off a press release handed down from American Petrofina. But as everyone suspected, the Belgians had simply grown tired of Tollett's drinking and sent word to sever ties. "Ray can make a better deal drunk than sober" is what they used to say, but Petrofina was too tidy an enterprise for loose cannons, regardless of their accuracy.

The story ran on the front page, top of the fold, along with a photo from a small, awkward reception thrown together at the Petroleum Building. In it, Tollett embraces the hands of Paul Meek, his second in command, who would take over the reins. Meek himself appears agitated. The look from Tollett, whose face is puffy and tired, is one of crushing surrender.

There was no sendoff at the refinery where his people labored, no opportunity to wave his hat under the lights while they bade him farewell. His only parting words, quoted in a later story, were aimed at them: "Of all the sins that may be counted against me, I do hope that I am never guilty of or accused of being ungrateful."

By the end of the summer, Tollett had opened a law practice with a local attorney, John Burgess, and hung his shingle on the fourth floor of the Permian Building. He still lunched at the Hotel Settles and held court on the downtown sidewalks, where people stopped to greet him. Impeccably dressed and radiating that same old magic, he never forgot a name.

For the better part of a year he put on a good face. He still visited old employees in the hospital and showed up to their funerals. But the life of a country lawyer was akin to death itself. His past threw an inescapable shadow. And for an alcoholic, the town itself must have loomed like a giant trigger — the company offices two blocks away, the *clang clang clang* of the Cosden filling station across the street, and every building he'd ever entered.

People told stories about Tollett's last few months, hushed ones that weren't meant for repeating. Like the time one of his former engineers found him passed out in his car on Main Street and had to drive him home. Or the night he fell down in Herman's new restaurant, before God and all of Big Spring, and couldn't get up. Not a fork moved as he lay on the floor. Herman finally ran over and helped him to his feet, then called a cab. The driver knew exactly where to go — they all did. Each time he fell down, the town was there to pick him up, to carry him home atop their able shoulders.

But then, one time they turned around and he was gone. On October 25, 1969, Tollett checked himself into the Starlite Village Hospital and Clinic, a rehab facility located in the Texas Hill Country. Perhaps he needed to get away from the old triggers, to dry out in a place where it might finally stick. Nobody knows exactly why he chose that place.

But the doctors at Starlite dried him out too fast. Two days into his treatment, away from the round-the-clock care of his trusted physician, Tollett went into convulsions and died. He was sixty-one years old.

His funeral, held two days later, was the biggest the city had ever seen. The refinery closed until noon so workers could attend. Over a thousand men and women who'd given Tollett the best years of their lives packed St. Mary's Episcopal Church, buried their faces, and wept. Joining them were oilmen from Fort Worth and Oklahoma, executives from New York and Chicago, along with so many others who'd visited the man at the top of the Petroleum Building and not walked away empty-handed. The church couldn't hold them all. They crowded the aisles and spilled onto the sidewalk, where the line to pay respects to the last king of Big Spring stretched around the block.

As for my family, Tollett's passing came as sad news, if nothing else. Aside from Bobby's cousin Granville, who developed patents at the refinery, our two orbits had remained mostly independent. It would take another eight years for them to cross.

3

Bobby grows up . . . meets Ronnie and Marie, battles the Lord . . .

Back when Tollett was still perched on his throne, in December 1966, Bobby turned thirteen. That winter a growth spurt seized him unawares and left him long and skinny, with two gangly arms that nearly reached his knees. At the same time, Bob finally concluded the boy would never dirty his nails beneath a Ford, much less drive one. The least he could do was impart a work ethic.

That summer, Bobby began sacking groceries at the air base commissary, where Zelda's Charlie was manager. From the start, he heeded his father's words and performed his work with pride. He took special care not to make the sacks too heavy, as most of his customers were women who'd be unloading them alone. Because he worked for tips, he learned early to spot the officers' wives and walk them to their cars, saying softly as they crossed the parking lot, "I'll bet that's a new dress you're wearing, ma'am," or "I was thinking how you remind me of the actress Audrey Hepburn." The women would smile and reach into their purses for change, sending Bobby home with around thirty dollars a shift.

It was the summer of 1967, but the tumultuous events taking place around the country unfolded without Bobby's knowledge. That July, as hippies protested Vietnam and paraded naked in the Haight-Ashbury, and as race riots raged in Newark, Detroit, Houston, and eight other American cities, Big Spring remained isolated by its geography, shielded by its sand and wind and locals bent on keeping the status quo. The 4-H Club held a "calf tour" through downtown and *The Alamo,* starring John Wayne, returned to the Jet Drive-In, back by popular demand. The Old Settlers held their forty-third reunion at the park to reminisce about the pioneer days and cook a batch of son-of-a-gun stew. And it rained, an honest-to-God soaker.

Aside from the Cosden engineers and air force officers who brought their modicum of high society, there was everyone else, who observed the rules of the frontier. For any longhair bold enough to express himself in the Summer of Love, there was a pickup full of cowboys ready to test his mettle.

While Big Spring didn't experience the same kind of racial strife as the rest of the country — the air base had long brought minorities to town and the schools had been integrated since the mid-fifties — most blacks and Hispanics still kept to the north side of the T&P tracks. Except for the times when the world's problems blindsided the town — like the dust and sand of the thirties and fifties, which came from above — things did not change. In fact, people still talked about the summer of 1963, when Dr. Martin Luther King rallied the march on Washington. All week long passenger trains full of blacks rolled through Big Spring from points west. None of them ever stopped, yet judging from the reaction to so much dark skin through the windows, you'd think each car cradled an atom bomb.

While the Summer of Love passed without incident, it marked Bobby's first awareness of girls. And much to his surprise, girls liked him, especially church girls. He was independent, which made him seem older, which suggested the possibility of mystery that girls reached for like a glimmer at the bottom of a pool. At fourteen he even got his own car, a used Galaxie 500 that he drove thanks to a hardship license his father had helped him get, more out of family tradition than anything.

Plus he could dress. With the money he earned at the commissary, in addition to mowing lawns, he shopped at the men's store downtown and pieced together a mod wardrobe of mock turtlenecks and cardigans, straight-cut jeans, and tapered slacks. And his gift of florid BS, which so enraged his brother Preston, had developed into something resembling confidence.

The ensemble was fully flexed the following summer when a new preacher arrived at First Assembly. Brother Farmer had a daughter named Marie, with long dark hair and big brown eyes that took in her new town. Big Spring was seven times the size of Muleshoe, whose only calling card was a life-size fiberglass mule named Pete that stood at the north end of Main Street. Marie had aspirations, which amounted to being popular and beautiful and driven around in muscle cars. She also desired a boyfriend right away, and Bobby seemed good enough.

He fawned over Marie and took great

pains to be the boyfriend he thought she wanted. He buried her in gifts and kitschy cards and kept his charm on overdrive. Marie gave him just enough in return: a brush of bare leg in church (while her father preached the sermon), a wet kiss at the end of a night, perhaps a note that suggested he was her only one. He'd never met anyone as pretty as the preacher's daughter or anyone harder to please. He would work for her love.

Marie had two older brothers, both of them infamous. The oldest boy, who now lived elsewhere, had been kicked out of Southwestern Assemblies of God College in Waxahachie for humping a mannequin in his bed, a prank that had clearly gone wrong. The next oldest, Ronnie, lived at home and was far worse. It was rumored that Ronnie was the reason his parents had to leave Muleshoe, on account of some trouble he'd gotten into on graduation night. He and some buddies got liquored up and painted Pete the Mule neon green, then spent the night in jail for it. And not long after arriving in Big Spring, he drove back to Muleshoe for a football game and got arrested with a trunk full of beer. His parents were trying to enroll him at Waxahachie, hoping it might straighten him out.

But until that happened, Ronnie was set loose on Big Spring, and right behind him was Bobby.

Bobby liked Ronnie even more than he liked his sister. For starters, Ronnie drove a two-tone Pontiac GTO. He was also handsome and sharply dressed, and girls paid attention. It frustrated Marie to no end when cheerleaders and class officers approached her and asked to hang out, then casually added, "So is your brother gonna be there?"

Ronnie dated several girls at a time. He filled his car with cheerleaders and drove to the top of Thrill Hill. Then he gunned the engine and flipped off his lights as they sailed over the rim into pure black, which encapsulated Ronnie's theory of everything. There was also the night when Ronnie and a bunch of girls filed into a local steakhouse and Ronnie ordered half the menu: T-bones and chops, fish and Sunday ham, every kind of dessert. When it came time to leave, Ronnie spotted an old man and his wife sitting at a nearby table.

"Hal Johnson, is that you?" Ronnie said.

No, the man replied, but Ronnie kept at it, just to keep him talking. As he walked out the door, he told the waitress, "Dad says he'll take the bill."

Ronnie did anything for money. "Ten

bucks says I won't drink this," he said once, holding a bottle of dish soap, then poured it down his throat. Years later up in Pampa, Ronnie consumed one hundred jalapeno peppers in fourteen minutes and nineteen seconds, and entered the *Guinness Book of World Records.* "My stomach felt like a small campfire," he told a reporter.

Ronnie drank beer and smoked cigarettes, so Bobby did too. And when Ronnie walked into a grocery store and slipped a pack of Winstons into his coat, Bobby did the same. Filching cigarettes became routine and the burden of sin clung like sediment to his soul.
  Lately, his relationship with God had gotten harder to maintain. It seemed that no matter how hard he tried, or how much he prayed, he could not follow the church's myriad rules that guaranteed salvation. Breaking just *one* knocked you out of grace. He tried keeping a tally of his sins, recounting each one when he prayed at night asking God's forgiveness. Aside from the beer and cigarettes and using foul language, there were the times when Marie took him to the brink of pleasure beneath the flicker of the Jet Drive-In, only to withhold her love. The sins kept mounting, and after a while Bobby concluded that he was too

weak to stop them. *I'm just not worthy,* he told himself.

Grandma Cora could see right through the dirty window of his soul. She even called him out for going to the movies once, saying, "What if Jesus came back while you were in that place?" He knew the answer, for it haunted him: empty sidewalks, cars without drivers, the plague of locusts with the faces of men. The great eye of heaven closing over the earth like a scroll. They'd taught him all about the Rapture.

He strove for salvation. When the preacher invoked the old story that everyone knew, how Billy and Suzy slipped out of church one night to go dancing instead of hitting the altar, and how they crashed and died on the way home and went to hell — well, Bobby was right up front, getting straight with the Lord.

Bobby was weak, and the fact that Ronnie's father was the pastor at First Assembly had worked like spiritual subterfuge. As soon as Bobby began to doubt his faith, Ronnie had appeared and led him into rebellion. Sinning with the preacher's boy not only seemed admissible, but it was fun.

"Hey, you know why it's a sin to screw while standing up?" Ronnie asked.

They were in his GTO, speeding home

from a youth rally in Lamesa and drinking beers from the trunk.

"No, why?"

"It might lead to dancing."

They flung their empties into the dark while Zappa screeched over the eight-track and faded into the plains. The gas flares out in the fields appeared like watchmen at their posts, on guard against those larger worldly fires. But their protection was not sufficient, and the problems of the world found their way past.

In April 1968, Preston was drafted for the war in Vietnam.

At the time, he and his wife, Linda, were living in a garage apartment on Twenty-third Street in Odessa. For two years, Preston had been trying to find his niche: first as a pipeliner, then a brief spell as a welder's helper. For a while he operated a forklift unloading boxcars, but the position was only temporary. He was out looking for another job when Linda opened the mailbox and saw the letter from the Selective Service System. "Uncle Sam Wants You," it said in big bold letters. She held it in her hand and wept.

The letter on the table, in addition to sending him off to fight someplace he could not picture in his mind, would also keep him unemployed. The government had classified him 1-A, fit for imminent service, rendering him untouchable. After he spent two weeks looking for a job, nobody would hire him.

His aunt Allie's new husband, Tom Henson, worked on a well-service unit based out of Odessa. When Tom heard about Preston, he called his boss and got him on a crew that serviced pump jacks. For three months Preston drove around with Tom, saying little as the dust boiled up in the cab. His mind was nine thousand mile away, running through the dark jungles of his future.

4

Frances moves right . . .

For Frances, life was entering its third chapter: Scottsdale, children, domesticity. When the judge gave Tommy custody of his two girls, part of the deal was that Frances had to stay home and care for them. She'd accepted this deal mainly so they'd be safe, but she also enjoyed being a mother to them and her own two girls. It was all she could do to try and right the wrong that had been done to her as a child.

Taking care of four girls was real work, and Frances ran the household according to a tight schedule. Each morning she got the girls out of bed and into the bathroom to get dressed, then tried to cook a real breakfast of eggs and bacon and pancakes. Once the house was empty she did her exercises: thirty toe touches and a few simple calisthenics (into her nineties, she would still be able to do splits). She showered, put on makeup, and if she had to leave the house to run an errand, she put on a pair of heels and a cute dress and styled her hair just so. Later, she tackled the usual mountain of laundry before the girls returned for lunch (the school was just across the street). In the afternoons she helped with homework and projects, and once the supper plates were cleared, she bathed the girls, scrubbed their heads, and set their hair into curlers. This was the routine, and time flew past.

Her one rule was that the family ate breakfast together every morning, and mostly they did. Those days, Tommy was often on the road and didn't come home until late. He and his salesmen were always out looking for new territory to claim, driving into towns such as Ajo and Eloy and going door-to-door, their customers mainly women like Frances who answered the bell at noon. If you'd asked her ten years ago what she'd be doing now, "the wife of a vacuum cleaner salesman" would be the absolute last thing she'd have told you. But Tommy loved it, and he made more money selling Kirbys than he ever had as a musician.

She had to admit she pined for the music now and then, the excitement of the life.

Dancing all night until her legs ached, the long bus rides with the Texas Playboys, and how the songs seemed to change from town to town as if they were living things, independent of the players. Most of all, though, she missed how much they used to laugh. It was strange, really, to think of herself now with those people — Bob Wills and Billy Jack, Ernest Tubb and the others. She could still see Hank Williams shooting dice on the floor of the tour bus as it rolled through the Valley, along with the note he gave to her that night in San Jose. She hadn't heard from Bob or Billy Jack in years. She still saw Hank Thompson whenever he toured through Phoenix, but that made Tommy jealous. Sometimes it felt as if the music had happened in a previous life and there was no way of going back. And what she had *in this life* was a responsibility to her girls and her husband.

In the early sixties, she found all the talk about women's lib annoying. When writer Betty Friedan came out with *The Feminine Mystique* and got everyone in a hissy about the "problem that has no name," it just set Frances off.

To her, staying home was part of the fifty-fifty deal. Tommy worked his butt off selling Kirbys and she kept the house, cared for the kids. God knows she didn't want it the other way around. If being a homemaker was a woman's role, then fine. She wasn't having an existential crisis about it, and she certainly wasn't whining like Friedan and these so-called feminists in the news.

Frances liked to listen to talk radio while she cleaned the lunch dishes and finished chores. At twelve fifteen she tuned into John Sage's call-in show on KWBY, where the discussion was always politics. She'd always considered herself patriotic — she'd led the war bond drive at National Biscuit Company in San Francisco — but she'd never voted in an election and didn't associate with any one party. But in the early fifties she became angry with President Harry Truman and the Democrats for bungling Korea, how they left it wide open for the communists to have their way, then let the United Nations tie our hands while our boys died by the thousands. The war became personal when it sucked in Tommy and Leamon and upended their lives.

So in 1952, when General Eisenhower, the most trusted man in America, declared in his campaign speech, "I shall go to Korea" and end the fiasco, Frances became a Republican. She'd veered to the right ever since, especially in the new decade, with

America coming apart at the seams.

Around 1962, she started calling into John Sage's show and picking fights. Not real fights, of course. Sage's program worked like a public forum where people called in and debated one another. Frances liked skewering the liberals, and even Sage himself, a gadfly who relished the back and forth. Arguing was something Frances was good at — plus she found it fun to sit in her kitchen amid a pile of ironing and feel her heart race in battle.

She set her crosshairs on feminists and left-leaning Republicans. But what made her blood boil were the civil rights protesters and everyone else bad-mouthing the South. After Kennedy was killed and Johnson — a Texan, of all things — started pushing the Civil Rights Act and welfare for all, she couldn't just sit in her house anymore and talk on the radio.

It's not that she didn't support rights for black people. They had a right to vote, be treated fairly, and not live in fear. But like many Americans, even Johnson himself, she was horrified by their marches and riots that began to play out in 1963 and 1964. The rioting was pure insubordination, she felt, a blade into the fabric of the country, and it was terrifying to watch it on television.

What's more, she hated the way people painted all Southerners as racists and bigots. These were hardworking folks, she would argue, most of whom had never done a thing in their lives to harm a person of color. And now their towns and neighborhoods were in chaos and their businesses ruined — all because outsiders came in and egged people on, outsiders whose own northern cities were no better to blacks than anywhere else.

Later, when LBJ rolled out his "war on poverty," it struck her as nothing but a big pity party sponsored by the government. *Poor me* — it was the new cry of the republic. Never mind that as a young girl in Big Spring, she'd stood in that same welfare line herself, and that check had kept her family from starving.

In early 1964 she began attending meetings of the Scottsdale Republican Women and was surprised to find that many of them recognized her voice from the radio, or had read her letters to the newspaper. Like Frances, most of the ladies identified as conservatives, having distinguished themselves from the more centrist members of the party who, at the time, were in step behind New York governor Nelson Rockefeller, whom they regarded as an East Coast elitist. The conservatives were a wily faction, still very much on the margins. Their ranks comprised radical elements such as the old America First isolationists, John Birchers, and McCarthyites, in addition to a growing number of Republicans who feared the country was swinging liberal for good.

In Scottsdale, their very own senator, Barry Goldwater, was the man carrying the torch for their swelling movement. The conservatives had wanted Goldwater in 1960 but got Nixon instead. They were mad when Barry refused to fight for the nomination on the convention floor, and even angrier when Nixon lost to Kennedy and flung the cause into obscurity. But Goldwater remained a conservative hero into JFK's term, "the favorite son of a state of mind," as *Fortune* put it.

Frances loved Goldwater and felt he was the only hope for a country facing its demise: from Kennedy bungling the Bay of Pigs and letting Castro make us look weak to the United States going head to head with Khrushchev and skidding to the brink of annihilation. Goldwater wouldn't pussyfoot around with communists and he took the Constitution at its word, believing the Civil Rights Act ran roughshod all over it.

But to back Barry meant you were part of the total movement. So that spring Frances helped the Scottsdale Republican Women get other conservatives on the ticket, including a local lawyer named John Conlan who was running for Congress.

Frances worked closely with Conlan, canvassing signatures to put him on the ballot, then manning the telephones to raise money and votes. The campaign put her in close contact with wealthy people in Phoenix who threw lavish fund-raisers. The fact that Conlan was a former army captain with a law degree from Harvard, and handsome to boot, made Tommy burn with jealousy. One evening Frances returned from a big gala to find Tommy piping mad. "Were you with Conlan?" he demanded to know. And when Frances answered yes, he smacked her so hard it left her face numb.

In November 1964, LBJ pummeled Goldwater in a landslide, but Conlan got elected. So did many other conservatives across the country, and that was enough to keep Frances pushing for the cause. In 1967, she became a delegate for Phyllis Schlafly's bid to lead the National Federation of Republican Women and traveled to Washington for the convention. Schlafly had written the book *A Choice Not an Echo,* which had galvanized the Goldwater movement and made her a cult hero among conservatives, including Frances. Schlafly ended up losing to a more moderate candidate, but she and Frances became friends. Later, Frances helped her mobilize housewives to take down the Equal Rights Amendment when it came up for congressional approval. In 1967, she also went to work for Goldwater's successful Senate reelection campaign. Frances was a demon on the telephones, so much so that Goldwater presented her with a gold pendant in the shape of a rotary phone, one of her most prized possessions.

But her biggest project came the following year. In reaction to the mounting protests over the Vietnam War, to seeing crowds booing returning servicemen and calling them names, Frances started writing letters to soldiers and recruited other housewives. "We are launching [this] campaign with the belief that our gratitude and support is much more effective than marching in the streets," she told the *Arizona Republic.*

After the story ran, hundreds of people began calling her house asking for names and addresses of soldiers, and Frances always had a ready list. Goldwater then encouraged her to start doing "talking letters" where people recorded messages on tape and sent them to boys overseas. Soon her kitchen table was a mess of wires and recording equipment, with strangers coming and going. It was during all of this that Tommy started staying out all night. "I slept at the office," he'd say, but Frances knew this was a lie because she'd hired a detective. Tommy not only had a mistress but he was keeping her in a rented house. The most devastating part of the news was that her brothers Leamon and John knew about it and never said a word.

Frances kicked Tommy out and filed for divorce. While their daughters reeled from the breakup — one requiring counseling and another plagued by violent nightmares — Frances circled her wagons and focused on being a good mother. After Goldwater sent a letter expressing sympathy and support, she pressed on with the yeoman's work of the cause — knocking on more doors, making more calls, registering voters. The cause is what kept her sane, and in the ground game there was but one objective: she was out to save America.

5

Hollywood comes to Big Spring . . . Preston goes to 'Nam . . . Bobby slips out of reach . . .

By 1968, the war in Vietnam had killed over a dozen young men from Big Spring. But if news of Preston being drafted had any impact on Bobby, who was fifteen, he doesn't remember. His brother barely acknowledged his existence, and the feeling was mutual. Besides, Bobby had problems of his own.

Marie was running around on him with another guy from church. His name was Steve, and people said he had a wild streak wider than Ronnie's. Steve had waited until the Wednesday night service when Bobby was working to ask Marie out, and she'd told him yes.

Already, she'd been giving off strange electricity. Bobby feared she might be drifting, so the previous weekend, he'd sprung for steaks at Herman's to try and see where he stood. Marie wore the necklace he'd given her for her birthday, which put him at ease, plus the expensive bracelet he'd bought for her. Half of every paycheck from the grocery store disappeared trying to make her happy. And how did she choose to thank him? By going out with Steve the very next week.

One of his friends saw them together that Saturday night. Bobby had driven around for hours trying to collect his wits, trying to decide if he ought to confront Steve and bust his nose. But that was too risky, he decided, because what if Steve really was wild?

He had to do *something,* so he drove to the cemetery and stole a big floral wreath off one of the tombstones, along with some flowers, and waited until the following night when Marie and her family were at church. He hung the wreath on their front door, scattered the flowers on the porch, and scrawled a note in big red letters that read STEVE'S FUNERAL!! When he asked Marie about it the next day, all she did was roll her eyes.

In the summer of 1968, Bob and Opal

found a two-bedroom house near the corner of Eighteenth and Main and for the first time in their lives signed a mortgage. From the front yard, they could look north along Main Street and see the town dip into the valley, where the Hotel Settles rose proudly. As a boy I would come to know the house's every crack and corner.

Unfortunately, it had sat vacant for years and needed a lot of work. Cardboard closets had to be ripped from the wall and framed out properly. The kitchen needed new cabinets and the whole house had to have a new ceiling. Opal assigned these tasks to Bob, without much success. However capable he was with dump trucks and drilling rigs, he couldn't seem to cut a straight line into wood and power tools ceased to function in his hands. To make matters worse, he was short on patience. He would fly into a rage and pitch his tools against the wall, screaming so loudly that Opal had to close the windows. Once when Bobby was helping him fix the washing machine, Bob grew so frustrated that he ripped it from its plumbing, pushed it all the way to the back door, then kicked it down the steps, shouting, "Dat-burn piece of garage sale crap!"

Opal then enlisted Zelda's Charlie for the home improvements, and slowly the work got accomplished. The house still needed a fresh coat of paint, but Bob wouldn't pay for it. Then, in late July, Opal saw a story in the *Herald* about a Hollywood film crew coming to town. They were looking to hire local talent, in particular, "rodeo-type performers, gospel singers, a 45–50 year old Negro male musician, boys 19–24 years old, and a blond man, six foot one or two." The movie was called *Midnight Cowboy*. Those interested needed to call Mr. Michael Childers at the Ramada Inn for an appointment.

Gospel singers, Opal thought, then picked up the telephone.

The paper described the film as "a late western," in which a young man leaves his small Texas town "in search of love and success" in New York City. British director John Schlesinger had driven five thousand miles across Texas looking for the ideal location for the character's childhood home before deciding on Big Spring. He also chose nearby Stanton for a few scenes.

For three weeks in August 1968, cameras and set crews, along with fleets of production and catering vehicles, took over the downtown streets. In the now-classic opening montage, a young Joe Buck, played by Jon Voight, leaves home dressed like Hopalong Cassidy and heads down Fourth Street clutching a cow-pattern suitcase. After quitting his dishwashing job at Miller's Pig Stand, he boards a bus for Manhattan and the town fades in the distance.

The production hired dozens of locals to play extras. The scene for which Opal was hired involved Joe Buck being baptized in a lake. Opal and about forty others spent the morning at Moss Creek, east of town, singing hymns along the bank. At the end of the day, she collected her twenty-five dollars and bought a trunk load of key-lime paint, feeling pretty proud of herself.

Of course, nobody told Opal or anyone else what really happened to Joe Buck once he reached New York City, how he became a male prostitute. That was revealed the following year when *Midnight Cowboy* hit theaters carrying an X rating, which the Motion Picture Association issued due to the film's "homosexual frame of reference." People in town were scandalized and felt betrayed.

Schlesinger had suspected as much. Shortly before the crew pulled out of West Texas, Voight found the director alone in his trailer. He was flushed and trembling all over, in the grips of an anxiety attack. Voight suspected he was dying.

"What have we done?" Schlesinger said. "What will they think of us?"

Thinking fast, Voight replied, "John, we will live the rest of our artistic lives in the shadow of this great masterpiece," and he was correct. That year the film dominated the Academy Awards, taking Best Picture and Best Director. Harry Nilsson, who sang "Everybody's Talkin' " over the opening montage, also won a Grammy.

But the accolades of the secular world did not impress Grandma Cora and the congregation at First Assembly, several of whom had been tricked into participating. It would take years for Opal to live down her role, even though she appears for only a split second. The choir's songs ended up muffled in the background anyway, denying the world her magnificent voice.

One of the people to come out publicly in support of *Midnight Cowboy* was Bobby's friend Grady Cunningham, who, at sixteen years old, worked as an usher at the downtown Ritz Theater. In November 1969, he penned an editorial in the *Herald* defending the R- and X-rated films the town's moral majority were in the habit of protesting, such as *Midnight Cowboy* and *Easy Rider,* pointing out that church people usually made up a good portion of the audiences.

Grady was taller now, yet pudgy still, and, like Bobby, wore his hair down to the shoulders. And also like Bobby, he'd become aware of the greater world outside of Big Spring and didn't much like his family's place in it. But it wasn't poverty that irritated Grady, because his parents weren't poor. It was their commonness. It was his father coming home day after day from the same job to sit in the same chair to be served the same suppers, forever until death, and his sisters marrying men he saw as no different. To him, his family seemed content with being invisible, and it made Grady want to live as loudly as possible. To be admired. To be known as a person of consequence.

Ever since starting at the Ritz, Grady had come to admire its owner, Ike Robb, whose family was revered throughout town. Since the early part of the century, the Robbs had built an empire of movie houses across New Mexico and West Texas and acquired vast holdings of property. In Big Spring alone, Ike's father, J. Y. Robb, had owned and operated over a half-dozen theaters over the years while keeping his hand in civic affairs.

During high school, young Ike had garnered his own fame as a standout defensive guard on the Big Spring Steers, lauded by the *Herald* as "fast, aggressive, and with a football heart." He later played for SMU, earning a spot on the All-Southwest Conference team before returning home to work the family business.

In the late fifties, as Big Spring's population soared from the air base and refinery, Ike's father ceded most of the theater operations to his son and turned his energies toward land development, in particular a project at the base of South Mountain, a large bluff on the southern end of town. Robb had purchased 960 acres from the T&P for a high-end subdivision he called Highland South — "Big Spring's Most Desirable Living Area," as described in its ads. The development would cater to Big Spring's new upper class: air force officers and Cosden executives, along with the bankers, lawyers, and wildcatters who'd grown rich in the recent boom. Highland South would feature spacious lots for family homes, plus apartments, a park for the kids, even an eighty-acre lake. But sadly, in February 1960, even before the bulldozers moved in, J. Y. Robb's health declined and he died.

Ike carried on his father's vision. By the

early sixties, giant homes were going up in Highland South and lots were in demand. Along one of its cul-de-sacs, at the bottom of a craggy cliff, Ike built a dream house of his own: a two-story ranch with a second-floor office that featured a built-in movie projector and screen where his staff could preview films before they played at the theater. As Grady sat watching *Love Story* and *The Hawaiians,* he knew what he wanted.

By then, people in Big Spring regarded Ike Robb — like they did Tollett — as a pillar of community life. He sat on the board at State National Bank, served as Chamber of Commerce president, and raised money for the YMCA, among other charities. In 1964, the city had honored him with its Outstanding Man of the Year award.

Grady deeply admired what Robb's family had built, and over time he began to see Ike as both a mentor and a person to emulate. The only thing was, Grady had an issue he didn't know how to resolve, one that could easily thwart his ambitions.

He was still attracted to men. Lately, he'd gotten into the habit of driving to the city park to meet them. He did this mostly at night, though he'd gone there in the daylight, too. He parked his car outside the stone bathrooms and waited alone inside, moonlight pouring through the window, until he heard the sound of car wheels and the crunch of footsteps. The consequences of being caught soliciting gay sex in a small town went unstated. At that same park, gay men had been dragged out and beaten. One man was pushed into the bushes and stabbed.

But Grady was young and careless, and during one of his visits, someone from school must have driven past and recognized his car. Now people were talking about him in the halls.

The one place where Grady found sanctuary was the Ritz. He strove to impress his boss, and his enthusiasm earned him a reputation as a little despot. Each shift, Grady patrolled the dark aisles like a mother superior, aiming his flashlight at anyone who violated his own code. Talking, even whispering, drew an instant reprimand. But he appeared to take special pleasure in punishing young couples drawn to the privacy of the balcony. He'd find them kissing and shoot his spotlight into their faces, then hiss for all to hear, "Up! Out!" The more popular the kids, the better: the sons of oilmen and bankers who spent their Christmases in Colorado, their summers at the lake; the football players and their cheerleader dates — the ones he suspected of mocking him behind his back. These people enjoyed no immunity on his watch.

Whenever Bobby heard rumors about Grady, he wanted to curl up and disappear. Part of him was nervous. What if people remembered that he and Grady were friends? What if they concocted some kind of lie? This made him terrified of being seen with Grady, made him want to smash his chubby face every time he shouted "Bobby Gaylon!" in the hallways like some deranged game show host. With Grady, there was always an act. If Bobby saw Grady first, he'd turn and walk the other way. His flimsy social standing would never survive that association.

But what people didn't know was that Grady still came around the house. He showed up on weekends dressed in his shirt and bow tie, acting like he and Bobby had never lost a step. Bob still called him Sister Grady, and Grady still did his preacher routine to make Opal laugh, grabbing Bobby by the forehead in a death grip of prayer.

During these visits, Grady rhapsodized about the prospect of being rich like Ike Robb, even saying things like, "One day when I'm rich . . ." But he never said how he planned on making his money. He'd started entering Publishers Clearing House and other sweepstakes, mailing away envelopes full of stamps for a chance to win a house, a new car, or tens of thousands of dollars. One afternoon he even switched on the electric adding machine that Bob kept on the breakfast bar, right next to his ledger, and started punching numbers. Opal asked what he was doing.

"Counting," Grady told her.

"Counting what?"

"The number of days I have to work at the Ritz until I'm a millionaire."

Like everyone at school, Bobby's girlfriend Marie was put off by Grady — by his ill-fitting clothes, his wild red hair. Marie didn't understand why Bobby stayed friends with him.

"He's so shabby," she said. "I don't know what on *earth* you see in him. You're just loyal, I guess."

He *was* loyal, and it was his cross to bear, especially with Marie. After he'd left the funeral wreath on her door, she'd apologized for trampling on his heart and he'd forgiven her, only to find out later that she'd kissed another guy.

But for the moment, anyway, things were good with Marie. Lately they'd started double-dating with her brother Ronnie and Merlee, who was Marie's best friend from church. The four of them piled into Ronnie's GTO and the girls rolled their skirts a little higher once they were out of their father's houses.

They dragged Third and Fourth Streets and stopped at the Wagon Wheel to have a hamburger. The night usually ended with them making out at South Mountain, Marie and Bobby in back, Ronnie and Merlee up front. Merlee was a good girl who didn't round the bases, not that it mattered to Ronnie. He knew plenty of girls who would. In fact, Ronnie had even started sleeping with a married woman. The problem was that she was a member of their church.

Marie knew about the affair, and so did Bobby and Merlee, and the stress of holding such a secret gave Marie an ulcer. How their father discovered the affair, nobody knows for sure, but by August 1969, Ronnie was gone. His parents put him on a bus to Waxahachie, hoping he'd meet a God-fearing woman who could set him straight. Within a year, he'd met his first wife.

By then, Preston was property of the United

States Army. The previous fall, a commercial bus had taken him from Odessa to Fort Polk, Louisiana. The morning he left, his wife, Linda, was so sick from strep throat she couldn't even get out of bed. Boot camp was hot and miserable. The swampy climate was supposed to help prepare them for the jungles. The drill sergeant was a lean black man who must've had some rabbit in him, Preston thought, because all they did was run. Mornings, after lunch, before bed. Run, run, run. After a month, Preston had dropped fifty pounds.

One day the drill sergeant came to him and said, "You should think about being a platoon guide. Lead these men." But Preston knew better. The men in his barracks were nothing but scared kids who dreamed at night of waking up alive in body bags. The draftees hated the army, and they took it out on the platoon guides — they beat them in the shower or threw blankets over them while they slept and pounded them with socks full of bar soap. Preston told the drill sergeant thanks, but no thanks.

Near the end of boot camp, they handed everyone a questionnaire that was designed to help place them in war. Nobody wanted infantry. They were asked things like, *Have you ever been hunting? Do you enjoy the outdoors?* Preston fudged every one. When it came time for assignments, the army made him a cook. Being married probably helped, too.

They assigned him to the Fourth Infantry and sent him to make bacon and eggs on the front lines. The Fourth moved like gypsies through the Central Highlands, setting up in forward-fire support bases carved out of a black mass of vines. Whenever the fighting moved, Preston's unit followed. They packed everything into a deuce and a half — a two-and-a-half-ton M35 truck — and chased the smoke. His kitchen was a sixteen-by-thirty-two-foot tent called a GP Medium, with a wall of sandbags all around it. There were five cooks in his unit, plus a crew of Vietnamese dishwashers, and together they served two hundred men at a time. Many of them were grunts fresh from combat who ate their meals and collapsed in pup tents with their boots poking out in the rain.

Preston was in charge of main courses. He'd crack six hundred eggs in the morning and flip three hundred burgers in the afternoon. His mess sergeant wanted everything on top shelf, meaning the soldiers ate plenty of beef, which Preston hauled through the jungles in "blood boxes" stacked on ice. On days off, he grew so bored and lonesome that he learned how to bake. From scratch, he made sweet breads and hamburger buns, biscuits, brownies, and other extras for the men. It gave him pride to fill their stomachs with good-tasting food, since each meal could be their last.

Each day he watched the choppers land in front of the mess tent and unload body bags into a line of deuce and a halfs. The walls of his kitchen breathed in and out with the blasts from the howitzers, and at night, artillery rumbled from the hills like muffled thunder.

It was hard for him to sleep at night because the bunkers were full of pot smoke. If Vietnam had a smell, he decided, it was weed. In order to get some shut-eye, he dragged his cot out under the stars, listened to the chatter of the jungle, and prayed, wishing he was home.

It was worse whenever they rotated back to transit base, located in Pleiku. The place radiated with insubordination and the army made you surrender your weapon. Some men were protesting the war and refused to maneuver, and racial tension was extreme. Men were beaten and stabbed and nobody dared venture outside their tents.

In Pleiku, Preston had a friend named Leo

who drove one of the deuce and a halfs. Leo's job was to transport grunts from the transit area to the fire-support base. One day Leo stopped by the kitchen before one of his missions, with his truck full of soldiers.

"Come along, keep me company," he said, but Preston couldn't go.

"Gotta work, buddy," he told him, and handed Leo a big bowl of vanilla ice cream.

On the way out, Leo hit an antitank mine and was blown in half. Many of the grunts were also killed.

Back in Big Spring, Opal and the family waged their own prayer war over Preston's safety, pleading with the Lord to bring him back alive. After ten months in theater, the army granted his wish to attend college and rotated him home early. In January 1970, the family met him at the Midland airport, then promptly whisked him home to fatten him up on coconut pie. Within two months, Linda was pregnant with their first child.

But while the family had been busy praying for the return of Big Boy Blue, Bobby was drifting from their reach. After his friend Ronnie left for college, Bobby started running with a guy named Mike Butler. He was several years older and, like Ronnie, cast a powerful spell. Butler's big brother Tony had been an all-state fullback for Coahoma, then played three seasons at Texas Tech. But Lubbock was a big city for small-town boys, and before long, the school booted him out for disciplinary reasons. Tony was now back in Big Spring working at the furniture store.

Mike Butler had played some football himself, but didn't have the coordination for it. Even so, he was big and loud and commanded whatever ground he stood on. He still lived with his parents out on the Snyder Highway and would roar into town demanding to party. The younger boys in his entourage called him "Der Butt" and had no choice but to obey. Being a wuss earned you a knock on the head from one of his meaty knuckles.

Der Butt was the kind of guy who swung by your house and yelled, "Let's go for a ride!" and then told you, once you were on the highway, that you were going to Mexico.

"*Mexico?* Man, I can't go to Mexico! I got a date!"

"Well, you ain't gonna make it."

Which is exactly what happened to Bobby one Friday afternoon. Four hours later he was shooting tequila in a scary Juarez bar, no idea how or when he was ever getting home. Hanging with Der Butt was like be-ing in a gang. There was no easy escape.

Der Butt was such the undisputed boss of good times that nobody, not even Opal, could stand in his way. One night he showed up at the house at 1 a.m., blasting his horn in the driveway, shouting, "Get yer ass out here! Let's go!" Then he did it again and again.

The first couple of times, Opal met Bobby at the door. "You aren't going out with him!" she said. He'd never seen her so angry.

But Bobby fought her. "Let me go, Mama," he demanded. I'm sure they both expected Bob to come out and put Bobby in his place, but that never happened. Opal finally moved aside, and out went her baby into a beer-soaked Tuesday night.

Bobby did wonder why his father never tried to stop him. The answer probably lay in the fact that Bob was an old rounder himself. Even now, Davey Jones would swing by the house in his dump truck and Bob would duck out for a while, never tell-ing Opal where exactly. Sometimes they just sat together in the cab, sipping whiskey from a paper sack.

In Bobby's mind, his father must've con-sidered him a man, and a man stayed out of another man's affairs. And *wasn't* he a man? Now in his senior year, he'd enrolled in work-study so that he left school each day at lunch. No longer working at the commissary, he'd taken a cashier job at a bigger grocery store in College Park. "Give me forty hours a week," he told his boss, and he got it. He graduated high school in January 1971, a semester early, having never taken algebra or geometry and having never read the classics. What was the point?

"When I was your age," he used to tell me, in a tone that implied a lesson was coming, "I always had money in my pockets."

And he did, enough to plunk down cash for a new car. But just not any car — a black Maverick, two door, straight six, with a fastback roof and a short deck. He loaded it with a Craig Pioneer stereo and a set of speakers that tickled the fillings in his teeth. And every other month, he walked into the men's shop downtown and dropped another two hundred dollars on clothes. Stepping out of the house in his new threads, he would spot his father sprawled beneath one of the trucks, his stained coveralls crusted in grime, and feel a sudden shame.

But he was young and unaware of his family's history. His own father had never sat him down and explained where they were from, what had happened in the years up to this point — never described the trail of indignities that had led him to the oasis of shade beneath the old Ford. His father hadn't told him about Julia dying, other than the obvious; about growing up on the road, motherless and hungry, and how it had turned him mean; about railroad bulls throwing him off the train, or the day they lowered his brother Bud into the ground. Did Bobby even know how that sadness still lingered to this day?

Bobby knew his uncles John and Leamon out in Arizona, but Flossie and Frances were virtually unknown, just names on Christmas cards his mother received. He knew nothing about the family farm they lost in Eastland; he'd never been told about the banker Frank Day or how people nearly starved to death back in the bitter days of '17. Perhaps it was because John Lewis had never told Bob about any of these things. And so in Bobby's selfish adolescence, knowing only the here and now, his father merely seemed pathetic.

Around that time, one of Bobby's buddies at the grocery store turned him on to weed, and he never looked back. After work they would hop into his Maverick and drive around town passing a joint, road beers tucked between their legs, blasting Zeppelin. Later he'd meet Marie and she'd gripe that he stank like cigarettes.

He was caring less and less about, well, everything, including Marie and her games with his emotions. It was sad, he thought, the way he'd let her do him. She made him so upset that he'd drive around for hours, then go to her friend Merlee's house and bang on her bedroom window. He'd sit outside in the flower bed and complain through the window screen about his love life while she tried to stay awake. Poor Merlee was caught between the two of them and strove to be impartial. At some point she'd cut him off, deliver one line that sliced to the quick, then send him home feeling better.

"I shoulda just gone with *you*," he told her once.

Finally, instead of venting to Merlee, Bobby exacted revenge. He'd met a girl at a youth rally and asked her out — a cute brunette from Snyder with big brown eyes. He drove fifty miles and picked her up, then took her to dinner and miniature golf, and at the end of the evening, even got a kiss. But their date never made it back to Marie, so what was the point of it?

The piano player at church had a cute niece from Stanton who visited one Sunday. After the service, Bobby walked right up and asked her out in full view of Marie. They went to Herman's for dinner, where everyone could see them, then to a movie at the Ritz. The damage to Marie was delicious. She came back to him sparkling like Sheba, wearing every piece of jewelry he'd ever given her. Up on South Mountain, looking out over the lights of Big Spring, she cradled her precious head in his arm, talked about college and marriage, and it was fine.

A few months later, Bobby drove to Midland and bought Marie a life-size teddy bear for her birthday. He was ready to surprise her when he discovered she'd been two-timing with a baseball player. So he gave the bear to Preston's wife, got drunk with Der Butt, and ended the night at Merlee's window, fooled by love again.

That May, both Marie and Merlee graduated with their class, and by August they were gone. Merlee enrolled at Evangel University in Missouri, while Marie went up to Waxahachie with Ronnie. After Marie left, her father resigned from First Assembly and left to pastor another church — some say because of the fallout from Ronnie's affair with the married woman.

Meanwhile, Bobby was left in Big Spring with Der Butt and his friends from the grocery store. But in March they'd found Der Butt's brother, Tony, dead in his apartment, shot five times in the chest by an acquaintance over a drug dispute. "Ex-Gridder Slain," read the headlines across the state. After that, Der Butt didn't come around much.

Still stewing over Marie, and with no plans, Bobby started using speed. One of the night stockers at work had a hookup for Black Beauties and other pills. Bobby favored L.A. Turnarounds because they delivered on their promise, though the farthest he ever went was around the Wagon Wheel and back, his heart thrumming like a straight six, the beer going down by the gallon.

If he sold the pills himself, he could get high for free and even make money, so he did. He bought them for a dollar and unloaded them for three, mostly to older guys he knew he could trust.

One of them was a hippie carpenter named Chuck who sold him weed. Chuck liked Pink Floyd and he also dropped acid. He and Bobby would gobble a hit on Friday night, then shoot the spectral galaxyway of Gregg Street on a soundtrack of *Ummagumma*. They uttered wild, half-baked truths from atop South Mountain while the skyline spun like a Ferris wheel on its side. Chuck was a cool guy, but Chuck had no idea how Bobby's spirit could rage against the flesh.

Zelda was having dreams about her baby brother. They were strong and vivid, heavy with meaning, and woke her up at night. In one dream, she saw Bobby in the back of a pickup along with others who were smoking, drinking, and shouting. Nobody in the pickup knew where it was going, but in the dream, Zelda could see that it was headed into a wall of darkness and only gaining speed. Whenever the dreams awoke her like this, she stayed up and prayed, her heart full of dread. In the mornings she called Opal and told her what the Lord had revealed.

"All we can do is lift him up," Opal said. And just like Little Opal had done for Homer, Bobby's mother took it to the altar, asking God to guide her boy toward the light.

Bobby's friend Doug had gotten hold of

some Orange Sunshine — the gold standard of LSD, which was making its way east from California. Big Spring was small, but the air base and oil field made it lucrative for trafficking. One Friday night Doug met up with Bobby and asked if he wanted to share.

"It's four-way," he said, "so I don't see why we can't each swallow two."

They ate the acid over at Doug's house, then smoked a joint to ease them into it. At some point Doug began to laugh and so did Bobby, and soon they were both weak from it, unable to stop, which is when Bobby looked up and saw that Doug had turned into the Devil.

Startled, Bobby wiped the tears from his eyes to see better, but he wasn't mistaken. Lucifer himself sat directly across from him, cloven hooves and all. But before he could flex his dark powers, Bobby jumped up and flew out the door.

He started his car and screeched out of the drive, just as the Devil appeared in the doorway waving his arms. He managed to go a couple of blocks before the windshield dissolved into patterns and forced him to stop. Up ahead he could see his friend Chuck's house, so he got out and ran.

He beat on the front door. When Chuck opened it up, he took one look at Bobby and said, "Holy smokes."

Bobby tried explaining the situation at Doug's house, but Chuck just laughed the same way Doug had, and then Chuck became the Devil, too. By the time Bobby reached his car again, he was convinced he was dying. His heart raced and he struggled to breathe. He'd never been so afraid. Suddenly a familiar voice penetrated the nightmare: *Billy and Suzy could have chosen heaven, but instead they chose hell.* It was just like the preacher described it: salvation presenting itself. Bobby knew: *Only one thing can save me now.*

He started his car and crept down the back roads, heading for the only safe place he could think of.

He drove to the preacher's house and rang the bell.

6

Bobby and Sharon . . . the Sea of Death
. . . Bobby makes a pledge . . . Bryan arrives . . .

For Bobby, the situation, as horrible as it
sounds, went down in the best possible way.

"I took LSD and I'm having a bad time,"
he told Brother Randall Ball, the new
preacher at First Assembly, who answered
the door and saw Bobby with his eyes
bugged out.

Rather than rebuking the boy, the preacher
invited him inside and sat him on the sofa.
And over the course of the night, they talked
and prayed together until Bobby was calm
and the drugs had left his system. When he
got home the next morning, Opal did not
ask where he'd been. If the preacher called
her, she never said.

The experience worked to turn him
around — at least for a while. He stopped

selling speed and even cooled out on the beer and weed. In fact, he stayed away from his old crowd for so long that they became suspicious — especially after Chuck got arrested for possession and went looking for the rat.

Bobby's brush with damnation had sent him back to the Lord — his head bowed, his soul laid bare. And when he returned to church, it wasn't a festive robe or a fatted calf that greeted him but a long-legged brunette named Tina Danko.

Her father had just married a woman who attended First Assembly and started bringing them, and Bobby asked her out that very first morning. She was a stunning beauty, in a league of her own. Bobby soon discovered that Tina had no real girlfriends, not like Marie or the others he'd dated, though plenty of guys called on her. And she'd gone out with a lot of them, not that he really cared. What mattered most was that now she was with him.

For Christmas that year, he went downtown and bought her a pair of red patent leather boots that hugged her calves just below her knees. She'd wear them with a miniskirt or pair of hot pants (which were no pants at all) and he'd have to stop and just watch her walk.

Mind you, the attraction was not purely physical. Pretty soon he was in love with Tina, or at least he thought he was. She was good to him, loyal and caring, and she didn't care whether he got high or not — so he did — which made her perfect by any definition he knew. And when his appendix burst, it was Tina who rushed him to the emergency room while he moaned and wailed on the floorboard of the car, rolled up in a ball.

But Tina's love could be intense, and sometimes it took him by surprise. She talked about getting married, and the ferocity of her plans went over his head like a stampede. "What do you think?" she'd ask, breathless, and Bobby would grin and nod his head, too chicken to tell her otherwise.

Then one night Tina's father called the house and Opal handed Bobby the phone.

"Tina says y'all are about to run off and get married — 's that true?" he asked.

And Bobby, finally put on the spot, denied it up and down.

That revelation must have broken Tina's heart and sent her running. Because two days later, one of Bobby's friends came up and said, "I just seen your gal down at the Wagon Wheel hanging all over Bucky Ford."

Bobby had no interest in tangling with

Bucky Ford. But he knew he had to do something, so he drove around until he found Bucky and confronted him like a man.

"Is it true?" he asked.

Bucky nodded. "It is."

"Well," Bobby said, feeling his emotions well up in his throat. "You can just have her then."

He bought some beer and drove around most of the night half crying. He knew Tina was a dead-end deal, but to have Bucky steal her from him? *That* stung.

To make matters worse, he'd just heard that Marie was about to get married to an aspiring preacher from Tennessee, who was helping her "to walk closer to the Lord." Not that he cared.

The sum of which meant that he was stuck in Big Spring, and alone. And the more he drove around, the more worthless that made him feel. The aloneness scared him, because by himself he could not resist the things designed to drag him down. In his heart he knew he needed God, but he could not reach out and touch the hand of God on his own, nor could he summon the courage to walk that rigorous road. Only bad things could happen when alone.

If only there was someone out there who

could do that for him, be a lifeguard for the times when the Devil spun his riptides. Someone pretty, someone who could love him.

He knew just the person.

It was Sharon Moore from Snyder, the girl who would one day become my mother. He'd dated Sharon on and off during his years with Marie, and just talking to her always made him feel better.

Sharon possessed something familiar to him, though he couldn't name it at the time. Later he recognized it as what his mother had, which was peace of the spirit. There were people like him who felt forced to walk the righteous road, as if gun-marched, for fear of the consequence. Sharon dwelled there instinctively, unburdened by hell. God dwelled in the calm center of her eyes, drawing people in.

Like Merlee, Sharon seemed wise beyond her age, and she met Bobby with honesty. She was dating other guys, but she did not sleep around. Back when they first dated, he'd understood that she wasn't rebound material. That and the fifty miles that separated them was enough for Bobby to turn back to Marie.

Yet all the while he was thinking of

Sharon, anticipating their next meeting. But whereas she was honest, Bobby was not. Whenever he'd see her, he'd swear that he and Marie were finished and that she was the one for him. "I could love you easy," he told her once, and she surprised herself by saying she could do the same.

She'd fallen for him way back in January 1969, at their first church youth rally together. It was in Snyder, and after the service Bobby had walked up and asked if she wanted to join their group for dinner. He drove her to the Rip Griffin truck stop in his black Maverick, which he'd just bought. She'd noted the red plaid seat covers and his mod wardrobe and wondered if he might be rich.

He called her a few weeks later. They played mini golf and he told her some sob story about Marie, and she kissed him then, not sure what to believe, but vanquished nonetheless by his charm. And ever since, he'd been writing her love letters. The letters were mostly silly ("Man alive do you have a set of hot lips, I dig 'em") but others were frightfully serious and revealed a vulnerability ("I don't have what I used to, I mean the experience with God. I want you to pray for me, please . . . I am falling fast.")

But he never truly declared himself, and

so she never took him seriously. His tendency toward backsliding, boozing, and dope made her nervous, although she did have this strange urge to care for him, to help guide him to the Lord. Perhaps God had sent Bobby for that very reason?

She dated someone else all senior year, a hulking tight end named Joe Bullard. But she ended the relationship after graduation. She was soon entering college and Joe was a year younger. She wanted no ties.

At Western Texas College, located in Snyder, she became head cheerleader for the basketball team, joined a sorority, and made new friends. Meanwhile Bobby kept sending her letters — each one a big event. She opened them in the kitchen and read them aloud to her mother, giggling at his stupid jokes, not even trying to conceal her joy. So of course, she felt a flutter of excitement when she learned that Marie had moved away and was seeing the Tennessee preacher. Yet the next thing she knew, Bobby was dating Tina — and every church girl from Hobbs to Sweetwater knew about her.

When she confronted him about Tina, he talked about her as if she were something that had befallen him unexpectedly, like a burdensome relative who appeared one day and wouldn't leave — something temporary, but unavoidable all the same.

"Do not worry about Tina," he wrote. "I shall take you back to Snyder and marry you and live happy after ever — or ever after — whichever you prefer. I will give you a choice. Be sweet, Baby, and remember I am nothing without your wonderful love."

She bought that line, though she was not proud of it. The sincerity in the letter was just enough to keep her excited about their next meeting — a youth rally the following weekend at the First Assembly of God in Big Spring. She arrived early with her friends and saved Bobby a seat. And when he entered the chapel moments later, grinning like he'd just won something, a hush swept the room. On his arm was Tina, wearing those killer red boots.

Sharon had been so angry, and so hurt. She dismissed him as a player and refused to answer his calls and letters. After a while, he stopped chasing her. But when she heard he was in the hospital several months later with appendicitis, she caved. She drove to Big Spring and visited him, then sent a sweet card to his house. The gesture surprised him and he seized the opening.

"I can barely remember you coming to visit me I was so doped up," he wrote in a letter dated March 9, 1972. "I can still remember those beautiful brown eyes. I think they are what pulled me through. Sharon, I still have your picture in my billfold and you are still one of the most beautiful people in it. Who are you going with? What is he like? Does your mother like him?"

He was right. She *was* going with someone else — a tall, blond golfer from Abilene named Tim, and they were getting serious. But Tim, unlike Bobby, was true marriage material. He was honest and attentive, and he was grounded — a college man on scholarship — without all the vice and spiritual torment. The only problem, if there was one, was that Tim was Catholic, and Tim's mother told Sharon that if they intended to marry, she would have to convert, which would mean renouncing the church she loved so dearly. Tim played it down, saying, "You don't have to listen to her." But still, she worried.

Somehow — not from her — Bobby had discovered Tim's identity, and despite the usual goofiness in his next letter, she sensed panic in his tone.

"I have prayed about this <u>four</u> times and I think it is God's will that you break up with Tim and start dating me again," he wrote in

May. "I will even learn how to play golf." He was losing weight, he told her. He threatened to end his life. "There is one thing I love more than myself and that is you, and if I can't have you, there will be no reason for me to live."

She read the letter to her mother, and they both rolled their eyes and had a good laugh. Then she sat down with a pen and paper and explained that she was committed to Tim. Yet even as she wrote the words, she wondered if it was really true.

His next letter arrived over a month later, during the second week of June. The return address was in Portland, Texas, along the Gulf of Mexico.

"I bet you are wondering why I am in Portland," he began. "It all started after I got your last letter. It almost killed me to know that you loved someone else. After five sleepless nights of worrying I came to the conclusion that I must really take my life. This was going to be very hard for me to do. Guns + knives were out because I hate the sight of blood, especially my own.

"Finally, I decided that I would move to Portland — from Portland I would drive to Corpus Christi, where the famous Harbor Bridge stretches across the Gulf of Mexico (but for me, the <u>Sea of Death</u>). Yes, I planned to jump off the bridge, and no one lives through that. But before I made such a big step I thought maybe I would give you another chance to break up with Tim."

The truth was he'd gone to Portland to lick his wounds. The situation with Tina and Bucky had embarrassed him so much, he thought it best to get out of town. His uncle Wendell had moved down in Portland with his family to take a job at the Reynolds Metal Company. This was Wendell Hahn, Granville's brother, whose daddy had doused him with boiling water in New London before walking out on him and his family. Wendell now had a son named Denny, who was Bobby's age, and the two of them had dug up a lot of mischief together as kids.

Denny knew a company in Portland that was hiring guys to run tests on oil field pipe. It was rugged, outside work, the kind Bobby abhorred. But Denny made it sound kind of romantic, and besides, Bobby was desperate for anything. His father, upon hearing this, even told him, "A man's job will do you some good."

He and Denny worked as a team, hoisting twenty-foot sections of steel pipe that were plugged on one end onto a rack, whereupon a guy sitting in a truck shot them full of pressurized water to test them for leaks. Most of the jobs were in soggy pastures, and the coastal heat and humidity left them wilted and tired. The work was also dangerous. Sometimes the plugs came loose and launched like mortar rounds from the pressure; other times the pipes just exploded.

He and Denny lasted two weeks before quitting, then applied for easier jobs in Corpus. Bobby interviewed at the local grocery store, and when the manager insisted he cut his hair, he bought a cheap wig instead, returned the next day, and was hired.

In exchange for his room and board at the Hahns' house, Bobby had to attend church with Denny and his family at the local Assembly of God, which his uncle had helped to build. That, plus the clean living under his uncle's roof, had delivered some clarity and made him pine for Sharon. One Sunday after church, he wrote her a serious letter, in which he apologized for his behavior and made one last appeal.

"I didn't want to live up to Christian standards and I knew that would be the only way it would work for you," he wrote. "I am ready to straighten up and live for God, but I need help. The kind of help that only you could give me."

Sharon grabbed the letter from the mail-box and took it to the kitchen, as she'd done all the others. But after hearing those sentences, her mother stopped her short, saying, "You better read that one by yourself." By the time she was finished, she was filled with confusion.

She spent the next week in a state of torment, pulled between two very different futures. Finally, her father sat her down and said, "You have to do what you know is right." And she did. That same afternoon, she drove to Tim's apartment and gave him the news.

"I can't let you go on loving me," she told him, then ran out of the room. She spent the whole night on her front porch, sobbing.

Down on the coast, Bobby received new life.

"I will straighten up and give you something to be proud of," he vowed. "I am tired of dope! I am tired of cheap girls! I am tired of running from God."

They were engaged by Christmas, during which time Bobby moved to Snyder and enrolled at the junior college. He worked at the Exxon filling station owned by Sharon's father, who taught him more about mechanics in three months than his own dad ever had.

Opal and the rest of the family couldn't believe his turnaround. The fact that he'd convinced such a prudent and godly (not to mention beautiful) young woman to marry him was not only a miracle, it was slightly suspicious. In fact, they made sure to give her the treatment the first time she visited for Sunday lunch. Opal and Norma Lou plied her with intrusive questions while Zelda simply stared at her throughout the meal. Finally Zelda asked, "What color eye shadow is that?"

Sharon froze. *She must think I'm a Jezebel.*

"It's Maybelline," she answered. "Electric Blue, I think."

"I love it," Zelda said.

They were married in Snyder in early June 1973, a week after receiving their junior college diplomas. The whole family was in attendance. They rented a small apartment in Odessa, which was cheaper than Midland and close to where Preston and Norma lived, plus his aunts Fannie, Allie, and Velva. Sharon took a receptionist job with a law firm while Bobby bounced between gigs, a married man searching for his niche. At first he tried the local Safeway, then a plumber named Bill hired him on as an apprentice.

"Two things," Bill told him right off. "Righty tighty, lefty loosey, and shit don't flow uphill. Remember that and you'll do fine." The two of them spent their days installing hot water heaters, digging ruptured pipe, and running closet augers through clogged toilets while housewives watched in their curlers.

For a while he painted houses with a couple of Big Spring guys, Mark Powell and Donny Janks, but their ways were nefarious and reacquainted him with old habits. Mainly they got high, thanks to another friend of Bobby's named Dale, who kept them in good supply.

Dale had landed on Sharon's bad side the day she came home and discovered he'd deseeded his marijuana at her kitchen table. He was a considerate dealer in that way. He was even kind enough to telephone from jail the day he got busted to alert her about the duffel bag he'd stashed under the house.

As someone who strived constantly to be a better Christian, how was she to handle these things? She did not smoke cigarettes or pot. She hated the taste of beer, and even today will only occasionally enjoy a glass of chardonnay. But that was *her,* and she didn't feel comfortable passing judgment on other people, particularly her husband.

She'd known his backsliding ways when she'd broken up with Tim and given her heart to Bobby. She'd signed on to be his lifeguard in the turbulent patches and his companion in their spiritual journey together. It wasn't her nature to nag — she was cool — but she also knew how widely her husband was capable of straying. The balance was hard to strike, especially since she was twenty years old and in the first months of marriage.

The answer, it turned out, presented itself on its own. In April the following year, she discovered she was pregnant.

She'd developed a kidney infection and the doctor took her off the pill, but with the assurance that she'd be safe for three months. He was wrong. She and Bobby hadn't planned on having children for a few years but made their peace with the news. Bobby was terrified, yet somehow he was also relieved. He needed walls to guide him, and between my mother and a baby, a fresh path revealed itself.

What he needed was a career, something he enjoyed. He was tired of the grocery business and didn't have the stomach for plumbing. He'd also sworn off the oil patch after his last experience. For a brief time, between Portland and moving to Snyder,

he'd tried working as a floor hand on a pulling unit south of Big Spring. He was one of
four guys on a small workover rig, extract-
ing sucker rods and steel pipe from old
pump-jacked wells they were trying to get
back on line.

Using two heavy wrenches, one in each
hand, his job was to unscrew the thin
sectional rods that acted like pistons to
bring the oil to the surface, then latch each
one to a traveling block and cable. The
operator then gunned a diesel engine that
lifted them up to the derrick man, who hung
them on a rack.

When they pulled the pipe itself, he and
another guy had to set the slips so the pipe
didn't fall back into the hole, then use a
pair of giant tongs to loosen each threaded
joint. Some of the pipe was decades old and
had to be beaten with a sledgehammer
before it gave, and even then, Bobby had to
push his whole body against the tongs to
get it unscrewed.

Oil and mud often blew up with the tub-
ing, and when it did, the other men, whom
Bobby found crude and intimidating,
shouted, "Hey worm, cup the hole!" And
Bobby would have to bend down and place
his body over the hole to keep the sludge
from spraying on them. It was the kind of

work his father had done for twenty-five years, and it was so punishing, so exhausting, that he passed out during lunch breaks and had to quit after two weeks.

Now, with a baby on the way and in need of a career, he remembered something his father once said, how back in the Depression, the only men with money in their pockets were the car salesmen.

Selling cars, he thought. That was more his speed.

He drove twenty miles to Midland, where the real money was, and pulled into the biggest dealership in town, Bill Rogers Ford. He walked inside and asked for the manager.

"I'd like to be a salesman," he told the man, who looked him up and down and said, "If you wanna sell cars for me, son, you're gonna have to cut that hair."

Bobby grinned. "Funny you say that, mister, 'cause I was just on my way to the barbershop."

He started on the showroom floor selling Thunderbirds, Gran Torinos, and LTDs. The pay was on commission, plus the dealership gave him a new demo to drive. Bobby chose a sleek black Mustang II, which happened to be sitting in the driveway the day Sharon went into labor.

Bobby was home for lunch when she came running out from the bathroom in a panic. At first he called Opal, screaming, "I think Sharon's water broke!"

"Baby, either it did or it didn't," his mother answered. "Now get her to the hospital."

He was so nervous about soiling the seats on the Mustang that he covered them in four layers of towels. Then, as an added precaution, he rolled up another and handed it to Sharon, saying, "Stick that one down your pants."

After three hours of labor at Odessa's Medical Center Hospital, I was born on January 23, 1975, at 2:37 in the morning. At first they named me Brett, then two days later, changed it to Bryan — after a tennis player they'd known at college.

"Bryan was just a nice guy," my mother would tell me.

They took a house in Midland so my father could be closer to work, then settled in as young parents. At twenty-two years old, neither parenting nor homemaking came easy for my mother. The day she brought me home from the hospital, she leaned over the crib and quietly apologized. Trying to cook dinner for Dad, she dropped a flaming skillet on the floor and burned a hole in the linoleum. Then one day a neighbor stopped by to say she'd seen the cloth diapers in the alley dumpster that Mom was throwing away instead of washing.

"Honey," she said, "you can't keep doing that."

Although I had been dedicated at Bethel Assembly of God in Odessa (as opposed to being baptized, which happens later in the Pentecostal church), Mom and Dad didn't belong to a church yet in Midland, and besides one other couple, they hadn't made many friends. Then one day Dad came home from work and said that Grady had just called.

"He's living here in Midland and has a big job," Dad said. "And get this — he's getting married."

"To who?" Mom asked, and Dad couldn't believe it himself.

"To Ann Tollett. Her daddy was president of Cosden."

7

Grady aims high . . . enters the Tollett family

After high school, and up until the mid-seventies, Grady lived along the margins and left little trace. From what I could find he had very few friends, and during this time even Dad lost touch. I do know that his mother, Louise, held him back in school, so that he graduated in May 1972, a year behind his class. He enrolled at Howard Community College the following autumn. On the cusp of adulthood, he still grappled with his desires.

He cruised the stone restrooms in the city park, where the smell of liquor and sweat filled the dark. He attended secret parties thrown in various men's homes. One man who had occasional relations with Grady described nights when he'd close his business and find Grady at the back door waiting. But he had no steady lovers, as far as anyone can remember.

Grady's desires weren't limited to sex. He wanted to live brashly, to experience extremes. It was during this period that he bought a plane ticket to London, passed through Dallas and New York, but was turned back at Heathrow Airport when he couldn't produce a passport.

Most of all, Grady wanted to be rich and he wanted it now, and the fastest course of action was through marriage.

According to Grady's cousin Selena, he first attempted to marry into money in 1969, when he began dating the daughter of a wealthy Odessa banker. Despite his sexual orientation, he managed to woo the girl to the point where he needed to close the deal, which he did one afternoon by stealing his father's checkbook. He bought a Cadillac de Ville at the lot in Big Spring, and before Luther's check could bounce and the dealership called the cops, Grady and his date were rolling into Fort Worth to pick out an engagement ring. By the time he returned home, the police were waiting in his driveway. Louise had also called a lawyer. To keep Grady from being arrested, she convinced the law that her son was unstable and required urgent care. She then drove him to the state hospital and had him admitted.

"Play crazy," she told him, and he did, though it's unclear how long he stayed or whatever happened to the girl and her ring. The state hospital no longer keeps records from that time.

By the summer of 1973, Grady had taken a break from college and was working full-time. A few years earlier, Ike Robb had refashioned the Ritz Theater to appeal to a more modern audience, and in the process, he'd done away with full-service ushers. For a while, Grady sold women's shoes at the Highland Shopping Center, then found a job at the Singer Sewing Shop as a clerk and courier.

One day the store received an order for a new sewing machine and sent Grady to make the delivery. He parked the car outside an imposing modernist home on Hillside Drive and lugged the machine up the long sidewalk. The name on the mailbox read, in bold white letters, R. L. TOLLETT.

Grady knew exactly who Tollett was, of course. He was the man whose long reach managed to touch everyone Grady knew, no matter where they worked or lived. Even before Ike Robb, Grady had admired Tollett, not for his money and influence, but for the way people loved him.

Grady had met Tollett once, back in high school. He walked into a downtown café and saw him seated at the counter, eating a bowl of chili. Gathering his nerve, he approached and asked if he could sit and talk.

"Go right ahead," the hunched figure said, barely turning from his food. "What's your name? What's on your mind?"

And although Grady's head was full of questions, he mumbled his name, then spent the rest of the time watching the man eat his lunch. Tollett died not long after that and Grady joined the rest of Big Spring in grieving.

It was Iris who answered the door when Grady rang the bell. She pointed him around back, saying, "Annie will let you into the basement."

A group of girls were sunbathing around a swimming pool when he opened the back gate. He recognized Ann instantly. They'd gone to junior high together, and in fact, Grady and my dad used to shoot paper wads into Ann's hair in Ms. Wally's class. She was prettier now, with short blond hair, and her two-piece bikini revealed a buxom figure.

"The sewing machine goes down here,"

she said, and motioned him toward the door.

Once in the basement, Ann instructed him to pull out the old Singer from its wooden cabinet and replace it with the new one, then she said, "I've gotta get back to my friends. Let me know when you're finished." She didn't even recognize him, he thought.

He wondered if Ann was dating anyone, but that was something he could find out. Even with Raymond Tollett gone and buried, everything that family did was news. After installing the new machine, he walked back upstairs and said good-bye, then showed himself out, making sure to register everything along the way.

For Ann, the memories of her father were rare — the good ones, anyway.

Like her brothers, she'd suppressed the most painful moments of her parents' alcoholic decline: her mother screaming under the wheel of the family car; the fights with her father, and her father transforming, becoming unrecognizable, in the months before he died.

The good memories stayed alive and vivid. Like the times she rustled out of sleep and felt someone in the room, then saw her father sitting in a chair in the dark, watching over her. Or the many nights when they sat together in his study reading books, him dressed in his smoking jacket and slippers.

In August 1961, Raymond had packed the family into the station wagon and driven all the way to Seattle. They could have easily flown first class, but her father insisted on taking a family road trip. The whole way up, Raymond and the children sang camp songs at the top of their lungs, driving Iris crazy. In Seattle they boarded a cruise ship to Alaska, then boarded a prop plane for Point Barrow, located on the North Slope along the steel-colored coast. Ann remembered there being little water or electricity and having to sleep in huts, and how one day, an Eskimo kid pelted her with rocks and ruined her new parka. It was the only vacation the whole family ever took together.

After Ann flunked seventh-grade math, her father pulled her out of junior high in Big Spring and sent her away to boarding school. The San Marcos Baptist Academy was three hundred miles south, near Austin, and just over an hour's drive from where, in three years' time, her father would die trying to get sober.

The girls' dorm where she lived was ancient, plagued by scorpions and rats, and had no air-conditioning. Her classmates were from places like New York and Michigan and did poorly in the Texas heat. Together they swore and learned to smoke, and on weekends they walked into town for Frito pies and movies. On Sunday mornings, Ann awoke early and walked to the Episcopal church, telling her roommates, "They're not making a Baptist out of *me*."

Once a month, her father appeared in the Heron or the Dove and sent a driver from the airport. They'd stop in Roswell, New Mexico, to pick up her brothers at military school, then fly home for the weekend, visits the *Herald* would record in Monday's paper.

The three kids would take trips with her father and his secretary, Helen Green, and Helen's two daughters. They went to the World's Fairs in New York and Seattle. Once, in Manhattan, Jason Blake got lost in the subway and they spent the whole day looking for him. But mostly Ann stayed in San Marcos, and when summer came around everyone went their separate ways, her brothers to Colorado, then abroad, and Ann to sleepaway camp in New Mexico. And this was how the family's years rolled past — separate and apart, and for Ann, marked by loneliness.

On one of her last visits home before her

father died, he picked her up in the car (no more company plane), and the sight of him actually frightened her. When they got home she asked her mother, "How come Daddy's so big?"

"Oh, he's just eating good," Iris answered.

It wasn't until she returned to school and spoke with her roommate that she realized her mother had lied, that her father was septic from alcohol.

It was Iris who called her late at night, several months later, to tell her that he was dead. She can't even remember if she cried, everything just went numb. Her brother picked her up the next morning, and for six hours they didn't say a word until they reached Big Spring. She remembered going to the funeral home and seeing him lying in the casket, and how his salt-and-pepper hair, which she'd watched him comb a thousand times in the mirror, now appeared bone white. She remembers the flowers and the mile of cars waiting to park outside the church.

Her mother sent her away again that following summer, just months after the funeral, when all she wanted to do was stay home.

"Being home will only bring back memories," Iris told her, and with both Ray and

Jason Blake away at college, she arranged for Ann to join a group of Dallas prep school girls on a long tour of Europe. Ann didn't know any of them.

She graduated high school in May 1971 and cried the entire day, missing her father terribly. That fall, she enrolled at Tyler Junior College, majoring in art and physical education, but soon grew discouraged. During her second year, she called Iris and said she was tired, ready to come home.

"Pack your bags and I'll come and get you," Iris said.

If anything positive came from her father's death, it was that her mother had eased up on her own drinking. Iris wasn't hammered anymore at two in the afternoon, stewing over mistresses. Raymond's dying helped to release her, unspooled her anger, and little by little, she stepped out into the world again.

She became a serious bridge player, and with Ann no longer in college, they embarked on tournament cruises that took them around the globe. They traveled to West Africa, Spain, Portugal, and India, and went exploring at each port of call. They visited the Soviet Union, where in Leningrad they watched the Kirov Ballet perform *Swan Lake* at the century-old Kirov Theatre, then capped the evening drinking iced vodka with their Russian guides.

Ann had no interest in bridge, but she embraced another aspect of ship life: men. There was the Swede who rowed her into port in a dinghy and couldn't keep his hands off her; the dark-haired Englishman from the ship's theater who snuck her into his cabin late at night. A hot romance with a Norwegian ship captain ended with a marriage proposal, which she rebuffed. A singer from Atlanta, a much older man, bought her a ring in the ship's boutique and proposed on deck.

"Come home with me," he said. "I have money, I can take care of you."

I've got a little money myself, she thought, and gently turned him down.

For someone who struggled with her weight, the attention did wonders for her confidence and helped ease her out of her shell. Iris not only approved of her daughter's exploits, she even put her on birth control.

Yet back at home, Ann had trouble finding love. The years away at boarding school had estranged her from the kids her age, and aside from a couple of girlfriends from childhood, her only companion was her mother, who until recently had stayed mostly drunk.

Now that Iris was getting older, she became keen on finding her daughter a mate. Yet she worried about Ann, the way she'd worried about her since she was a girl. Part of it was her appearance. She'd fussed over Ann's weight for years, at one point taking her to endocrinologists in Atlanta to determine why she kept gaining. But doctors could never figure out the problem (years later they discovered a faulty thyroid).

"I want you to find someone," Iris told her daughter. "I don't want you to be alone."

"I'm looking, Mama," Ann replied. "But how will I ever find him?"

In January 1974, Iris got a call from a friend who worked at the *Herald*. It was Jo Bright, the society columnist and women's editor, who wanted to know if Annie was dating anyone. When Iris told her no, Bright said that she had someone in mind, a young man who had recently started working at the paper. In fact, his birthday was coming up, which was the perfect excuse to have dinner down at the country club and introduce them.

"I'll tell him to swing by at seven and pick her up," she said.

Ann recognized the name when Iris told her, but she couldn't quite place him. It wasn't until Grady appeared at the door that it finally hit her. She couldn't explain what made her do this, but she walked over and immediately wrapped him in a hug. Both Grady and Iris were taken aback.

"Do you know each other?" Iris asked.

"Mama, I went to junior high with Grady, and he's the one who delivered the sewing machine. I knew you looked familiar!"

Iris asked Grady what he did at the *Herald,* and Grady said that he worked in circulation, assembling the sections of each day's edition to send to the carriers. It was a better-paying job than the Singer shop, and he hoped for a promotion soon.

They drove to the country club in Grady's car, an old green Bonneville, the color of which reminded Ann of vomit. She also noticed how Grady's clothes didn't quite match. He'd paired a brown checkered shirt with blue pin-striped pants. But he was nice, gentlemanly, and she felt at ease in his company.

Jo Bright had a table reserved when they arrived. She'd brought along another writer named Gene and after a round of cocktails and a quick toast to Grady's birthday, they settled into dinner.

Jo and Gene hadn't seen Ann for months and wanted to hear all about her and Iris's latest cruise. *Where in Europe?* they asked. *And where did you stay?* So Ann told them everything, and still they asked more questions. At one point she realized Grady hadn't said a word, but what could she do? Before she knew it, dessert was over and it was almost time to go.

On the drive home, Grady seemed to sulk. Finally he said, "You talked the whole time. Don't you think that was rude?"

"I'm sorry," Ann said.

Back at the house, he opened her car door and walked her to the porch, then stuck out his hand and said good-bye. Not even a peck on the cheek.

"Maybe next time you'll keep your mouth shut," Iris told her.

A week went by and he didn't call. *Mama's right,* she thought. Then one day, when she was ready to give up hope, it was Grady on the phone, asking for another date.

She discovered years later why it had taken him so long to call, and it had nothing to do with her talking too much at dinner. He'd been engaged to another girl in Odessa and that relationship had ended. She didn't know who the woman was or which of them had broken the engagement.

All she knew was that the second time Grady appeared at her door, he surprised her by wearing a suit. He then drove her to Midland and bought her a steak.

As the new boyfriend of Ann Tollett, Grady approached each date like a job interview. His manner became formal, and so did his dress. He now rotated between two leisure suits, one green and the other brown, each with oversized lapels and flared legs, and accented with a garish tie and pointed shoes.

And he insisted they eat in what he called "the finest restaurants" in Midland and Odessa, which included Steak and Ale, the Shrimp Boat, and Jay's Barn Door. Dating the daughter of Raymond Tollett gave him a new role to play, one he felt he needed to practice even when Ann wasn't around. At Kimo's Palace in Big Spring, the Chinese restaurant where the *Herald* staff drank beer, Grady ordered rounds of pitchers for the table, then waved his hand at the waiter, saying, "Put it on my tab."

But playing that role required real money, and often Grady found himself overextended and had to borrow from Ann. "Why can't we just go for a burger?" she would ask, and Grady replied, "Because I want you to have the very best."

Then there was Grady's car. It was probably their third date when the old Bonneville sputtered and left them stranded on the interstate. Grady had to walk several miles to call a wrecker, leaving Ann exposed to whatever oil field trash happened past. The Patty Hearst story was still big news, and already Iris had forbidden them from parking out front, lest Ann be kidnapped for ransom. Anything they wanted to do with each other could be done in the poolroom, she said.

One night Ann didn't come home until 3 a.m. and Iris was frantic. Turned out Grady's car had broken down again on the way back from Odessa, and they'd waited hours for a tow.

Grady perplexed Iris, and his behavior left her with many questions. In addition to hating his car and his clothes ("Why does he wear those ridiculous suits?" she asked Ann), she grew disturbed by the rumors that Grady was gay. But Iris never discussed the subject with Ann, who led her mother to believe that she and Grady were intimate (and they were). And besides, for the first time, Ann seemed so happy.

"Who else is going to be there to love her?" Iris would say.

After nearly two years of dating, Grady

dropped to his knee one night in the living room and proposed. Although he didn't have a ring, Ann said yes, then started planning their wedding.

They married on a Tuesday morning in March 1976 at St. Mary's Episcopal Church in Big Spring. Ann wore "a princess-style gown of silk organza and Alencon lace," the *Herald* reported, and carried a French bouquet with blue streamers. Standing at the same altar where her father had come to rest, her smile radiated through the chapel.

Grady was so ill he could barely stand. He'd come down with something the night before and spent half the night in the bathroom. That morning he told Ann there was no way he could travel, so they'd canceled their honeymoon in Ruidoso, New Mexico. Down at the altar he wobbled in his rented tux, his skin pasty, but managed to repeat his vows.

From the front pew, Iris watched the ceremony as if it were a car accident in slow motion. She squeezed her niece Judi's hand and Judi could feel her whole body trembling. As Grady and Ann exchanged their vows, Judi looked at Iris's face and saw tears pouring down her cheeks.

"Oh, Mama, you were crying," Ann said

afterward.

"I was just . . ." Iris said, forcing a smile. "I was just so *happy* for you."

8

Dad survives the storm . . . oil returns to Texas . . . a new adventure begins . . .

Dad served as best man in Grady's wedding. The request came as a surprise, considering he hadn't heard from Grady since his own wedding in 1973, to which Grady had been invited only because Opal had insisted. Ann told him later that when it came time for Grady to choose a best man, he was the only friend Grady could think of.

Another surprise was that Grady had landed a job in Midland with the American Diabetes Association — as its regional director. He had an office on West Wall Street, a small staff of associates, plus a cute blonde who answered his phone with her long legs up on her desk. The *Herald* even ran a story about the appointment, along with a photo of Grady dressed in one of his leisure suits. It noted that he'd recently organized a seminar at the hospital entitled "Podiatry in Chronic Disease and Diabetes."

It all seemed strange, because whenever Dad thought about Grady, he saw only the awkward guy in the hallways. And, of course, how could he forget that day at the old man's house?

He'd never spoken to Grady about any of that stuff, and he wouldn't know how to, anyway. So what if Grady was suddenly married? *People change,* he thought. Sometimes it took a good woman to turn a man around, something Dad could attest to personally. Ann Tollett made Grady seem more approachable and somehow less of a freak. If anything, Dad was rather proud of Grady. He'd always talked about getting rich, and from what everyone said about the Tolletts, Grady had done well.

In fact, it wasn't long after the wedding that Grady called one night asking if Mom and Dad wanted to go to Lubbock for supper.

"Lubbock's two and a half hours away," Dad told him.

"Well, not if you take a *plane,*" Grady said. A limousine would pick them up in an hour.

Sure enough, Grady had chartered a little single-engine Cessna for the night. Mom and Dad had probably taken one plane ride in their entire lives, and neither had ever seen the inside of a limousine. The Cessna had four seats and no bathroom, but they were in the air for only thirty minutes, during which Grady sat up with the pilot, headset on, pointing to the instruments like some kind of boss.

Another black car sat waiting in Lubbock to take them to Smuggler's Inn for steaks and seafood, all of which Grady paid for. They flew back to Midland in time to relieve the high school girl whom Mom had called to babysit me. When the limo pulled up to the house, Grady walked inside and handed the girl a hundred-dollar bill. After that, she was always available.

By the time Grady was chartering Cessnas, the energy crisis was on.

Since the fifties, limits on domestic production had opened the door for cheap crude from the Middle East and Africa, which we consumed without limit. At the close of the sixties, America was using more oil than ever before, while producing less. The once abundant fields in East Texas and the Permian Basin, at one time so mighty that world markets swung on their output alone, couldn't begin to satisfy our thirst, no matter how much we drilled. In fact, Texas reached peak production in 1972, but with little effect. The country needed more oil, and by then we were almost out (or so we thought).

That year saw the first long lines at filling stations from coast to coast. People seemed gripped with an end-times panic as they sat idling in the cold. Dallas nearly ran out of gas in early 1973, and in May, the mayor of San Antonio said the city was ten days from shutting down. Arab producers in the OPEC nations took advantage of their new power and doubled prices, and still we paid — especially the refineries.

Texans were so desperate for oil, the pipeline from West Texas to the Gulf was reversed. In May 1973, the Big Spring refinery received its first shipment of Middle Eastern crude — 354,854 barrels from Iraq. By that summer the nation was importing 6.2 million barrels per day, twice as much as in 1970.

Then, in October, Egypt attacked Israel. OPEC raised crude prices by 70 percent, then refused to sell the United States anything because we supported the Jewish state. They also cut their own production, a power move that sent the global economy into a tailspin recession. Everything went up, from interest rates to the price of a gallon of milk. Gas prices went up 40 percent and the long lines continued, people fuming as Watergate played over the radio.

Although high oil prices crippled the rest of the country, they came as a boon to states that still could pump crude out of the ground, chiefly Texas. Better, by late 1976, in an effort to curtail imports, the government began its slow easing of price controls and restrictions on domestic drilling. The gates wouldn't fully open for another few years, but rig counts across the state steadily began to climb. The Permian Basin, at long last, was headed for another boom, and Midland would shine at its tawdry center.

By the mid-seventies, the twenty miles between Midland and Odessa marked a chasm of civilization. Both cities had experienced a slump in the sixties, when foreign crude squeezed the oil patch, but Midland had fared much better. Midland was still home to dozens of oil companies where executives in hand-tooled boots sat in glass buildings with ornamental pump jacks on their desks. It was where the big-gambling contractors lived, along with wildcatters blessed with knack and fortune, and the men of science: the sober geologists and petroleum engineers dressed in slacks and dusty oxfords. Midland was where men made money without dirtying their hands. And as the pendulum of the oil market began its upward swing, it was a hell of a place to sell cars.

What Dad loved selling most were pickups. His first year at the dealership, Ford introduced the F-150, which combined the dumb ruggedness of a truck with custom interior and trim options. The result was a pickup that a midlevel accountant could get away with driving ("hauling air," as they said) and one that quickly became the standard fleet of the oil field.

Selling new cars and pickups required more than a good line, and for a time, Dad studied the other salesmen. First, he noticed, you had to look for signs that a person was serious about buying. The best sign was if a man brought along his wife. There were also a few questions you could ask to determine the potential of a sale, the most basic being "What is it you do for a living?" It wasn't too intrusive, but the answer would tell you plenty.

Dad learned as a general rule that he shouldn't waste time on roughnecks and oil field hands unless they had the price in cash, or at least enough credit through their wives to sway the finance team. No need to be rude, his coworkers told him, but he shouldn't get a sunburn for a man who might not have a job in two months.

Now, if the guy's an energy loan officer at First National? An engineer for Consolidated Petroleum? Well, give that man a card, get him behind the wheel of one of those new Explorers — limited edition, with red Styleside stripes, Cruise-O-Matic, and plush-carpeted interior. Let him drive down West Wall Street and up Marienfeld, see through the windshield the great city their money was building.

"You look good in this truck," Dad told these men. "It's about time you bought the vehicle that's made just for you."

He also learned that there was no cause for dishonesty or flimflam. The trick wasn't to rob a man of his dignity, but to make him feel like he was walking out with a deal. What you wanted was *volume.* Move the units, get the volume, and the gross will come — along with a thousand-dollar bonus if you're sharp.

For the customers that Dad just couldn't close, Bill Rogers sent in his muscle — a man named Dick Bratcher. Dick was tall and kind of bald, with eyes like cut glass. "What would you buy it for? Come on, name a price," he'd say, just friendly enough not to rattle anyone. Dad learned from Dick Bratcher how to bear down with an easy touch, and after a year under his tutelage, he'd sold enough units to win Salesman of the Month. His name was added to a plaque on the wall.

Dad worked long days at the car lot, usually twelve-hour shifts, and got close with the other salesmen. There was Schroeder, a former baseball player from Portland, Oregon. He was handsome and fit and spoke with a cute northern accent. When friends of the boss came to buy cars, they always asked for Schroeder, and so did their wives.

There was Lying Larry, "the Most Cheatin'est Car Salesman Alive," who seemed honest but played the stereotype to the hilt. He liked to walk the showroom floor and pretend to greet customers by saying, "Welcome to Rogers Ford. Lying Larry's the name, how can I jew ya, screw ya, and tattoo ya?" The guys standing around just died every time. Larry was also strong, and to demonstrate his strength, he'd grab a pole with both hands and hoist himself sideways in the air, then hang there like a flag.

After a while Dad started hanging out with some of the used-car salesmen who worked the adjacent lot, and he realized he could be making a lot more money. Unlike new units, which carried the manufacturer's sticker price, used cars had a mystery component that a good salesman could exploit. No one knew how much the dealership had paid for the car, either at auction or on the trade-in. As long as you didn't veer too far over the Blue Book, the profit margin was usually bigger. But you had to know how to sell it.

The used-car salesmen were older, grizzled men, but they liked Dad because he was. cool and could score them weed. They took him drinking after work, but never to bars. Most nights they drove around town in one of the hot-rod demos, passing around a bottle of whiskey and pouring it into their beers.

They also threw high-stakes poker games where Dad was out of his league. A few mornings he woke up to realize he'd lost several hundred dollars the night before, then had to hustle all day to try to earn it back before Mom discovered it was gone. On those mornings, hungover and fragile, the Lord tugged at his conscience and brought him down a peg, leaving him feeling small in that vast car lot.

He was twenty-three years old with a family to support, and he'd made Mom a promise to walk a Christian path. Between Opal and Mom's mother, there was already pressure to find a local church, and they'd talked for months about looking for one. Finally we joined an Assembly of God congregation on East Pennsylvania, but after going a few Sundays, we never really went back.

On Saturday nights, Dad invited Mark and Donny, his buddies from Big Spring, over to play cards and drink beer. Mom put up with the late nights. She said nothing about the pot smoke that wafted into the nursery while she put me to bed. One day she walked out into our small backyard, discovered a marijuana plant growing as tall as the fence, and nearly had a heart attack.

"Honest, I never thought it would get that big," Dad told her. She handed him a pair of scissors and sent him to cut it down.

Mom was busy trying to establish herself, to put down roots and feel settled.

Around that time, she answered an ad for an interior designer position at Sears, and they hired her. Sears even arranged a two-week training course in Dallas so Mom could learn to sell custom drapes and upholstery. When it came time for Mom to leave, she left me with her mother in Snyder while Dad stayed behind to sell cars.

After a week of being alone, he started missing her. That Friday he called her hotel and said he was driving to Dallas after his shift. He didn't tell her that first he planned to meet his buddies for drinks. He was drunk and swerving between the lines at 40 mph when the state trooper pulled him over outside of Weatherford. They charged him with a DWI and let him spend the night in jail.

The next morning, the judge summoned Dad to the bench, saying, "How much money you got in your billfold, son?" Dad told him three hundred dollars. The judge fined him $295, leaving him just enough to buy breakfast and think of all the ways to explain it to Mom.

Of course, she was angry, but as always, she forgave him. A few weeks later, however, she snapped. One night Mark and Donny were over playing cards and Mark was drunk. Mom didn't like Mark. He had the stink of crazy on him, plus a pair of eyes like an owl that caught you when you weren't looking. Just being in the same room with Mark gave Mom the creeps.

Mom was in the back room changing my diaper when she heard a loud crash. When she ran into the living room, she saw her antique tea set — the one Dad had given her before their wedding — shattered on the floor. Mark stood over the ruins, laughing.

Mom curled her lip, then did something she'd never done before. "Damn you, Mark Powell," she shouted, "you get the *hell* out of my house!"

The sudden vulgarity coming from Mom's mouth was so shocking that Dad stood there slack-jawed. Even Donny ran out the door.

Gambling, the DWI, and Mom being forced out of her character — Dad could read the signs. The Lord was telling him something.

"I feel like I'm slipping," he told her one night. "But I can't seem to stop."

Then one morning Bill Rogers sent a few salesmen to Houston on a dealer transfer to deliver some pickups. It was Dad's day off, the first he'd taken in months, otherwise he would have gone. When he returned the following morning, his manager told him there'd been an accident. On the way back to Midland, Lying Larry had fallen asleep and his car had veered off into a ditch, killing him instantly. Larry left behind a wife and a baby girl, who wasn't much older than me.

The tragedy shook Dad deeply and plunged him into a fog, one he was still trying to escape when he and Mom nearly died themselves.

It happened on one of Grady's airplanes. One Saturday night, they left me with the babysitter and flew to Lubbock for dinner. As they sat in the Smuggler's Inn, one of those epic sandstorms rolled in with thunder and lightning on its back.

The pilot, at Grady's urging, had ignored the weather forecast earlier that evening and gone ahead with the trip. Now they were stuck at the Lubbock airport, the wind howling and the sky flickering a foreboding pink and brown. The pilot refused to budge. "No way I'm going up in that," he said.

But Grady was drunk and insisted they leave. "We're gonna be stuck here all night. Just take us up high, get above the sand." He then offered the pilot more money, and away they went.

It took minutes for the storm to swallow the Cessna somewhere above the plains. The wind flipped it sideways, then into a tumble. Then lightning struck the tail with a crash and sent them lurching even more. The pilot fumbled with the controls, trying to wrench the plane from nature, while Grady screamed and panicked, his face the color of skim milk.

At the rear of the plane, Mom, Dad, and Ann held each other, shouting *"Jesus! Jesus! Jesus!"* into the swirling maw. After what seemed like an eternity, the storm opened its jaws and let them through.

At the airport in Big Spring, they crawled down the plane's ladder and lay flat on the tarmac until the earth stopped spinning. At home, Mom scooped me up from my crib and clutched me tight. She was furious with Grady and furious with herself for taking such a risk. "That night, you came close to being an orphan," she would tell me, as if reminding herself of their misjudgment.

Not long after, Dad put in his two-week notice at Bill Rogers Ford and called his sister Zelda. She and Charlie had recently moved to Albuquerque, where Charlie managed the commissary at Kirtland Air Force Base. Dad didn't have to say a word because his sister already knew. She'd been having those dreams.

"Something's been telling me to pray for you, Bobby," she said. "You've been heavy on my heart."

Zelda and Charlie knew someone at

church who sold cars at one of the big dealerships in town, so they put in a good word for Dad. We put a for sale sign in the front yard of the house on Amigo Drive, and within days, thanks to the latest boom, it was sold. In October 1976, we packed a U-Haul truck with everything we owned and pointed west for New Mexico.

9

Adrift in a foreign land . . . the Texas miracle . . . Preston's plea . . . the family comes home . . .

My first-ever memory comes from this time, and there at center stage is my grandmother Opal. Mom believes it was probably a Sunday, about a week before we pulled out of Midland, and I was staying at my grand-parents' house in Big Spring, as I sometimes did.

She was still plump, or "big-boned" (her own description). And her laugh, which sounded as if it came from some bottom-less, untroubled place, had not changed. The singing voice that soothed Uncle Bud during his final days had lifted an octave higher with age, but remained her greatest power. As I got older, that voice carved its own beautiful groove in my memory.

Gone were the oversized, matronly dresses

432

and the lace-up oxfords. The sixties and seventies had brought her and Bob out of black-and-white poverty and into modern color. She now wore bright silk dresses accentuated with bling: dazzling brooches, big necklaces, and rings the size of little gumballs. For their recent anniversary, Bob had dipped into his savings and bought her a white Coupe de Ville, previously owned.

By now Opal worked at the dress shop that her brother Herman had bought for his daughter Evelyn. The shop, Miss Royale, was located at the Highland Shopping Center, where most days she and Evelyn wore high heels on the thick carpet while dressing ranchers' wives in Nardis of Dallas and other designers. On weekends Opal worked the cash register at Herman's restaurant.

Herman's wife, Little Opal, had experienced a similar transformation. Her outfits sparkled and her makeup was perfect, and the sixties had taught her a few things about pantsuits. Her old churchiness had also relaxed, something her son Homer discovered in the mid-sixties. He was eating at the restaurant when his parents appeared dressed to the nines.

"Where are y'all going?" Homer asked.

"To the ball game," Little Opal said.

"The ball game? You never let me go to the ball game. What, did God change his mind?"

Homer let it slide, no longer the angry boy. Time and fatherhood had softened his edges.

But that was the state of things when I finally entered the world: salad days all around. No one had to jump boxcars like Bob during the Depression; we weren't grieving over dead children the way we'd done with Mary Lou, Orville, and Jack; no mad drives out west to save us from the Devil and ourselves, and no having to burn tires to stay warm. At that moment, no one in my immediate family even worked on an oil rig. The day my memory came alive, our house in Midland had just sold for twelve grand over market, a white Cadillac was sitting under Opal's carport, and my grandmother smelled of sweet Estée Lauder.

She was putting me to bed in the back room, and her face was close to mine. Nearly two years old, I was already talking and had learned the name of my soon-to-be home. We were playing a game.

"Mee-mee!" I said. This is what I called her. "Babel-kirky."

She threw her hands over her face and pretended to cry. "No, no, no! Don't go to

Babel-kirky!"

"Mee-mee," I said again. "Babel-kirky."
And on we went.

I loved my Memie and was sad to leave
her and Papaw, and I'm sure she was sad to
see me go. But she knew something that I
didn't, that Albuquerque — more than West
Texas — was the best place for my daddy to
save himself from the Devil and his own
flesh.

Albuquerque was an adventure. Neither
Mom nor Dad had ever lived outside of
Texas, and the clean break from Midland
was a fresh start.

In Albuquerque, we bought a little adobe
house on the east side of town. And thanks
to Zelda's friend, Dad began selling cars at
a giant dealership off the Coronado Freeway
called Rich Ford, while Mom took an
administrative job at the Bureau of Land
Management. Most days, I stayed with Aunt
Zelda, who taught me Sunday school songs
and spoiled me with Fig Newtons.

Under the guidance of his oldest sister,
Dad became strong again in the Lord. He
quit partying and started reading his Bible
in the quiet mornings before breakfast. We
joined Zelda and Charlie's congregation at
First Assembly of God. After church, Zelda

cooked a pot roast and the men gathered in the living room to watch the Dallas Cowboys. These were the team's glorious years, 1977–79, when Roger Staubach, Tony Dorsett, and Randy White led America's Team to back-to-back Super Bowls.

My memories from Albuquerque are few, yet it's funny what stuck. I remember that a neighbor kept a giant model train set like the one on *Mister Rogers* that filled his basement from wall to wall. He'd designed little towns, forests, and tunnels, and one of the locomotives even blew steam as it chugged up the hills. And I remember the day Mom and I were driving in the car and heard on the radio that Elvis Presley was dead. The memory locked only because of the sound my mother made in response, a kind of gasp.

I don't remember Mom ever being pregnant, but nearly nine months after Elvis died, my sister was born. My cousin Tammy — Zelda's oldest — came to stay with me while Mom and Dad were at the hospital. She brought along her fiancé, a former state-champ discus thrower from Eldorado High School named Mark Longerot. At three years old, I told Mark how I'd wanted a brother, but would still lobby my parents to name the girl after my favorite Dallas Cowboy, Tony Dorsett. Mark promised he would help, but it was no use. When my baby sister came home, her name was Marci.

As much as Dad enjoyed his new little girl, she arrived just as he started questioning our big adventure. He couldn't say exactly what it was, but Albuquerque was tough.

Away from the oil boom, the city of 330,000 was like most everywhere else in the country, still suffering under high inflation and fuel prices and limping through a prolonged recession. Kirtland Air Force Base and Sandia National Laboratories kept many people employed, but the economy was nothing compared to Midland. No big-swinging oilmen were paying cash for a new stepsider pickup.

To Dad, Albuquerque felt like another country. It had Indians, which Dad had never encountered before. The food was different, people drove too fast, and Dad found these big-city folks to be mean and cutthroat. One day he said to Mom, "They must hate Texans here."

He learned this the hard way at work. Rich Ford was twice as big as Bill Rogers in Midland. Thirty salesmen prowled the lot, compared to four or five, and according to Dad, these men didn't care if you lived to eat breakfast the next morning. Dad's first week on the lot, he spent three hours with a customer, who left to go and get his wife. But when they returned later and asked for Dad, one of the salesmen lied and said he'd gone home, then closed the deal himself. That was called skating, and those sharks in Albuquerque would skate you with a smile, even while you stood and watched.

Four hundred miles from home, Dad was in a foreign land, where his slow, sidewinding flattery was met with hardened stares, his sharp wardrobe screamed overkill, and friends were hard to come by. The old mojo was bootless, and his sunny confidence began to wane. For once, he went dark.

Through the dealership, he met two men who'd opened their own used-car lot, a "tote-the-note" operation that catered to ex-cons and people with bad credit, and they were looking for a third partner. They'd leased a weedy stretch of asphalt along Central Avenue, the old Route 66, on the blighted edge of downtown. Outside they hung a banner that read: BUY HERE, PAY HERE. NO CREDIT? NO PROBLEM! WE FINANCE! The sign rippled against the traffic noise and the songs of drunks looking for the bus. Dad took in the scene, his own empire of dirt, then returned to Rich Ford

438

and told them he quit.

He leased a corner of the lot and sold his own cars, mostly high-mileage trade-ins the dealerships had shucked off to wholesalers: mid-sixties Chevelles and Bonnevilles, Vega GTs, a Falcon Coupe. They were old, but they ran fine — as far as Dad could tell. His partners, however, didn't pretend to care. They sold clunkers to Indians, even coined their own warranty:

As Long as the Waters Flow
As Long as the Grass Is Green
Or to the End of the Block

It worked like this: Dad bought a car from the wholesaler for five hundred dollars, then advertised it for nine hundred. The buyer put five hundred dollars down, then financed the rest under the maximum interest allowed. Payments were made by the week, and in person. Fear the repo man if you fell behind.

"The fastest way to get back on your feet," his partners told people, "is to miss a payment."

Dad managed to move thirty cars his first couple of months. His customers were punctual, dropping by each Friday afternoon with handfuls of payday cash. But when his wholesaler ran low on inventory, Dad had nothing to sell. So he started buying dirt bikes, little Kawasakis he moved fairly fast but with little profit.

Then one morning he came to work and saw that thieves had busted open the trailer where he stored the bikes and cleaned him out. He didn't have insurance and the police were no help.

From a guy who hung around the lot, Dad bought a little Smith & Wesson .38 for a hundred dollars cash, only because the man had papers proving it was clean. Dad tucked the pistol into his jeans and tried to feel at ease, but he couldn't. The whole neighborhood had turned toxic in his mind, and he started feeling paranoid. The drunks and low riders cruising past suddenly beamed with menace.

Eventually his conscience wrestled free, and he saw his customers for what they were — poor folks with few options. Hapless suckers not even aware of how much they were getting screwed because they'd been getting screwed their entire lives.

"Can't you get a cosigner?" Dad asked a guy one day, a customer who just couldn't qualify. The man shook his head. "Last cosigner I had, he never made *one payment.*"

Dad knew he'd made a mistake, as if the pistol pressed into his back wasn't already a sign. There had to be a different way.

Turned out there was, and it came from Big Boy Blue.

Back in Odessa, his brother Preston had nearly worked himself into an early grave. After returning from Vietnam, he probably could have used some time to readjust, but with a baby on the way, he had to earn a living. The only job he could find was selling insurance for Metropolitan Life. He sold door-to-door, cold canvassing, like he'd done as a boy hawking manure in a tow sack. The work was torturous and every bit against his nature, but he did it for a year because there were bills to pay. Then for a while he tried selling vacuum cleaners — Filter Queens — but that job ended the day he flipped the wrong switch and blew dirt onto a woman's dress.

His father-in-law worked for El Paso Natural Gas, which had a plant in Odessa, and managed to get Preston hired. The pay was solid and he could choose his shifts, which made time for other things. Growing up around his uncle Herman, Preston had always wanted to start his own restaurant, and his tour in Vietnam had fostered in him a love of cooking. He and Linda prayed about the decision for several months. Then Preston borrowed the money from friends and family.

He bought a little place on the corner of Tenth and Whitaker, spruced it up, and opened for business. Preston's Place advertised "Good Home Cooking" and Preston cooked most of it himself, usually after graveyards at the plant, coming in early to bake biscuits and desserts from scratch. For the breakfast and lunch crowds, he served the standard bacon and eggs, hamburgers, and usually a blue plate special, like meatloaf or chicken-fried steak. Standing over a grill was tiring, but the work was wholly fulfilling. And better, word was spreading about how delicious the food was. Most days, the dining room was packed.

But his job at the plant sucked away time from the restaurant, which proved his downfall. It turned out one of the women he hired was stealing food. Preston kicked himself for not noticing this earlier, but for whatever reason he didn't. By the time he put the pieces together, he was in trouble. After only a year in business, Preston's Place had to close.

The failed restaurant buried him in debt, and now he was desperate. When a second child arrived, he picked up more shifts at the plant, and when that wasn't enough, he took a night job unloading boxcars at the freight docks — all while taking college courses in mechanics and carpentry before his GI Bill expired. Days and nights bled together and he was rarely at home. His body and mind grew so tired that one evening at the plant he broke down in the yard, draped himself across one of the big steam lines, and begged the Lord for mercy.

Get me out of this, he prayed.

The next week, he got word that the plant was closing. But before he could register the bad news, recruiters from Monsanto and Amaco were there snatching up men for two facilities near Houston. The boom was on, and chemical companies had to compete with the oil patch for workers, even if it meant having to poach them. Dozens of guys applied for the jobs and were hired right away, including Preston. But still, the companies needed more. So Preston put in a good word for Dad, who hopped on a flight the next week for an interview. When he returned home, Mom looked at the grin on his face and saw another nine hundred miles of road. And she was right — we were moving back to Texas.

■ ■ ■

The Monsanto plant was located thirty
miles south of Houston, in a little town
called Alvin, where we moved just after
Christmas 1978. Dad worked in the hydro-
carbon unit as a board man, walking the
pipe alleys where chemicals thrummed
overhead into giant distillation columns,
checking the pumps and gauges for any
leaks or changes in pressure. It was an easy
job and it paid well, even if it meant pulling
graveyards.

The transplants from Odessa called themselves the Sandblasters, and they were a
tight and rowdy group. The night Dad
started work, he wore blue jeans and a
monogrammed belt buckle. One of the
cowboys singled him out.

"What's the B stand for on that buckle,
hoss?"

"It stands for Big Bucks," Dad said,
without missing a beat. From that day
forward he was Big Bucks Bob.

Alvin was flat and swampy and thick with
mosquitoes, but we enjoyed it because we
were close to family. In addition to Dad,
Preston also got Mark Longerot, the discus
champ from Albuquerque, hired at the

plant, and he and my cousin Tammy now rented a small apartment in town. Opal's sister Veda and her husband Bill lived twelve miles south in Dickinson. Down the street from them lived Opal's other sister, Dorothy.

This unlikely family outpost along the Gulf marked the first time in decades that so many of us had lived together. Already, it had been forty years since Dorothy took a bus to New London and found her sister Agnes and her kids nearly starved after her husband walked out on them. Thirty years since Veda insisted on playing preacher, standing at the bedstead where time and again she sent poor Gloria Jean to hell. Agnes was long dead now, having passed suddenly in 1968 at the age of fifty-eight while undergoing gallbladder surgery ("I won't survive," she told Homer a week before the operation, and she was right).

All the children of Clem and Cora — who died in 1970 — now had grandkids of their own. Their hair was turning gray and their bodies were starting their slow march of revolt. On holidays we stood around Aunt Veda's dining-room table, where Uncle Bill or Uncle Charlie offered up the *Oh Heavenly Father* before the meal. Afterwards, we all pressed into the living room to watch

445

football and home movies of family reunions past. The old films brought back the dead and for a brief moment, they were among us, back with the choir. Each movie ended with the whole gang singing around a piano.

Preston and Linda rented a house on the outskirts of Alvin that backed up to a bayou. On Saturdays, their two boys, Matt and Brandon, and I stalked the marshy banks catching frogs and hunting crawfish. We'd find their mud chimneys and lure them out with bacon tied to a string, then make them fight one another. The frogs we tied to bottle rockets.

In late July 1979, a storm blew in that took us by surprise. That day, Mom's parents arrived from Snyder to stay the weekend, and that's when it started to rain. By ten o'clock that evening it was still coming down, but even harder. I was lying in bed listening to the wind whip and howl when I heard a loud crash and ran into the living room, where my grandfather was staring out the window. Half the neighbor's roof had just landed in our front yard, missing his brand-new Chevy Impala by a foot.

By the next day, Tropical Storm Claudette had dumped forty-three inches of rain on Alvin — the largest twenty-four-hour rainfall ever recorded in the United States. I looked out the window and saw a man rowing a johnboat down our street, loaded with furniture. Luckily our house was on a small hill and was spared.

Much of Alvin, however, sat under ten feet of water, including Preston's house. The bayou had jumped its banks and rushed under their doors. Dad waited a day for the water to recede, then waded into the flood to help his brother salvage what was left. But everything they owned was gone, including their family photos and Preston's letters from Vietnam.

Earlier that day, Dad had hoisted me onto his shoulders to walk down our street. Dad stood six foot three, and the water was all the way up to his chest, just below my feet. At the end of the block, men were guiding boats between the houses, loading them with coat racks and toasters and other random belongings. A man in a canoe rowed by and shouted something to us, then pointed to the water, where a poisonous cottonmouth slithered past my toes.

The entire coastal region was paralyzed by flooding, including much of Houston. Yet at the time, it seemed no amount of water could drown the rising totem of the Texas miracle. Midland may have bustled, but Houston was benefiting from the high oil prices more than any other city. As many as a thousand people like us were moving to the region each day to work in the plants and refineries, and to help build the offices and apartments that were shooting up overnight. Office space in Houston was so limited that companies were leasing it eighteen months in advance.

"This is not a city," *U.S. News & World Report* had written the previous year. "It's a phenomenon — an explosive, churning, roaring juggernaut that's shattering tradition as it expands outward and upward with an energy that stuns even its residents."

The Texas miracle, long before our politicians began throwing it around on campaign stops, started here. Between 1973 and 1981, while the rest of the nation suffocated under double-digit inflation, rising fuel prices, and stagnant growth, and while mills and factories seemed to close each month in Michigan, Pennsylvania, and eastern Ohio, Texas added 2.2 million jobs and personal incomes tripled. In just three years between 1979 and 1982, when the country as a whole shed two hundred thousand jobs and $26 million in gross domestic product, Texas hired eight hundred thousand people and boasted a $43 million GDP.

For the men and women in northern cities ravaged by the recession, who lost work in the plants and factories and watched their communities unravel while crime, addiction, and suicides increased, Texas seemed to be the only working part of America. They looked toward Texas the way John Lewis had nearly a hundred years earlier from the claustrophobic hollow, and the way the sodbusters saw California when the bankers and dusters drove them off the plains.

Before long, the Texas dream was broadcast on prime time. The television show *Dallas* premiered in April 1978 and gave us the oilman J. R. Ewing, who embodied the wealth and glamour our nation desired and the greed it would embrace in the coming decade. At the height of *Dallas* madness — when ninety million people around the world tuned in to see who shot J.R. — the Southfork ranch north of town drew more tourists than the grassy knoll.

While the Dallas Cowboys were known as America's Team, the Houston Oilers were the mascots of the Texas miracle. If the boom had a color, it was Columbia blue and white. The Oilers had a derrick on their helmet and a coach who looked like a tool pusher. Bum Phillips was potbellied, wore boots and white cowboy hats, and gnawed on a giant cigar. In 1978, the Oilers drafted the locomotive running back Earl Campbell. Two years later they acquired Kenny "the Snake" Stabler from the Oakland Raiders. Stabler's long hair and beard made him look like a roadie in Waylon Jennings's band, and he partied like one too, walking into bars around Houston, yelling, "Strike, lightning!" as if to let everyone know what was in store. Like most people in Alvin, our family burned with Oiler fever. Dad and I wore Snake jerseys and searched the stores for Snake Venom Cola, while Mom had a pair of tight Vanderbilt jeans with "Luv Ya Blue" embroidered on the butt.

With so many men pouring into the region, bars and honky-tonks sprang up in the dingy towns like League City and Pasadena, with Gilley's being the biggest. In fact, *The Guinness Book of World Records* listed Mickey Gilley's dance club as the largest honky-tonk in the world, capable of holding six thousand people. It had pool tables, a mechanical bull, punching bags, and, of course, a giant dance floor and stage where top-billed country acts played seven nights a week. Its slogan was "We Doze but We Never Close." They filmed the movie *Urban Cowboy* in Gilley's in 1979, and when it came out a year later, around the same time as *Honeysuckle Rose,* starring Willie Nelson, Texas chic became the rage. People everywhere started wearing Stetsons and Tony Lamas. Bolo ties were popular, along with rhinestones, cow chip jewelry, and anything with armadillos.

Gilley's was only twenty miles up the Dixie Farm Road from our house, yet its carnival of temptation had no power over Dad. Between the graveyard shifts and a buffer of family all around, our time in Alvin was proving to be his most stable. The friends he hung around with were from our Assembly of God church, and together with Mom, they helped keep him on the path of righteousness. What also helped was the arrival of another child — my sister Melissa — born in June 1980 in nearby Texas City. It wasn't long after she arrived that a telephone call late one night brought Dad's old life roaring up from behind, flashing its high beams.

Earlier that day, his old friend Mark Powell, whom Mom had cursed and kicked out of her house, chose to take revenge on his ex-wife for their recent divorce. Mark walked into the Odessa Chamber of Commerce, where his former mother-in-law worked, and shot her with a twelve-gauge while she sat at her desk. He then stole a car, kidnapped three teenagers at gunpoint, and embarked on a twenty-four-hour shooting spree across Texas.

When the news reached Big Spring that morning, Opal nearly fainted. Just the previous day, Mark had stopped by the Miss Royale dress shop and asked for Mom and Dad's address in Alvin. "Well, it's at home," Opal said, lying. She could tell something was amiss.

After killing his ex-mother-in-law, the police reported that Mark was headed for Houston, and that's when the phone started ringing. "He might be coming after y'all," friends warned, harking back to Mom's episode over the broken tea set.

But Mark was after his ex-father-in-law, whom he also wanted dead. He managed to find him at a trailer park and get off a few pop shots, then turned back toward Odessa, firing at random cars along the highway and injuring four people. Sheriff's deputies finally caught up with him in a little town south of Midland. A high-speed shootout ensued and Mark was fatally wounded. His car then hit a guardrail and flipped.

"The subject was terminated one mile east of Rankin," the sheriff's office reported, adding that the hostages, who'd been in the trunk, had survived unharmed.

The whole ordeal, the family believed, was the Lord's way of affirming Dad's choice to leave his old life behind, for the wages of sin was truly death. Dad believed it too. He and his brother had even started making plans together. Preston had a dream of starting his own construction company — Mealer Homes — to take advantage of the current housing crunch. To that end, Dad started taking classes in heating, ventilation, and air-conditioning at the local community college. Air-conditioning was a steady trade, he figured. In good times or bad, Texas was always hot.

At twenty-seven years old, my father finally had it together. He had the confidence and support of his church and extended family, a strong wife and three healthy kids at home, a recent promotion, and a career goal within reach. By his own description, life was grand. That is, of course, until Grady called.

"Bobby Gaylon!" he shouted. "Answer me one question: How'd you like to be a millionaire?"

Within a month, I was riding in another U-Haul on the bench seat next to Dad. We were on our way to Big Spring, headlong into the boom.

10

Grady seizes the kingdom . . . Dad takes
a leap . . .

Around the time we moved to Albuquerque,
Grady lost his job with the American Diabetes Association. For budgetary reasons, the
ADA had decided to close its Midland office and not transfer his position as regional
director. Suddenly in need of a new job,
Grady turned to his wife, who asked her
mother for help.

By this time, Iris's displeasure with Grady
bristled on the surface. She disapproved of
his charter planes, his thousand-dollar dinners, and especially the way he threw Raymond's name around town. And foremost,
it made her sick to hear that Grady was
leaving Ann alone at night so he could
carouse with homosexuals, if the rumors
were true. If the police ever arrested him in
one of those clubs in north Odessa, how

would Ann survive the shame?

So when Ann asked her mother for help finding Grady a job — at the refinery, of all places — Iris honored her daughter and made the call. But the favor came with a pleasurable twist of the knife. If the son-in-law of Raymond Tollett expected a corner office and a long-legged secretary, he had another think coming.

"You want a job at Cosden?" Iris said to him. "Then you start at the bottom. The very bottom."

Grady's title was yardman. On his first day of work, someone handed him a shovel and assigned him to a crew digging a ditch. Other duties entailed making coffee twice a day and chauffeuring engineers around the plant. The sight of Grady laboring in the heat, flushed and sweaty, provided endless entertainment for the other men.

"Say, ain't you supposed to be runnin' this place?"

"Hey, if the boss man likes your coffee, he might let you take out the trash."

Grady gritted his teeth and forged on, determined not to crack. In order to take the job, he and Ann even gave up their apartment in Midland and moved back to Big Spring. They rented a small house that was old and drafty, and one morning Ann awoke to find a giant snake curled up on the kitchen floor.

As a reward for moving home, Iris bought them a new washer and dryer, along with a bedroom set. And evenings when Grady came home demoralized and covered in filth, he sometimes found her sitting in his living room, as if to remind him who was boss. Whatever path to easy street her son-in-law thought he'd chosen had just taken a turn straight into a wall.

But Grady was biding his time, no one's fool. The man in charge of the machine shop, Joe Faulkner, felt sorry for him and put him inside cleaning tools. He still made coffee — once in the morning and once in the afternoon — and drove the men where they asked to go, saying nothing even while they insulted him. And he remained civil — even charming! — when he and Ann drove Iris to Midland once a week for lobster night at the Steak and Ale.

The truth was, he actually liked his mother-in-law. He recognized that despite the dragon scowl and vicious tongue, she could be generous and kind. She'd continued some of her husband's charities, and every Christmas mailed out checks of three thousand dollars to members of her family.

But even more, he wanted Iris to like *him,* as she was living proxy to the man whose legacy he was trying to internalize.

But Iris showed no signs of thawing, so Grady endured. About six months later, in December 1976, his wait came to an end. Iris grew very ill.

She'd been plagued with circulation problems for years, which caused her constant pain and sent her back and forth to doctors for treatment. Then, a week before Christmas, she was having trouble breathing and called Dr. Thomas, her husband's former doctor, who admitted her to the hospital. It turned out that Iris had suffered a stroke. For two days she kept getting worse, then finally slipped into a coma. Around three in the afternoon on December 19, she opened her eyes and took her last breath. She was fifty-nine years old.

Ann phoned her husband at the refinery, sobbing, and broke the news. Grady's brother-in-law, Roe, happened to be working in the machine shop when Grady flung open the door, screaming.

"Roe! Roe! You'll never believe it! Mrs. Tollett is dead!"

Roe was stunned. "My god, that's terrible," he said.

Before he could inquire how, Grady had

climbed out of his coveralls and flung them into a trash can. "I quit!" he shouted and ran for the door, stopping just short of leaving.

"Don't you know what this means, Roe? Do you know what this means?"

A look of ecstasy washed over his face. "It means I'm a millionaire!"

The transformation was rapid. In July 1977, Grady and Ann bought a thirty-five-hundred-square-foot ranch home in Ike Robb's Highland South — "Big Spring's Most Desirable Living Area." The developer and theater magnate remained Grady's idol.

After the house, the couple flung themselves into local causes. Ann became secretary of the Big Spring Garden Club and hosted events in the home, where the society page noted her décor and wardrobe as it had her mother's. They became members of the Heritage Museum, gave generously to the Bible Fund, and were lead sponsors for a celebrity fund-raiser to benefit the Dora Roberts Rehabilitation Center, donating a trip to Las Vegas. The event featured the New Christy Minstrels and comedian Foster Brooks, famous for his role as the Lovable Drunk on *The Dean Martin Show*. The *Herald* ran a photo of Grady and Ann posing at the rehab center with little Christy Clifford, a girl with cerebral palsy. Grady stood rigid. There was talk of him running for local office.

But as Grady embarked on his long-awaited ascent, the plates beneath the town began to shift. The first jolt was cataclysmic: In 1977, after twenty-five years in Big Spring, the federal government decommissioned Webb Air Force Base and closed it down. Nearly three thousand jobs vanished in a matter of months. The overall loss of population was even greater, as families of airmen and civilian employees fled in droves, leaving nearly a thousand empty houses and a public school system in crisis. Simply maintaining the empty base cost the town forty thousand dollars per month.

More bad news arrived two years later, when Cabot Corporation, which had made carbon black in Big Spring since 1950, announced it was a victim of the fuel shortage and shuttered its plant. Gone were a hundred jobs and $2 million in payroll. But most painful was the announcement that the Cosden Oil and Chemical Company was vacating the Petroleum Building and moving its headquarters to Dallas, where it would merge with its parent company,

American Petrofina.

While the refinery itself would remain, gone were the scores of executives who'd sat on school boards and directed charity drives — Tollett's "corporate citizens" who'd put some skin in the game. The Cosden Country Club had long closed its doors on the lake. For those still around — the bankers, newspapermen, and ranchers who'd mingled with that crowd — all that remained were memories: Duke Ellington and Harry James and wild nights in the Blue Room; Prince Farman and delegations of boots-wearing Japanese; Mr. T. toasting war heroes and cops at the policeman's ball.

When the tremors finally quieted, much of Big Spring's educated upper class was gone. Aside from the few bankers' and ranchers' kids, no more students returned in the fall with tales of Vienna or Paris, or arrived with previous addresses on the other side of the globe.

The mighty T&P had ferried its last passenger more than a decade earlier. The iron rail gave way to the interstate that now shuttled tourists and traveling salesmen around the edges of the city as if it were invisible. They slept at the Holiday Inn and waited to spend their money in the shopping malls of Midland and Lubbock. Downtown stores began to close, their windows papered over.

Finally, to everyone's disbelief, the Hotel Settles locked its doors. Competition from the interstate chains had lowered its standards, leaving it to the winos and circuit prostitutes who rented rooms by the month, then moved on to other towns. Workers pried away the mahogany pancling and marble stairs and dismantled the wrought-iron railing. Everything that could be removed — beds, stoves, telephone booths, room keys, even the plumbing pipes — was carted out and sold at auction. Built in 1929 as a monument to Big Spring's first oil discovery and a symbol of its soaring ambition, the Settles was reduced to an empty shell.

The little town with boundless vision that had lured my grandfathers and uncles, the oasis of the plains that Tollett championed from Fifth Avenue to Tehran, was folding in on itself, threatening to collapse. Holding it afloat for the moment, and numbing some of the pain, were the champagne bubbles of another rising oil boom. And folks, as they always do, mistook them for progress.

Soon, contractors were buying up the discarded drilling rigs that were mothballed on back lots and roadsides, sanding off the rust, and adding a fresh coat of paint. The clerk's office filled with landmen looking for leases, white service trucks stacked up at the red lights, and suddenly there were strangers in the restaurants. To Grady, this presented his biggest stage.

Because no matter how much he loved Ann, or tried to love Ann, the town would always regard him as a gold digger, the theater usher of ill repute who hoodwinked the beloved family. Already, the old man's legacy mocked him from the grave: Tollett, the "four-career man" who dropped out of school and taught himself law, who busted gangsters for the FBI and turned a scrap heap into an empire. If Grady were to step out of that shadow, he had to build something on his own, and he had no interest in growing cotton or buying cattle. But the oil business — *that* was the arena of immortal men. The question was how.

The answer, remarkably, arrived in late 1979, when Grady won the lottery.

Starting in 1960, the U.S. Bureau of Land Management began holding oil and gas lotteries to award drilling privileges on public land in the American West. Most of it was located along a pay zone called the Overthrust Belt, which covered seven states in the Rocky Mountain range. Wyoming and Idaho alone were thought to hold 1.5 billion barrels of crude and over 7 trillion cubic feet of natural gas.

Each month, the BLM sent out a list of available tracts. Applications cost ten dollars each and were open to both individuals and private companies. (Wyoming's state-run lotteries had already proved lucrative for con shops, who targeted retirees with letters that read, "Dear Potential Oil Baron" while tripling the entry fee.) Winners could lease up to 2,650 acres for an annual rate of a dollar per acre. The odds of the bureau's computer pulling your name were around four hundred to one.

There was no limit to how many times a person could apply, so Grady had entered dozens of applications in the years after Iris's death, devoting the same furious energy he had to his Publishers Clearing House Sweepstakes. And lo and behold, he finally struck pay dirt: Grady won an oil well in Wyoming.

The news arrived while Grady and Ann were driving through California on vacation. Grady telephoned from the hotel and spoke to his gardener, who described an important-looking envelope that had arrived in the mail. "Don't move a thing," Grady told him, then drove straight home.

He flew to Cheyenne to claim his prize, which was a small tract near Thermopolis. But a few telephone calls were enough to tell him how deep the water was. Even though he'd married into money, he didn't have the ready capital, much less the savvy, to organize the men and tools to spud a well beneath the sagebrush. Since he didn't know the first thing about finding oil, he'd have to hire an engineer and geologist. And say they did find some — they weren't popping Spindletops on these plains. The taxes on royalties, after the feds took their one-eighth, probably wouldn't justify the headache. Not with one little lease.

Luckily, one of the loopholes in the government's rules was that no one was legally obligated to drill on the land they won. And considering all of the above, most people ended up selling their leases to oil companies — which is exactly what Grady did, unloading his acreage to a Louisiana driller for a cool sixty-five thousand dollars. It wasn't the way he'd envisioned becoming an oilman, but it would do.

Elated, Grady called his cousin Selena, who was living in Denver, and said he was stopping in town on his way home. He wanted to celebrate.

"We'll have dinner at Lafitte's," he announced, meaning the erstwhile gem of Larimer Square whose clientele had been reduced, according to one review, to "tourists and the tastelessly rich."

"Grady, they're booked for weeks," Selena said.

"Tell them I'll pay five hundred dollars for the reservation."

He then told her to wear a black dress.

Selena sold real estate in Denver and knew some high rollers, one of whom offered to call Lafitte's and got her a table. She was seated in one of the red-leather booths, dressed in all black, when Grady entered through the heavy wooden doors. He'd come straight from Perkins Shearer, the men's store, and had shucked his shabby duds for a stunning pinstripe suit and Italian leather shoes. He brushed aside the hostess and, before reaching the wide stairway that led up to the oyster bar, stopped for a brief moment, as if to present his new persona to the room.

With the money from selling his Wyoming lease, Grady started Cunningham Oil Company and set out to make his name. He first partnered with a wildcatter named Adkins on a series of wells being drilled around

Sweetwater — putting in the standard third of the operating expense for a quarter of any royalties. And with proper oil transactions under way, Grady acquired his headquarters.

There was little question of where. That fall, he announced that his company was taking over the top suite of the Cosden. Petroleum Building, the very offices from which Raymond Tollett had steered the great refinery. He hired a decorator to fill the rooms with expensive carpeting and Henredon furniture. For his desk he bought an executive pen set and decorative pump jack, along with an extra-large checkbook, monogrammed and bound in leather.

He hired an actual secretary to answer his telephones, one of middle age who chain-smoked cigarettes and swore like a floor hand. And over a game of golf one afternoon at the Big Spring Country Club, where Grady was now a member, he met Hugh Porter. Hugh was fresh out of Texas Tech, where he'd majored in biology and chemistry. Not only was he sharp and good-looking, he also came across as steady and self-possessed. Right away Grady offered him a salary of thirty thousand dollars to come sell oil investments, then dispatched him to Midland for a seminar on how to do it.

Next, Grady sought legitimacy. Using his father-in-law's name, he managed to convince one of the Permian Basin's most respected geologist-and-engineer teams to come aboard. Orville Phelps and Bob Wilson had been finding oil since the fifties and had employed just about every method to get it. They were shrewd, sober men who spoke a language of origin and deep time. After listening to them explain porosities, anticlinal structures, and stratigraphic traps, Grady threw up his hands, said, "Hell, just go find me some!" and gave the men whatever they asked for.

It was early 1980. Ronald Reagan would soon campaign for president on a platform of rescuing the sluggish economy, and part of that plan involved further decontrol of oil and gas. But in West Texas, the boom was already in full swing. Oil prices would soon reach their highest level yet, forty dollars a barrel, thanks again to the kingmakers half a planet away, whose names most Texans wouldn't know how to pronounce.

In early 1979, revolutionaries in Iran had ended the dynasty of Mohammad Reza Shah Pahlavi and installed their supreme leader, Ayatollah Khomeini. A second global panic ensued that drove crude prices even higher than OPEC's embargo ever had. The gas lines returned, forcing President Carter — after delivering his soul-searching "malaise speech" on television — to finally start lifting price controls to spur domestic drilling. Suddenly all the abandoned "stripper wells" that produced less than thirty-five barrels a day — most of them remnants from the old booms — were worth bringing back online. Texas had thousands of those wells, which were soon producing millions of barrels of oil.

Farmers and ranchers who'd come to ignore the pump jacks on their land now saw royalties for the first time in years. In January 1980, deposits into Midland banks surpassed $1 billion for the first time ever — despite one of the worst droughts in twenty years that decimated the season's cotton crop. Banks across the state whose foundations had been built on cattle and agriculture made quick decisions to diversify into energy loans. They dispatched officers to the oil and gas seminars, then sent them running after tomorrow men short on equipment and cash.

By then Grady's company, Cunningham Oil, was officially in the race. Two of the Adkins wells near Sweetwater came in strong. In April 1980, Bob Wilson and Orville Phelps zeroed in on some acreage a hundred miles east of Big Spring, near San Angelo, where they wanted to drill. Hugh then drove out and charmed the rancher to secure the lease. While the team pitched investors and waited for drilling permits, Grady — by another stroke of fortune — acquired yet another piece of his kingdom.

On April 9, Ike Robb, his old mentor at the Ritz Theater, collapsed while exercising at the YMCA and died of a massive heart attack. He was forty-nine. Just four months earlier, Robb had sold off his movie theaters to focus on real estate — in particular, to finish Highland South and realize his father's vision. Now, with father and son both dead, Robb's wife Betty put the undeveloped land up for sale. By year's end, Grady had snatched up seventy-three acres at the foot of South Mountain.

It was right after buying Highland South that Grady called Dad, offering a seat beside the throne.

Grady was typically drunk when he called the house, but that night Dad heard something different. Grady carefully laid out what sounded like an actual business plan. And it was impressive. He told Dad about Adkins and the wildcat wells, about hiring

Wilson and Phelps, and his excitement about landing their first lease.

"Bob Wilson says each well on that ranch is worth twenty thousand barrels," Grady said, quoting the engineer's review, while not precisely. "And we're gonna drill forty, fifty of those wells. With prices how they are, you're gonna be stuffing pillows with all the money we make."

But what sounded most appealing to Dad was what Grady called the Cunningham Development Corporation, which would carry on Ike Robb's legacy at Highland South. Back when we lived in Midland, Dad had actually taken real estate courses, mainly on a lark, and gotten his license. Grady asked if it was still valid. Dad told him yes.

"Then I can't let you pass up this kind of opportunity," he said, then lowered his voice. "Bobby, I'm asking you to come back to Big Spring and be my vice president."

"Well . . ." Dad replied.

"You'll be my right-hand man — just the two of us."

Dad's job would be to sell investments in the oil wells, and with each deal he closed, Grady would give him a percentage of the royalties. He'd also be in charge of selling the luxury home sites to all the people getting rich in the boom. "And if you're gonna sell 'em, you might as well live on one," Grady said. "First thing we'll do is pick out your own lot." Grady then offered Dad a salary of thirty-five thousand dollars plus perks, which was a lot more than he was making.

The next morning, when Dad got home from his night shift, he and Mom talked it over. They agreed that if even half of what Grady said was true, it was still a heck of a deal. The boom was making millionaires out of fools left and right, no one could deny that.

"This is that *one* chance of a lifetime," he concluded, "that chance that you'll kick yourself later for turning down." The next day, he called Grady and said he'd take the job — as soon as we could sell our house in Alvin.

Yet when Dad broke the news to the family, he was surprised by their disappointment. Preston, especially, seemed hurt.

"You got a great job down at the plant," his brother argued. "All the guys love you. Your family is happy here, and besides, what about the homes *we* were gonna build? You were gonna be my AC man."

"Sorry, Pres," Dad said. "I'm going to get rich."

Preston begged him not to go. He knew all about Grady Cunningham, and the whole enterprise Dad described made him feel sick to his stomach. But Dad wouldn't listen. Even Zelda couldn't talk sense into him. And it was no use appealing to Opal, whose only thought was having us grandkids back home.

Bob was even less help. "I never had that kind of opportunity," he told his son. "I say go for it. It's about time one of us made money off them dat-burn rigs."

11

Dad returns home, enters the fast lane . . . Grady's world . . . an accident and a choice . . .

While Dad and Grady worked out the details for our lot in Highland South, we needed a place to live. Dad had heard stories about all the empty houses left over from the base closure and assumed we'd have easy pickings. But once we arrived in Big Spring, we discovered the boom had devoured everything. No rentals, not even mobile homes. Every hotel was full and charging double rates. People blamed the Yankees from Ohio and Michigan who'd driven down to work on the rigs, but many of them were sleeping in their cars out at the truck stop.

After two days of looking, we found only one house that was available. It was a tiny three-bedroom on West Seventeenth Street with peeling yellow paint, ratty carpet, and no air-conditioning. We took it right away. Mom managed to fix it up as nice as possible while Dad used what he'd learned in HVAC classes to install central air and heat. Weeds and sticker burrs covered the tiny backyard, but near the porch was a dead pecan tree with a wooden fort in it. All that remained was its rickety platform, which my sister Marci and I quickly claimed as our lookout. Not long after we arrived, we sat eating popsicles and watched a sandstorm roll off the distant Caprock. It approached quietly, like a brown fog, before the wind grew dirty and stung our bare skin. We dropped our ice cream and raced inside, our mouths full of grit.

At work, Dad was getting his own taste of the landscape, and it was more crowded than he'd pictured. In all of their conversations about the company, Grady never said anything about Hugh Porter. The day Dad arrived, Hugh joined him and Grady at the Branding Iron for a welcome lunch that stretched long into the afternoon and left Dad overwhelmed. From the start, it was clear that Hugh, not Dad, was in charge of putting together oil deals, and Grady treated him like a second in command. Grady even introduced Hugh as "operations manager."

Grady had also failed to mention anything about the half dozen other guys who joined them at the table, each one resembling a *Tiger Beat* model. They didn't seem old enough to drink the liquor they ordered, and in fact, most of them were still in high school. There was Jacques, a farm kid from Indiana whose family had recently arrived in town. He'd been working as a waiter at the Branding Iron when Grady flagged him down one night and offered him a job.

"A job doing what?" Jacques asked.

"Oh, hell," Grady said, "we'll find something!"

Grady put Jacques in charge of the company's new IBM and gave him the title of assistant comptroller. But his job mostly entailed driving Grady to the country club. The same went for David, and Kenny, who was Grady's nephew. Another guy, Buddy, seemed to be in charge of this group and held special sway with the boss. Buddy was basically an orphan — his father dead, his mother absent — when he'd first come to work for Grady. Like Hugh, he was bright but also streetwise and shrewd. When Grady learned that Buddy was living alone in the Barcelona Apartments, unable to pay rent and trying to go to school, he asked Ann if Buddy could live with them. Buddy now had his own bedroom on Glenwick Cove and a rack of nice suits in the closet.

There was also the company pilot, a twenty-three-year-old Norwegian named Torstein. He and his fiancée, who was from Michigan, had found their way to Big Spring when the municipal airport needed a safety instructor. They later got married in one of the big hangars, and for the reception, one of the crop-duster pilots filled a wooden coffin full of ice and Coors.

Coincidentally, Torstein had served as assistant pilot the night Grady chartered a Cessna 206 to fly to Dallas for a convention. Grady was so drunk by the time he returned to the plane that he promptly passed out while Torstein and the pilot headed back to Big Spring. Halfway home, they ran into a fog that paralyzed the western half of the state. Every airport within a hundred miles was closed and wouldn't let them land, forcing them to circle five hours in a panic as the fuel gauge dropped to red. Grady happened to wake up just as Torstein found a way to get them to Roswell, where they landed on vapors. The following week, to express his gratitude, Grady invited Torstein to dinner at his home.

"I told Ann, 'This is the young man who

saved my life,' " Grady said, and raised his whiskey glass to duplicate the moment. "And that's when I asked him to be my company pilot. You know what he said? He said, 'But Grady, you don't even have a plane.' I told him, 'What the hell, Torstein, I'll just buy one!' And damn it, I did, too."

Grady bought a Piper Seneca II, which he painted in Dallas Cowboys colors — even though he hated football — before upgrading to a ten-seater Navajo. When Torstein wasn't flying Grady and his crew to the "company cabin" in Ruidoso or the ski slopes in Taos, he helped Hugh knock on farmers' doors and pitch oil investments. Grady even gave Torstein his own desk and a company car, a slick Ford Thunderbird, which Torstein posed with for photos he sent back to Norway — proof of the jumbo Texas dream.

Not long after Torstein joined the company, Grady asked him if he spoke any foreign languages. When Torstein replied that he knew a little French, Grady invited him and his wife on a grand tour of Europe. A whole group from Big Spring went, making stops in Amsterdam, Paris, Austria, and Switzerland. It was in Switzerland that Grady bought a floor-length fur coat made from a wolf. In fact, he bought two.

Torstein's wife, Linda, worked at a new bookstore that Grady opened called The Book Inn: An Overnight Home for Books. It was located in the Highland Shopping Center, next to where Grady had sold women's shoes. Grady had even paid a local decorator twenty thousand dollars to design the store's interior.

Upon hearing this along with everything else, Dad's head was swimming. "What are you doing buying a bookstore, Grady?"

"Well, Bobby, Ann likes books. She needs something to keep her busy while I work all the time."

"Is it making any money?"

"It will, it will," he said, and patted Dad on the back. "That ain't for you to worry about anyway, sugar, not on your first day." He lit one of his long Carlton 100s, then motioned the waitress for another tray of drinks.

It turned out that Grady really was hard at work, mainly trying to fashion himself after Ike Robb and Raymond Tollett. A couple of days later, Dad was shocked when he opened the *Herald* and saw that Grady was running for the school board.

He'd taken out a campaign ad that stated: "As a businessman, I have learned about necessity of sound management, and of conserving and wisely using our resources." A photo showed Grady in his new suit sitting behind his desk with a pen in hand, ready to sign a stack of documents. "My door is always open to you," he added.

The election was held that week, and it was a slaughter. Out of six candidates, Grady came in second to last in the polls, garnering little more than four hundred votes. Ann was secretly relieved. "You'll never be like Daddy," she told him, and the comment made Grady furious. He never spoke of the election again.

Ann was the real owner of the bookstore — it was her money that bought it, anyway — but she let Grady's sister Nancy manage the place, with help from Torstein's wife. After Grady won his oil well, Ann and Nancy had gone up to Portland, Oregon, for a seminar on how to succeed in the book business. And ever since opening the place, Nancy had done a fine job of stocking the latest bestselling novels, romance and mystery paperbacks, plus a wide variety of Bibles — which was their bread and butter. No one got rich off selling books — that's what they told them up in Portland. The store had to be a labor of love, and theirs would survive as long as Nancy could keep Grady's hands out of it.

Ann visited the store nearly every day, usually in the mornings when it was quiet. She complained to Nancy and Linda how Grady was never home, how he and the boys would leave and be gone for two or three days, never telling her where, and spending money on God knows what. One day, a woman named Bobbi Joe showed up at the door and said she was the new cook. Another time, after a trip to Houston, Grady had come home with a stretch limousine and a chauffeur, who now waited outside their house to drive Ann to the grocery store. "I don't need somebody driving me around Big Spring. Not unless I'm sick!" she said. Already she refused to get in that silly car.

Grady had started leaving Buddy at the house when he went away, along with another high school kid named Kirby who'd also moved in. "Buddy can look after you," Grady told Ann. "Just tell Buddy if you need anything." Most nights it was just her, Buddy, and Kirby, and sometimes Buddy's girlfriend, playing cards at the kitchen table. For supper they endured whatever concoction Bobbi Joe pulled out of *Mastering the Art of French Cooking.* "That woman can't cook worth a plugged nickel," she complained to Grady. It was clear to Ann what

Buddy had been hired for, and Bobbi Joe, too — to be her guardians.

What Ann really wanted was a family. She wanted to raise kids and give them the grand happy childhood she never had. What she wanted was to love someone and to be loved in return. She told Nancy and Linda how she and Grady had tried to get pregnant but hadn't been successful. The doctor said it might be her thyroid. She was also taking diet pills to keep down her weight, hoping that being thinner might keep Grady at home. The truth was that Ann couldn't care less about her money. She would trade every dollar her father had ever made to have a happy family of her own.

"Why don't you just cut him off?" Nancy would say. But for reasons she didn't know, perhaps loneliness, Ann enabled her husband, which she would regret for the rest of her days. She enabled him with the money she had no use for herself. And under its influence, she was watching him turn into a monster.

The shock of Grady's spending propelled Dad into overdrive during his first weeks on the job. There was plenty of work to get done: forty-five lots out in Highland South needed to be supplied with water, sewage, and electricity for the first phases of development. Permits needed getting, streets needed paving, and most of all, the lots needed to sell.

On the oil side, despite feeling sidestepped by Hugh (and having to share his royalties), Dad realized they made a good pair. Hugh was a real go-getter, plus he had a head for numbers when it came to drafting promotional deals for investors. In fact, for their first well, he structured the deal so the company wouldn't pay a dime until after the hole was completed, which in turn freed up capital to put them in what Hugh described as "a multiple-well position."

It was Dad's job to find the people who might have ten or twenty thousand dollars to play with. He called bankers and car dealers in town, doctors and lawyers in Midland and Odessa, a dentist in Dallas, people who might be eager to drop an oil well into their cocktail conversation (a 70 percent tax bracket for high earners meant that people were investing for the write-off, which only fueled the reckless speculation). Others were just hoping for one lucky strike that would guarantee some go-to-hell money when they grew old.

Sitting in his office, with his own brass pump jack on the desk, Dad was never sure if it was Ike Robb, Tollett, or J. R. Ewing whom Grady saw each time he looked in the mirror. That seemed to change by the hour. But like Tollett, he spent the morning greeting visitors who came seeking favors.

"Come in, come in," Grady would say, half rising to his feet. The visitor's eyes would take in the rich draperies and stuffed leather chairs, the liquor cart with its crystal decanters of whiskey, the glass jar of complimentary cigarettes. The sweeping vista through the window was like its own trophy on the wall. And perhaps for a second, their minds recalled an image from a secondhand story, told to them by their fathers and uncles, of the old man who once sat at the top of the Petroleum Building and how his office was always open. But a young man sat behind the desk now, one who'd managed to get his hands on the old man's money. They'd ask for a loan anyway, for an investment in the business they dreamed of starting, because they heard the old man had cared about the community and his money still coursed like a river beneath it. Grady listened patiently, then thanked them for coming in, the way he knew the old man had done when he'd occupied that same seat of power. And then he'd open that big checkbook and put his fancy pen to work.

By the afternoon, the Seagram's VO was on the desk, the room was hazy from smoke, and Grady was loose and restless. The boys arrived from their half day at school — the same work-study program that Dad had done. Grady would stick his head out and find one of them loafing with his feet up on the desk, dipping snuff, and send him out to wash Ann's Mercedes or the cream-colored Rolls-Royce he'd recently bought up in Dallas, previously owned. Or he'd ask one of them to gas up the limousine. Grady had granted Ann's wish and fired the chauffeur, mainly because the Houstonian couldn't orient himself on the flatland and kept getting lost. They'd pile into the limo and race down I-20, bound for Midland, like a gang of highwaymen, then blow into the Petroleum Club and order half the menu.

It didn't take long for Dad to acclimate to the new lifestyle. Like most men when they first entered the oil game, he went into Bob Brock Ford, walked past the Mustangs and Thunderbirds of his youth, and picked out a new F-150.

The thousand-dollar dinner bills no longer fazed him, and he got accustomed to a life lived by the minute. Like the afternoon when one of the boys in the office said something like, "You know, I've never been on an airplane," whereupon Grady stuck his head out the door and told his secretary to call Southwest — they were going to Vegas. Soon everyone who'd been in the office was crammed into two limousines, speeding to the airport for the six o'clock flight. Dad barely had time to get Mom and take us kids to Bob and Opal's, much less pack a bag. Grady even said, "Don't worry about clothes, Bobby. We'll buy new ones there." But Mom said that was silly and packed a suitcase anyway.

When they returned three days later, Mom told us about staying at Caesars Palace and playing the slot machines, how one night they'd gone to see David Copperfield's magic show at the Hilton. "He made an entire Cadillac disappear," she said, and I couldn't even imagine.

Dad came in wearing a brand-new leather jacket, a pair of snake-skin boots, and a crisp black Stetson. And with this ensemble, I noticed something else for the first time. Dad was wearing a western belt that his coworkers in Alvin — the Sandblasters — had given him as a farewell gift, their own talisman for the boom. On the back, in bold, black letters, it read: BIG BUCKS BOB.

■ ■ ■ ■

I was still thinking about the disappearing Cadillac when, a few days later, Dad pulled up to the house in a long white Lincoln, a gift from the boss. We all ran outside to behold the thing, then stood in the yard gawking as Dad stepped out to show it off. Aside from my grandmother's de Ville, it was the prettiest car I'd ever seen. Best of all, it was parked in my driveway.

"Well, how do like your new car?" he said to Mom. "It belonged to Ann, but she hardly drives it anymore. Grady said I might as well give it to you."

Mom giggled. "I think it's beautiful. But shouldn't this be your company car?"

"Those lease roads would tear it apart," Dad said. "And I've got my pickup." The wife of the vice president needed something better than an old Torino.

When Dad saw my sister Marci and me peering through the window on our tippy-toes, he said, "Look here," and opened the back door and motioned us in. "Go on, have a look." I climbed into the backseat and marveled at the luxury: burgundy carpets, chrome-plated instruments, wood paneling. The white leather seats were as long and wide as my own bed and carried the familiar traces of liquor and cigarettes. I took a deep breath and ran my hand along the cool surface, not knowing that this beautiful car would be the first casualty of the boom.

It happened at the tail end of that long, wild summer of 1981. In early August, Mom received word that her grandmother had died in Snyder. The funeral for Mrs. Curtis was being held that coming Saturday morning and Dad was scheduled to be a pallbearer. Mom's Lincoln would lead the procession, since everyone agreed it was the nicest car in the family. Mom was distraught by the news, naturally, so the evening before the service Dad hired a babysitter and took Mom on a date to the Branding Iron, hoping to raise her spirits.

For Mom, it was a much-needed night out. Dad had been working late, staying out until 3 a.m. with Grady and the boys and leaving her at home with us kids. She got Marci and me ready for school in the morning and picked us up in the afternoon, then fed and bathed us and got us into bed — all while tending to my baby sister. Many nights we ate dinner at Bob and Opal's house because it was easier. Now that it was summer, Mom kept us busy all day with swim lessons and trips to the supermarket, maybe lunch at Uncle Herman's restaurant to see him and Little Opal. At the same time, Mom was studying to get her real estate license. Each night after we were asleep, she sat up studying at the kitchen table, then crawled out of bed on Saturday and Sunday mornings to attend classes in Midland — forty miles away.

That night at the Branding Iron, Dad promised he'd talk with Grady about getting off earlier and helping more at home. And he was sincere, for he knew how hard Mom worked to take care of us. They ordered steaks and shared a bottle of wine, which went down so easy that Dad ordered a second one, which he drank on his own. By the time they left the restaurant, he was feeling pretty good.

He drove the babysitter home in the Lincoln. On the way back, Mom asked him to pick up some Pampers. But as he left the 7-Eleven, the warm August night beckoned, and he got a notion to keep the party going. Figuring Mom was already asleep, he threw a dime in the pay phone and called his buddy Chuck.

It was good to see his old friend. They hadn't hung out since Dad had eaten Orange Sunshine and thought Chuck was the

Devil. Later, of course, Chuck had said those mean things about Dad being a narc. But all of that was behind them now. Dad told Chuck about driving to the preacher's house that night and knocking on the door, how he was convinced he was dying, and they both had a laugh. Then Chuck rolled up a fat number and put on a Jethro Tull record.

Before long, Dad started feeling sleepy and said he had to leave. He told his friend good-bye and walked out to the car, and that's the last thing he remembered.

It was his own blood that woke him up — blood running down his face and pooling in the cracks of his lips. He opened his eyes and felt a pain in his head. He saw broken glass covering his lap, the windshield spiderwebbed from end to end, and the hole where his head had momentarily left the vehicle. Outside, the hood was crushed like a beer can where it had hit a telephone pole. His first thought was to run.

The DWI from Weatherford still hung on his record. Another one would cost him his license, or worse. Dad's mind went into red alert. He cut the engine and took his keys, then reached over and popped the trunk. As he pushed open the door, he took a quick read of his bearings: he was on Goliad

Street, heading south — or, on the sidewalk, anyway, and surrounded by houses. It was only a matter of minutes before somebody called the police.

He ran around and opened the trunk. Down inside sat the flotsam of his life — a baby stroller, diaper bag, and extra car seat, all of it highly incriminating. He scooped them up one by one, then ran across the street to a small vacant lot. A little mesquite bush was the best cover he could find, so he dumped it all there, then camouflaged the pile with handfuls of dead grass. He was out of his mind. Whenever the police found the car, he figured, it would be registered to Cunningham Oil.

The accident occurred less than a mile from our house, and only a few blocks from Bob and Opal's. But rather than go there and call Mom, Dad turned toward his office downtown. He weaved through empty streets and past darkened windows, his boot heels clicking in the morning silence. At every intersection, he looked for a place to hide — a tree, the shadow of a house — before sprinting across. Whenever a car approached, he dove onto his belly like an infantryman and waited for the headlights to pass.

He reached downtown in minutes and

kept on running, his lungs on fire, his head throbbing with pain. At one point, looking out for cars, he failed to see an open manhole and stepped right into it, badly scraping his leg. He was hobbling by the time he reached Fourth Street and saw the silhouette of First Assembly, standing in high relief against yet another Friday night. For a brief second he considered turning, just as the parapet lights of the Petroleum Building came into view. He quickened his pace.

He used his key to get in the door, then took the elevator up to the Cunningham suite. In the bathroom, he washed the blood and bits of glass from his hair and examined the scrape on his forehead. The wound was still seeping blood, but it wouldn't need stitches. He rinsed it off with some paper towels and cleaned up his leg, which was its own mess. He then thought about calling Mom, but decided it was best to stay there and sober up.

Thinking about Mom, he remembered: *her grandmother's funeral.* The service was eight hours away, and her Lincoln — the one they'd promised to lead the procession — sat totaled on Goliad Street. Dad walked into his office and sat behind his desk. There, alone with his brass pump jack, he felt himself well up with shame.

The old church came to mind again like a haunting. He was taken back to those Sunday mornings, straight-backed in the pews, watching his mother and sisters lead the chorus, listening to how their voices weaved into one another like thread. He saw himself as a young boy during one of the sing-alongs that erupted after supper, all of the family pressed into Grandma and Grandpa's living room to harmonize on "Just a Little Talk with Jesus." How the dry summer breeze came through the open window smelling like cinnamon, and how the neighbors pulled up lawn chairs and listened to the music. And he remembered how one night, he glanced up and saw his uncle Herman not singing at all but just sitting there, smiling. The most amazing expression of peace had settled on his face. Tears rolled down his cheeks, and Dad knew at that moment that his uncle beheld the very Kingdom of God.

How far he'd strayed from that place, and how many times. There had been moments when Jesus had confronted Dad and said, *Give up everything and follow me.* But Dad was like the rich man described in the Gospels, who was almost ready to commit, but could never surrender it all. He wondered now if it would be up there, at the top of the Petroleum Building, where his testimony would begin. He could hear himself telling the congregation, "I was drunk, stoned, and covered in blood, and still I was running." And this time, he was running from the law while his wife and kids slept at home. Finally, he'd reached the end of the line. There was nowhere else to run.

At that moment Dad became very afraid. The police were probably looking for him now, making calls, knocking on his door. Mom would see them, and for an instant she would fear that he was dead. He could see her face as she absorbed that wound, and the image made him physically ill.

He needed to reach her before the police did, tell her what had happened, perhaps even try and find another car for the funeral. *That* he could fix. But even if she did forgive him, he still had the law to contend with.

He picked up the telephone to dial the house, then stopped when he heard a noise in the hallway, near the elevator. He walked out to the reception desk to get a better listen, then heard it again — the sound of men talking. *The cops,* he thought. He ducked behind a file cabinet just as a key slid into the lock and the door swung open. Dad held his breath and waited. But it wasn't the police he saw. Instead, standing in the doorway, silhouetted against the light, was Grady — wrapped in the hard embrace of a young cowboy.

When Dad stepped from behind the file cabinet like a blood-soaked ghoul, the men wheeled around in fright.

"What the hell are y'all doing?" he shouted.

He pointed to the cowboy, whom he did not recognize. *"I said what the hell are you doing?"*

Grady stammered, but Dad cut him off. "Tell your friend he needs to leave. *Now.*"

Grady looked at the man and nodded, and the man walked back into the hallway and disappeared. Embarrassed, Grady tried explaining the situation, but it was too much for Dad to process, especially in his condition. They would revisit it another day, or probably never. Dad stopped Grady short and motioned for the door. "I need you to take me home," he said.

The police had no trouble finding Dad. A lady across the street had heard the crash and called 911. She then gave the dispatcher a play-by-play of a man who looked just like Bobby Mealer dumping a baby stroller into a vacant field, then running off into the night. Big Spring was a small town. When

494

the police called Grady's house and Ann picked up the phone, she confirmed everything they needed to know.

But *because* Big Spring was a small town, there were escape hatches. There were friends. A squad car parked in front of our house the next morning and an officer knocked on the door. I was up watching cartoons when Dad staggered out of his bedroom, his head wrapped in a bandage. The officer was a guy named George whom Dad used to run with in high school. George told Dad how they'd traced his car and how the lady watched him flee the scene. He asked what happened.

"I must've fallen asleep and hit the pole," Dad said. "And I hit my head."

George nodded, scribbled some notes, then looked up at the bandage. "So you're saying you left the scene in a *confused state*?"

Dad smiled. "That's exactly what happened."

The police dismissed the case as an accident. And although Dad was off the hook with the law, Mom was still furious. That night, he'd called her from the office before coming home, and she was waiting for him when he walked through the door. The sight of his bloody head and the threat of the police had pushed her into action. But after George left, she killed Dad with silence. That morning, they borrowed my grandmother's Cadillac and drove to Snyder for the funeral, and Dad carried Mrs. Curtis's casket, looking half dead himself.

But Mom never stayed mad for long. Dad knew this better than anyone. After a few weeks, he was singing "Brown Eyed Girl" to her again and easing back into her graces.

As for Dad and the Lord, well, he figured there was always time for forgiveness. And besides, it was hard to stay focused on Jesus when you were busy drilling for oil.

12

The gusher . . . life at the top . . .

Cunningham Oil drilled its maiden well, the Guinn No. 1, on a warm day in September 1981, on a little ranch of low-slung hills covered with juniper, runty oak, and mesquite. The land, located in southwest Tom Green County, had belonged to a now-dead doctor, John Guinn, for whom the well was named, although a burly rancher named Bowman lived there now, leasing it for cattle. Up in the rock house by the road, Bowman's beautiful wife gave Grady's boys plenty to talk about.

The oil lease was a farm-out from Amoco, which had drilled the acreage for years and resorted to flooding old wells with water to get them producing again. Although it was a heavily trodden field, the geologist, Orville Phelps, saw potential in a shallow formation called the San Angelo Sand.

Next to the drill site, Grady parked his mother's travel trailer to serve as a doghouse and place for Orville and Bob to examine the cuttings. Dad, Grady, and the boys were gathered round, beers in hand, ready to party, when the driller arrived at the gate. Whether Dad realized it or not, this marked a family milestone. After fifty years in the oil fields, we'd finally climbed down off the trucks and derrick floors and into the soft shoes of investment and return. Drinking a can of Coors while other men did the work was something akin to evolution.

But the moment was somewhat spoiled when, to everyone's disappointment, the driller pulled in carrying only two rough-necks. Furthermore, on the back of his International was a puny drilling rig that lay folded like a fireman's ladder. Grady looked crestfallen.

"Isn't that thing supposed to be bigger?" he said, turning to Bob.

"It's plenty big," the engineer replied. "We're only going eleven hundred feet."

The rig was a Failing 1500, used mostly to drill for water. Its derrick stood maybe forty feet when erect and remained mounted to the truck, requiring no elaborate plat-form. It looked nothing like the towering rigs Dad and Grady were used to seeing from the highway, the ones that tunneled deep into the Ordovician and beyond, painted red and white and festooned with pretty lights at night — the kind of big iron my grandfather had worked for half his life.

"Hell," Grady said, "this thing looks like a peashooter."

The driller parked behind a line of trees, then lowered the hydraulic jacks and stabilizer to keep the truck from tipping over. A diesel motor thundered to life and the long pageant of drilling for oil began, a procession that in most places carried on for days and weeks, regardless of weather, the way it had been carried out for a hundred years through sand and deepwater, into mountain and pasture, and now here behind a quiet patch of juniper.

A large drilling bit chewed into the soft ground about three hundred feet, spitting a gray cloud of sand and limestone. This was the starting hole, which the men filled with casing pipe that would guide the drill stem and keep the hole from collapsing as they bored deeper. A hopper truck arrived and poured cement down the hole to set the pipe in place, and once it was firm, they attached a smaller cone bit and prepared to go to depth. The driller revved the engine to three hundred rpms, the rotary table atop the International started to spin, and the process of drilling and casing repeated all the way down the hole — this one in the name of Grady Cunningham.

By that time it was the middle of the night. Dad and Grady had gone back to the hotel in San Angelo while the others had stayed behind to help. Since the well was shallow and mostly through sand, the driller used compressed air to push the debris out from the hole, rather than fluid or "mud." The sediment billowed back to the surface through a pipe called a blooey and into a shallow sump pit, where Buddy and Jacques caught samples in cloth bags and labeled them according to depth. Once the driller reached six hundred feet, he telephoned Bob at home, as instructed, and the engineer climbed out of bed and drove over. Inside the doghouse now, he and Orville examined the rocks for signs of hydrocarbons. Some oily film appeared, but not much. The logging tools they'd sent down the hole — which measured the density and porosity of the formation, among other things — hadn't convinced him, either.

"Probably not," he told Orville, shaking his head.

But the geologist stuck to his hunch. "It's there all right."

So Bob told the driller to go deeper — past a formation called the No. 5 San Angelo Sand and into another one known as the Clear Fork. At some point the next afternoon, Bob walked by the blooey and saw it running dark, then motioned for the machinery to stop. His nose told him everything.

They lined the well to twelve hundred feet, then lowered frac charges to blast holes in the casing. Some acid was poured down to melt the sediment, mostly dolomite and siltstone, which freed the oil to bleed its way in. To get the well flowing, a swabbing unit arrived to suction the hole and pull the liquid to the surface. When that happened, the whole gang was ready.

They were perched over a small storage tank when it first appeared, pulsing through the pipe. It was a gorgeous light crude, a perfect forty gravity, and shone dark green under the sun.

"Look at it!" Grady said. "Will you just look at it!"

He let the oil run over his hand and the others followed, weak in the face of its power, the way Billy Sunday had done over sixty years before in Ranger.

"It's still warm!"

"Yeah! It feels like . . . like milk!"

By now the swabbing crew had a decent flow, and the casing head gas rising off the tank was so strong that it soured the air. About the time Bob's nose registered the smell, he saw something that filled him with terror. It was Grady, one hand under the black fountain while the other waved a lit cigarette. It took him two seconds to reach the tank and pull Grady off, otherwise they would have been blown to bits.

The Guinn No. 1 delivered twenty barrels a day, thanks to a second-hand pump jack, and because the well was shallow and cheap to drill — around eighty thousand dollars — it began paying for itself in no time. Dad and Hugh typed out the numbers and took them to investors, and within weeks, they'd raised enough money to drill two more successful wells. Just like that, Grady Cunningham was a bona fide oilman.

The desired persona was now complete. For the first time, Grady stood in the same arena with Joshua Cosden and Raymond Tollett, along with every historic man in Witch-Elk boots who'd ever harnessed the land to his will. The maiden strike, however average, set things in motion. Twenty barrels a day, it turned out, was plenty of fuel for Grady to punch the business into warp speed.

He had a lot of catching up to do. Over in Midland, men were flying in Bulgari reps from Manhattan to sell jewelry to their wives. In Big Spring, Grady had to contend with a pair of men who'd married into the Dora Roberts family, whose ranch was giving oil again like the days of '27. Their Ferraris whizzed down Gregg Street, and one of them even flew around town in a helicopter.

From his perch in the Petroleum Building, Grady aimed for the stars.

The whole gang flew to Dallas to buy gold. By now they were taking chartered King Airs. Gone were the Piper Navajo and the sweet-faced Norwegian pilot, along with his friendly wife. No one seemed to know where they'd gone, or why, and no one seemed to care. One shiny thing was replaced by another and the ride continued to move.

This time to Bachendorf's. Its owner, Harry Bock, was a Lithuanian immigrant who'd survived Dachau and now operated one of the most exclusive jewelry stores in the country. The private collection was housed at its well-guarded office in downtown Dallas, where select clients were greeted with open bottles of champagne to pique their spirits. Leaving Bachendorf's that night, Ann wore a ten-carat diamond as big as her thumbnail, while Mom wore a gold nugget necklace pendant the size of a thick half-dollar. For Dad, Bock designed a gold nugget ring in the shape of Texas, with a diamond in the place of Big Spring. On his wrist was a gold-plated Rolex.

That Christmas morning, I was lying on the floor watching cartoons when a big moving truck pulled up to the curb. Before I knew it, two men were walking through our front door carrying the largest television I'd ever seen. It was cased in polished wood and its weight caused them to waddle. They set it down in front of me and carted off our old TV, which suddenly looked pathetic.

"Watch your cartoons in style now, kiddo," one of them said, and winked. A minute later, a beige Rolls-Royce appeared in the drive. Grady stepped out wearing his giant wolf coat, its hem nearly dragging the dirt. He didn't even bother to knock and was shouting before he stepped inside.

"Let's see that new Magnavox," he said to Mom, gesturing with his cigarette as he marched toward the living room. "Those sonsabitches wanted an extra hundred dollars to deliver it on Christmas, so I wanna see it."

He stopped at the doorway and his face brightened at the sight of his gift. It was the first time I'd ever seen the coat, and instantly I thought of Luke Skywalker stuffed inside the cavity of the Tauntaun.

"Sugar," he said to me, exhaling a big cloud of smoke, "I want you to always remember something: your uncle Grady takes care of your family. You hear that?"

"Yessir."

"Now," he said, spinning around, "where's that Bobby Gaylon? We got work to do."

I watched Dad follow Grady out to the Rolls-Royce, and Grady opened the trunk. Down inside were dozens of Christmas hams and turkeys he was delivering to old people and the poor — to retirees living in the Canterbury apartments down the street, and to elderly parents of friends and employees (Bob and Opal got a ham, and so did Uncle Herman and Little Opal). Families on the north side opened their doors that morning to find Grady standing there in his wolf coat like Daddy Warbucks, the Rolls billowing steam by the curb, and Dad lugging a frozen bird up the sidewalk. Meanwhile, Buddy was driving around doing the same.

That afternoon, Opal was pulling the ham out of the oven when we arrived for Christmas lunch. Zelda and Charles were visiting from Albuquerque, Norma Lou had driven over from Odessa, and Preston's family had come from Alvin. They were seated around the living room when Mom and Dad walked in wearing their own fur coats, which Grady had given to them that morning (I'd been too preoccupied with the new television to notice). Mom had on a frothy white fox and Dad's was a velvety chocolate rabbit. Standing in Bob and Opal's tiny living room, they looked like a couple of Hollywood stars who'd lost their way.

"My lands!" someone said.

"Will ya look at that!"

"I told ya he's Richie Rich!" Bob shouted, and beamed at the sight.

Preston looked at the coat and the fat Rolex on Dad's wrist, then shook his head. "I guess you've done pretty good for yourself, little brother," he said, and there was a tinge of sadness in his voice. He knew Dad was never coming back.

The high life suited us kids just fine. My sister Marci and I loved riding in Uncle Grady's fancy cars. We especially liked the Rolls-Royce and the long white Excalibur. The company dinners at the Branding Iron and the Brass Nail club and restaurant were huge events. I sat next to Grady while he waved his cigarettes at Dad and the boys and shouted orders at the poor waitress: "Bring this kid whatever he wants!" he demanded. I got a chicken-fried steak with onion rings, ate half, then announced, "I'll have chocolate cake!" making sure to wave my arm with the same bravado before Mom took me down a peg. With Grady, every boy was a king! Every man a millionaire!

For show-and-tell that year, I brought a mason jar full of oil that Dad had skimmed off one of the tanks. Standing in front of the class, I popped open the lid and dipped my finger into the green-black liquid. As the oil streaked down my hand, the room filled with its sulfury vapor.

"Y'all smell that?" I asked, my accent as thick as the crude. "My daddy says that's the smell of money."

Out at the company's oil lease, Bob Wilson and his colleague Orville Phelps were trying to gauge the extent of the field, looking for the oil that had eluded them in the San Angelo Sand. First and foremost, they wanted to explore the property across the fence, which belonged to a rancher named Jones. So one afternoon, Dad drove up the caliche road and knocked on his door, a case of Coors under his arm. By evening, the two men were backslapping on the porch and Jones was wearing a Cunningham Oil cap, a few of which Dad kept out in the truck.

The Jones No. 1 hit the seam they were looking for. Their second attempt missed it completely and came up dry, and so did their third. But the next four produced nicely. With the wells across the fence, the company now had greater cachet when pitching investors. Bowman was so happy about getting his small percentage that he threw a cookout at the rock house with fried catfish and homemade ice cream. He even let the boys dance with his pretty wife. For the first time, Dad and Hugh started receiving royalties. With every check came a tiny stake in the jumbo Texas dream.

Sometimes the boys came to the house for dinner when Dad grilled ribs, and I'd hear things. Like how Grady flew twenty-five people to Aspen in two King Airs for St. Patrick's Day, running up to strangers in his wolf coat trying to kiss them, shouting, "Are you Irish?" Or how Grady was talking about opening offices in Denver and Midland, even buying a big farm.

"What's he doing buying a farm?" one of

them asked.

"Hell if I know," another said.

"Hey, is it true he just ordered twenty-two Cadillacs?"

"Probably, crazy sonofabitch."

"Hell, you should've been with us in Denver," one of them said, then described eating at the Chateau Pyrenees, where down in the basement was a bowling-ball-looking thing with a spout that poured wine down your throat, "some VIP deal." And how on the way home, someone said, *Shit, Grady, we ain't far from Vegas!* So Grady told the pilot to turn the plane around and they spent three days at Caesars Palace.

For me, Grady's boys epitomized cool, with their long feathered hair, tight jeans, and leather jackets. What's more, their language was profane and intoxicating and reminded me of Bud Davis from *Urban Cowboy,* which I watched on our big new television. When Mom and Dad went out, our babysitters were too busy with my sisters to worry about what I was doing. And what I was doing, at seven years old, was killing time in a Pasadena honky-tonk. Never mind that John Travolta was an Irish-Italian from New Jersey, the people in Gilley's were familiar to me — their clothes, the longnecks they drank, the way they talked, and how they stood half-cocked with their thumbs down in their jeans. That was how Dad carried himself, with his weight shifted to one leg, and how I too saw myself standing against the intemperate world.

One night the boys took me on a beer run, but instead of going straight to the store, they peeled off the main road and raced down the side streets along the dark arroyo. Sitting in the backseat, the rush of wind from the open windows roared in my ears, and the speakers — cranked with Van Halen — seemed to pulse in my stomach. I began to panic, and when I opened my eyes the boys were hanging out the window hurling beer bottles into street signs. Each time one exploded into a puff of glass, they whooped and hollered, causing my whole body to flinch.

Once Mom finished with her real estate classes, she started taking us to First Assembly of God, on the corner of Fourth and Lancaster. Our history hung heavy under the roof my great-grandfather Clem had built a half century earlier, mixing with the smell of old hymnals and wood polish. And there was living history, too. Little Opal and Herman occupied their usual places toward the front, along with other aunts and uncles.

And after years of resistance, my grandfather Bob was finally in regular attendance.

It helped that Papaw had taken a liking to the pastor, Brother Rick Jones. Shortly after being hired, the preacher had dropped by the house to introduce himself, and he and Bob became fast friends. Jones hailed from West Texas and his father was a driller who still worked the rigs. "A man I can listen to," Bob said. The preacher started stopping by regularly for coffee and even helped Bob tinker with the trucks. In turn, Papaw put his fly-killing days behind him and donned a three-piece suit. He sat in a pew and listened to Brother Jones, digging the grease from his nails with his Old Timer knife.

With Mom and Opal as my examples, I began to form my own relationship with God, whom I loved without question. It would be another two years before I gathered the courage to kneel at the altar and accept him as my savior — it would happen at a summer church camp — but until then, I became a half-pint soldier in service of the Cross. That year our youth group, the Royal Rangers, asked us to raise money to help missionaries in the godless reaches of the globe. They gave each kid a little Buddy Barrel to fill with coins, but I knew I could

511

do better. The first chance I got, I went with Dad to the Petroleum Building and had Grady's boys stuff mine with cash.

Dad attended church sometimes. But for him, the old tabernacle where he was raised was too redolent with history. It was like every sin he'd ever committed still hung at the foot of the altar in a cloud, always there to remind him, along with church ladies who gave him the same sidelong stare as when he and Ronnie used to come in buzzed. Brother Jones may have been an easy touch, but hell was still hot and sin was the only ticket there. Most Sundays Dad stayed home.

"If those blue-haired women wanna treat me like I'm sixteen and running wild, I won't disappoint 'em," he said.

And how could he? He was having too much fun. Or at least he told himself as much. Never in his life had he been so well dressed or had as much money in the bank. And wasn't the price of oil only going up? It was true the recession was pummeling the market and people were starting to sweat. But when analysts were still forecasting eighty-five-dollar crude ("Eighty-five in '85!" the oilmen chanted), who was Dad to doubt them? The local economy was so strong that the new prison was training inmates how to roughneck so they could join a crew upon release. "There's not enough qualified men," the instructor told the *Herald.*

But for Dad, it wasn't easy keeping up with Grady, especially with three kids and a wife, and with Grady showing no signs of slowing down. The last few trips with the boys, Dad had managed to stay behind — *someone* had to run the business — but Grady still found ways to steal him away.

A morning trip to the Guinn lease turned into three days in San Angelo, spent mostly in the hotel bar with eight of the boys and whoever else happened past. They must have gone through a case of whiskey, and one of those nights Grady kept trying to buy the marble statue in the hotel lobby. Or was that in Midland? Then it was over to Hobbs, New Mexico, to look at Grady's farm, which he was buying off his uncle Alvie. Standing hungover in a cotton field had proved disorienting, to say the least. A weekend ski trip to Ruidoso had turned into . . . how many days? He couldn't say.

He remembered being at the horse track and Grady talking in fake Spanish to people and how everyone thought it was funny, then after dinner he and his cousin Selena tangoed in the street and backed up traffic.

Grady must've lost several thousand dollars on horses that day and there he was, dancing. Then to Dallas and back again, the blur of tarmac lights. Then, of course, there were all the late nights at the Brass Nail, Dad's Rolex pushing one more midnight while a bunch of strangers plundered the company's tab.

By early 1982, Mom was working full-time selling houses and had to wake up early. Whenever Dad left for three days and stuck her with us kids, she grew furious. "Just tell him you have to leave," she implored, but Dad said he couldn't. Grady was like Der Butt, the leader of a gang he just couldn't quit.

Truth was, they were flying too high to even see the ground. To help Grady maintain that kind of velocity, he started buying help. It seemed like everybody was on something, whether it was coke, speed, or Tennessee whiskey. Even the *Herald* ran a big series about the amount of drugs flooding Big Spring because of the boom. "Almost all the workers smoke and eat speed," a roughneck told the paper. Men were doing anything to pack in the long hours, not knowing how long the bonanza was going to last. In fact, the driller on the Jones lease had to send his crew home one night because they were so high. Buddy, who lived with Grady and Ann, wound up running casing for them. He was nearly killed when a rabbit line shot out of the hole and knocked him off the pipe rack, breaking three of his fingers.

Roughnecks could buy speed for a dollar a hit, while "prellies" were ten and guaranteed sixteen hours of wide-eye. But cocaine wasn't for oil hands. At three thousand dollars an ounce, it ran like pipeline across cherrywood desks where the toy pump jacks bobbed up and down to "Deep in the Heart of Texas." It was railed out in restroom stalls at the Petroleum Club, in dark-windowed limousines that snaked between the tanker trucks blasting "Bloody Mary Morning." It was at the Brass Nail, where the waitress left it in cellophane baggies in the change slot of the cigarette machine. And it ate like cancer at the great families that had built Big Spring.

The Dora Roberts fortune melted away not only on Ferraris and helicopters, but famous "Elephant Hunt" parties thrown at one of the biggest ranch houses in town, where bowls heaped with cocaine sat out like bean dip for all to share. As for Raymond Tollett's money, it walked from an idling limousine to a dealer's door on West

Eighteenth Street, with Buddy's one good hand sweating against the wheel.

Buddy's role as Ann's watchman, and his place under their roof, meant that he was the one Grady awoke at 2 a.m. when he needed to score. But that was the extent of Buddy's involvement, as Grady was careful to keep the cocaine from the boys. Like his affairs with men, the habit remained separate and tidy, contained in its own little world, and each little world had its own landscape and cast of players. The world of Cunningham Oil was Grady's most carefully designed, with an all-star ensemble. He'd even tapped the only friend he'd ever had, the only person he ever trusted, to be its true-north leader. And for a while, these worlds remained exclusive to their own orbits. But the more cocaine Grady put up his nose, the more he upset the gravity, until the bodies in this delicate balance spun out of their paths and came barreling toward each other.

The cocaine moved into the Petroleum Building, along the conference room table and in smudged powder traces atop the liquor-tray glass, and it moved into his home. As luck would have it, Ike Robb's widow went up north to be with her kids and put the dream house in Highland South

516

up for sale. Grady bought it for cash, plus the house next door. He put his office in the old upstairs theater where, as an usher, he'd once sat at the foot of the master, screening *Love Story* and *The Hawaiians,* and said to the boys, "Can you believe it? Can you believe I own Ike Robb's house?" He then hired a butler named Riggins, a tall black man he'd met on the north side, to wait on Ann, Buddy, and Kirby while upstairs he got high with town trash who congregated like pilot fish. *Every man a king! Every man a millionaire!* The next week, those same people were filing past the secretary, asking Grady for money.

When mixed with Seagram's, the cocaine turned him into a beast, out to devour all that passed by. That included waitresses, whom he tormented on a nightly basis to try and impress the boys, making them all but bleed for their tips while fetching his VO and water. Some had seen much worse and were game, rolling their eyes at his weird misogyny, while a few — to everyone's relief — even gave it back. He loved them the most. But then one night his hand would find a thigh or an ass. When the women twirled around, eyes sparking with rage, it usually fell on Dad to make the peace and get everyone out. Ann would yell at Grady in the parking lot, and Mom would get quiet and upset, and once home, she'd vow never to go out in public with him again. "But of course," she'd say, "he did leave that girl a three-hundred-dollar tip."

Even the boys learned to watch their backs when Grady was on a binge. They'd surrendered a piece of themselves already to his appetites, back when they first came aboard, gold struck and eager to impress. Grady had taken a few of them down to Melba's on Third Street for prostitutes and let them pick out a girl. And while they were going at it, one of the gals pointed to a peephole in the wall and said, "He likes to watch, you know," and the boys had had to live with that information.

Over time, though, they learned they could still draw honey from the hive without getting stung. In Vegas, when the knock on the door came at three in the morning, they pretended to be asleep. Or in the limo, if his wandering hand brushed their thigh, they could scold him like a naughty child and that would be the end of it. And if necessary, they could go even further. One bleary night in a shared hotel room in Dallas, Grady's desire boiled over on Jacques, who put a finger in his eye.

"Try that again," he warned, "and I'll kill you."

With Grady, each person had to ask himself the same question: How much were the good times really worth?

13

The unraveling . . . Dad finds his wings . . .

In July 1981 Congress approved President Reagan's Economic Recovery Tax Act. The legislation sought to boost the economy by slashing corporate and personal taxes along with government spending. While the wealthy benefited most from an across-the-board tax cut, the poor took a hit on programs and social services. And when the Federal Reserve jacked the interest rates to drive down inflation, the economy plunged into the worst slump since the Great Depression. Over seventeen thousand businesses across the country failed, unemployment spiked to 10 percent, and homelessness surged as families who'd lost jobs in the mills and factories now saw their safety net disappear.

By the summer of 1982, the recession caught the Texas miracle and gave it a hammering. People stopped buying as much gasoline, which resulted in a surplus of oil, and OPEC couldn't keep the prices from tumbling. Before anyone knew it, the rig count had fallen by half and jobs began to follow. All the Yankees from Pennsylvania and Ohio found themselves out of work and with nowhere to live. In Houston, they stood with their wives and children at soup kitchens while police pulled over cars with northern plates and pointed them out of town. Many left on trains, just like in the old days, headed west, where they heard there were jobs and the dream was still alive. Fort Worth alone had fifty hobo jungles full of hundreds of men.

In Big Spring, unemployment that summer reached its highest level since the air base left town. Meanwhile, the banks were reporting a sharp increase in loans from the previous year, while deposits were spiraling down.

For Dad and the boys, it was difficult to determine their true coordinates. They still had their jobs, which could only mean they were safe in that privileged patch of airspace. The company still had plans to drill seven more wells in the coming months, plus they'd partnered again with the wildcatter Adkins for a deep well out in Haskell

County — a potential gusher if the geology proved legitimate. And the sagging market hadn't spooked investors, not yet, anyway. Dad and Hugh had little trouble raising the money for the drilling expense.

But there was indeed trouble. One morning, a service company rep called the office asking to be paid for a well they'd drilled months ago. Someone alerted Grady, who pulled out his big checkbook and settled up, then apologized for the oversight. But a few weeks later, the same thing happened again.

For every dollar Cunningham Oil earned, Grady turned around and spent fifty. The burn rate was alarming, yet Grady always assured Dad, "The trust is covering it." The Tollett trust paid the seven-thousand-dollar bills for charter planes and covered the payments on the bookstore and cotton farm. It paid cash for Ike Robb's house and the house next door ("I liked the swimming pool," Grady said), and the twenty-seven thousand dollars' worth of carpeting, not to mention all the cars, jewelry, and whatever hedonistic whimsy crossed his mind. And so Dad suspended all logic and took Grady at his word, assuming that it was Ann's money also buying his cocaine.

But after examining the books, Dad discovered that Grady had nearly cleaned out the company. The investment capital, which was supposed to sit in escrow until it came time to drill, was all gone — used to cover payroll and past debts, as well as expenses. As it stood, they were behind in payments on three wells.

Dad found Grady in his office. "The heck are you doing?" he said. "You've nearly bled us dry."

"I got overextended, Bobby. It's normal in this business."

"We've got seven wells to drill and no way to pay for them. When these companies find out we're broke, nobody's gonna work with us."

"Don't worry," Grady said. "Bill Read will take care of it."

Bill Read was president of Coahoma State Bank, ten miles east of Big Spring. Coahoma was a tiny oil-patch town, but Read and his senior loan officer gambled like the big boys in Midland and Dallas when it came to chasing deals.

Read took care of the problem, and then some. The huge loan he gave Grady not only got the company out of trouble, but kept the circus on the move. The next thing Dad and the boys knew, Grady was donating twenty thousand dollars to the Catholic

Church, and he wasn't even Catholic. "It's what Mister Tollett would have done," he explained.

At the same time, Grady started a scholarship fund at Howard College for local high school students. This was in addition to already covering the tuition for dozens of other kids, including a niece he was putting through medical school. As Dad and Hugh tried to make sense of it all, Grady announced he was buying the Brass Nail restaurant.

Furthermore, he'd also bought a football team which he named the Cunningham Oilers. It was part of the semi-professional Dixie League, which included teams from Houston, Oklahoma City, and as far north as Wichita, Kansas. Most of the players had high school or college experience and now worked in the oil fields. For their road games, Grady gave them first-class treatment, chartered fancy buses so they could ride like Nashville stars and fed them prime rib. He even hired local girls to be cheerleaders. At their first home game, in mid-November 1982, held at Memorial Stadium, we all watched as Grady descended onto the fifty-yard line in a helicopter. The door flung open and out he jumped in his wolf coat, arms raised in victory, a cigarette in one hand and a tumbler of VO in the other.

That fall, oil prices continued to drop. Investors started getting spooked and stopped returning calls. The phone rang instead with contractors asking about their money, which Grady moved around with such mystery it merited its own Vegas act. Worse, the company had just drilled three consecutive dry wells on the Jones and Guinn leases. Sixty grand down the tube. And there was Grady leaping out of a helicopter. The check he gave the referee after the game — for $150 — bounced at the bank on Monday.

"Grady," Dad told him, "you got no business being in business."

But somehow Grady found the cash to stay airborne, the whole operation hovering just above the rocks. A month later, he threw a Christmas party at the Brass Nail club that people still talk about, with music and dancing and thousands of dollars' worth of food and booze. Inside, a parade of oil chic bounced along as Hoyle Nix and His West Texas Cowboys fiddled down on "Big Ball's in Cowtown" and "Comin' Down the Pecos." Snapshots of that night show Opal wrapped in a fabulous mink shawl and scanning the crowd through a pair of huge tinted glasses. Dad posed with Ann, who wore a big smile and a wine-colored muumuu, the kind of dress Grady insisted she wear lest her figure attract a man who'd treat her better. Dad himself wore a mustache and a sharp black suit. And like most local men, his expensive leather shoes were covered in dust.

On Christmas Day, over at Bob and Opal's, our family had just finished lunch when Grady and Ann stopped by for dessert. Halfway through his pecan pie, Grady declared a fishing trip to Florida to celebrate New Year's.

"Just the men," he insisted. "I'll call a plane. We'll leave tomorrow."

Ann said, "Well, if the men are going fishing, the women are going to the Bahamas."

By that evening, nearly twenty people were booked on two chartered jets, including Mom, most of the boys, and our local state representative. Even Zelda and Charles, who were visiting from Albuquerque, got invited. The group stopped first in New Orleans, where Grady tipped two hundred dollars to a carriage driver, who danced a jig in the street, before splitting off to Fort Lauderdale and Nassau. While Dad, Grady, and the boys went fishing, the others enjoyed scuba diving, shopping, caviar and champagne. The planes alone cost forty thousand dollars.

As 1983 got under way and the sky continued to fall, Grady doubled down and aimed even higher. In January, he reopened the Brass Nail restaurant, which he'd remodeled into a five-star establishment, hiring the maître d' from Il Sorrento in Dallas, along with a top-flight chef and manager.

"It Began with One Man and a Dream," the announcement in the *Herald* read. "Grady Cunningham Founded a Restaurant Where Dining Equals an Art Form."

A selection of menu items followed: oysters Rockefeller, escargot in mushroom caps, consommé Celestine, prime rib, milk-fed baby veal, Dover sole, and a dish called "Poulet Ann."

A few weeks later, Dad's coworker Hugh got married. He told his bride to expect something extravagant from Grady — a trip to Hawaii, a new car. But after the wedding Grady summoned Hugh to the house, ripped a brass lamp from the wall, handed it to him, and told him congratulations. When Hugh returned from his honeymoon, his paycheck bounced, along with everyone else's.

Meanwhile, we'd moved into a new house.

It was not in Highland South, as Grady had promised; that commitment, like many others, had long been forgotten. Mom and Dad bought this house on their own, and it was special: the very three-bedroom ranch that Zelda and Charles had built back in 1966. The red bricks had come straight from the old T&P station after they'd torn it down. There were towering pines in the front and back yards and a long covered porch painted with a shuffleboard court. The day we moved in, I stood on the flagstone steps of the big sunken living room and knew it meant one thing: we were finally rich like Uncle Grady.

But the very sight of the house filled my father with terror. He was now saddled with a fat mortgage, and his paychecks were worthless. He began considering other careers, escape routes. He enrolled in a fourteen-week Dale Carnegie course at the junior college, which he attended at night, convinced his speaking skills were lacking. But really, it was just his old reliable confidence that was battered.

Mornings, I'd go into the kitchen for a bowl of cereal and find him at the table, hunched over a three-ring binder of Carnegie's boiled-down principles for self-improvement. How to become a leader.

How to gain influence and overcome worry. The last part seemed particularly poignant. "Ask yourself, 'What is the worst that could possibly happen?' " Carnegie advised. "Prepare to accept the worst . . . Remind yourself of the exorbitant price you can pay for worry in terms of your health."

So Dad tried doing as instructed. But each morning he pulled out of the driveway, he had to wonder which method of slow suicide Grady would attempt next.

It was around that time that Grady walked into his office.

"We have a situation," he said.

He explained they still owed money for the casing they'd used in the deep well out in Haskell County, back in November. This was the partnership with Adkins, where Grady had agreed to cover half the operating costs. He'd held off the pipe company as long as he could, but their lawyer had telephoned that morning, threatening legal action if they didn't get paid. The bust was squeezing everybody.

"They need thirty thousand dollars," Grady said.

Dad was indignant. "We already raised that money, Grady. Where'd it go?"

"You know where it went, Bobby!" he shouted. "The fact is, it's gone now and if we don't pay these people, they're gonna file a lien on that well or else take us to court — you, me, Adkins, and anyone else who invested with us. Then we'll really be in trouble."

Grady had temporarily exhausted the trust. Most of the bankers in Big Spring and Midland no longer took his calls. Even Bill Read in Coahoma had turned him down, Grady said.

"But State National says they can arrange something with *you.*"

He was looking straight at Dad. Out of options, Grady had come to the only person who could save him.

"I can pay you back in a month."

So Dad climbed into the Rolls-Royce and they drove up to Ninth and Main, where inside he scribbled his signature that gave Grady the money, an amount that nearly equaled his own salary. As he put down the pen, the awareness of everything he'd just signed away, plus all he'd given up, bored in deep like steel against stone. They were down in that hole together.

Grady missed the first payment on the loan. When the ninety-day period was up, he kicked State National some interest money, persuaded them to give Dad another extension, then missed that payment, too.

Finally, in August 1983, so stressed he could hardly sleep, Dad walked into Grady's office and quit.

If the catalyst to change his life hadn't come the previous year, on the top floor of the Petroleum Building, filthy with his own blood and covered in broken glass, it came now at the dawn of the great Texas oil bust, when Dad found himself suddenly jobless and thirty grand in debt. The loan, not the car crash, finally forced Dad to find his way back. But this time, the road was too narrow for Mom or Zelda to guide him home as they'd done so many times before. There was only room for him and God, and so Dad accepted who he was as a man, acknowledged the choices he'd made, then followed that road.

He went deep into the Word. Mornings, I'd find him at the kitchen table reading his Bible, still in his pajamas, and know that he'd been sitting there for hours. If I was quiet, I could listen from the hallway and hear him praying. It was during these moments that I understood that something had changed. No longer was my father the honky-tonk hero with the Rolex watch and alligator boots, but a man sitting in his underwear in the dark, humbling himself and admitting weakness. Big Bucks Bob was nowhere to be found. In fact, I never saw that belt again.

We started attending church regularly — not First Assembly, but First Baptist, where Dad's cousin Homer and his family went. It carried none of the baggage of Dad's youth, and unlike the Assemblies of God, the Baptists believed that once saved, always saved. "This way I can still sin and get into heaven," Dad said with a wink.

The church had a much wealthier congregation than First Assembly, comprised of retired airmen and Cosden executives, doctors, lawyers, and ranchers who'd invested generously in the church building. First Baptist had a real bowling alley, plus a skating rink in the gymnasium. But best of all, it had a small library where volumes of classic literature shared shelves with the Christian books and study Bibles. While the rest of the family went bowling, Mom let me roam the stacks under the supervision of the librarian, Ms. Edna Ames, who I'm sure regarded me as a curiosity, since what eight-year-old kid would rather read books than bowl or roller-skate?

"What kind of books are you looking for?" Ms. Ames asked.

"Adventures," I told her, and she went to

work, introducing me to the Hardy Boys and Encyclopedia Brown series, *Treasure Island, Swiss Family Robinson,* and the novels of C. S. Lewis. There in the confines of a Southern Baptist church, my love of reading began.

At First Baptist, Dad also reunited with an old childhood friend, Rex Rainey, who was emerging from his own rowdy years in the boom and seeking a change. Mom became close with Rex's wife, Wanda, and their boys were the same ages as my sisters, so our two families began spending a lot of Friday and Saturday nights together.

After supper the kids played while the four of them sat around the kitchen table and talked. Almost always the conversation turned toward church and faith and how they wanted to do things differently than their parents, to forge their own relationships with God. For Mom and Dad, it was the first time they'd ever discussed such things with friends, as adults anyway, and it felt as if they were inventing a new kind of language. Oftentimes the discussions went so late into the night that Rex and Wanda put the boys to bed with us, then racked out on the sofa.

Back when we lived in Midland, Dad had bought an acoustic Alvarez guitar that he pulled out occasionally, but he'd never been serious about learning more than a few chords. But Rex played, and some nights over beers, the two of them went back and forth on a melody and jotted down lyrics.

Before long, Dad had a notebook full of his own songs. The first one he wrote was called "Rolex," about the life he was leaving behind and where it was likely to lead him if he kept on running. *"Rolex on my arm, Cadillac in the drive, a brand-new home, I'm all alone, isn't this the life?"* He and Rex played that one a lot, along with a heart-break number they penned together called "Adios, My Love." But Dad's second song was the one he was most proud of, about the change he was undergoing and how he strove to live. It was called "Eagle Saints," inspired by a verse in Isaiah that says, "But they that wait upon the Lord shall renew their strength; they shall mount up with wings as eagles . . ."

As Dad understood it, the eagle could fly above a storm. With that in mind, he embraced the eagle as a kind of spirit animal of his new self. Soon I started noticing eagle things around our house, including a brass eagle statue on our coffee table next to the pump jack. Whereas Grady and the boys occupied the altitude of man, Dad could

mount up with mighty wings and rise above
the noise, into heavenly space, into glory.

Here I am, Lord.
Mold me and mend me.
I'm all yours, Lord, wash me and cleanse
    me.
As your spirit changes me and the new
    man comes to life,
I can soar the highest heights,
Eagle Saints arise!

14

The Texas Miracle goes bust . . . Bob's last load . . . Dad finds new life . . .

Outside the Permian Basin, the fallout from the oil bust actually started the previous year, in July 1982. As oil prices dropped, banks caught up in the go-go lending craze found themselves in deep trouble. The royalty checks customers used as primary means of repaying their loans suddenly shrank or disappeared. At the same time, the value of proven oil reserves that customers used as collateral — the secondary source of repayment — was also heavily diminished. When cash flow stopped, the regulators swooped down on the banks and started charging off delinquent loans, which sparked a run on deposits as people feared the banks would close. When that happened, banks did begin to fail, starting spectacularly with Penn Square Bank in

Oklahoma City, whose collapse sent tremors through the financial sector.

Penn Square was a tiny commercial bank headquartered in a shopping mall, but its executives never let size deter them — not during an oil boom. When oil prices were at their peak, its energy-loan officers — led by the eccentric Bill Patterson, who famously sipped amaretto from his Gucci loafer — fluttered away over $2.5 billion in handshake oil loans.

Penn Square sold many of those dubious loans to major banks, such as Continental Illinois of Chicago, Chase Manhattan, and Seafirst Seattle (now part of Bank of America), which were eager for their own stake in the bonanza. Continental was the largest commercial bank in the Midwest and itself an aggressive energy lender. When prices finally dropped, the loans from Penn Square were enough to push Continental off the cliff, making it the largest American bank to go under (until Washington Mutual in 2008). Its $4.5 billion bailout from the Federal Deposit Insurance Corporation became, at the time, the biggest in American history, and introduced the term "too big to fail" into the lexicon.

The collapse reached the Permian Basin in July 1983, when Metro Bank in Midland went under, followed by National Bank of Odessa. Each reopened the next day under different names. Then, on October 14, 1983, Texans turned on the evening news and discovered that the unthinkable had taken place. First National Bank of Midland, the true king of the oil patch, had also gone down. The near century-old lifeline to cattlemen and dry landers, benefactor to civic groups and Little Leaguers alike, was gone.

First National's president, Charles Fraser, hadn't squandered a single moment in the boom. He'd brought in every wildcatter, oil company honcho, and man off the street who thought they could make a buck. At First National, people said, any one banker could approve an oil loan, but it took two to turn one down. And with pump jacks dancing, it seemed like a sound policy. As a result, First National swelled to become not only the largest independent bank in Texas, but in all the United States, with assets totaling $1.3 billion, and its collapse would stand as the second-biggest in American history.

Fraser wasn't a showboat like Patterson, and many argued that his freewheeling handouts were justified in order to sustain the country's fastest-growing city. But like so many others, Fraser got greedy, and by the time it was too late, hundreds of millions in loans were in jeopardy and the feds were in the books, carving up his portfolio. The bank sold its own glass tower to stay above water, but over $600 million in lost deposits finally put it under. The Comptroller of the Currency finally declared the bank insolvent on October 14, a day people still refer to as Black Friday. At the same time, the FDIC arranged a humiliating takeover by RepublicBank of Dallas.

Republic assumed only $250 million of First National's $1.3 billion debt portfolio, leaving the FDIC to try and collect on the rest. A slick New Yorker in lizard boots stepped in — Thomas Procopio, an FDIC agent who became known around the region as the "Liquidator-in-Charge." Almost overnight, his agency became one of the biggest creditors in the Permian Basin; within two years, it would be one of its largest employers. Among the holdings it secured from First National was the huge tract of land where the bank had planned to build two more skyscrapers. It also held condominiums, office buildings, scores of homes, oil wells, luxury cars, even rights to books and movies.

Procopio's office would come to staff over

three hundred people charged with recovering the bank's debt. The yellow FOR SALE — FDIC signs became a common sight across the oil patch, but in order to sell the properties, the government needed to know what each one was worth.

To help with this enormous task, they hired a man in Big Spring named Jeff Brown — and with him, my father.

Jeff and his wife, Sue, owned Home Real Estate, the firm where Mom worked. During World War II Jeff had flown B-24 bombers and still wore his hair in an airman's buzz cut. A fat cigar usually sat clenched between his teeth. Jeff ran his firm's appraisal division, and lately Mom had noticed that he could use an extra hand. The week after Dad quit Grady he called Jeff, who hired him after a brief interview.

Originally, Dad's job was to conduct research on sales comparisons in and around Big Spring. This entailed photographing properties, pulling deeds and tax cards at the courthouse, then using that information to determine the value. Jeff paid $125 a day, plus Dad could sell his own real estate for a straight commission. But two months into the job, the FDIC called Jeff, asking his help. Within days he and Dad had more work than they could manage.

Each morning, Dad drove the forty miles to Midland and appraised the fallout of the bust: half-built condominiums frozen in midair; acres of vacant office space and miles of metal buildings, once busy with tomorrow men, now padlocked and quiet, their white caliche yards full of weeds; tracts of raw land, now relinquished back to horny toads and tumbleweed; warehouses and apartment buildings; a convenience store.

With each property, Dad snapped photographs and toured the surrounding area for comps. He pulled deeds and copied tax cards, then drove back to Big Spring. With every trip, he passed the stacks of discarded drill pipe along the highway growing bigger and slowly turning to rust.

Dad was only an apprentice, but the amount of work became so great that Jeff taught him how to do more, then let him keep the commissions. Soon Dad was able to earn the same money as he had with Grady, and this irony did not escape him. Much like his father selling blowdirt in the drought, or his uncle and cousin hauling starved cattle off the pastures, he too had found his niche on the bottom, like a catfish in a muddy river. But there was still that $30,000 loan.

■ ■ ■ ■

No one seemed more worried about Dad's debts than Bob. Each time Dad stopped by the house, Bob tried giving him money. "Just take this here," he'd say, pushing a wad of bills into his hand. But Dad always refused.

"The Lord'll take care of it," he'd say, and Opal would nod her head.

"That's right, baby. You give it to Jesus."

"The Lord will provide," he repeated, trying to ignore the clawing in his gut.

Ever since quitting Grady, Dad had started having breakfast with his father on his way to work. By the time he arrived at seven thirty, Bob was standing over the stove finishing the bacon and eggs. They sat at the bar with cups of black coffee and drizzled honey into hot buttered biscuits, and Bob would say, "If there were canned biscuits when I was young, I never would have married," causing Opal to roll her eyes. After she left for work, in a streak of green silk and perfume, Dad and his father stayed and talked.

As a grown man with children of his own and some recent disappointments, Dad had come to see his father differently. No longer did he feel ashamed of the dump trucks and piles of dirt, or the family backstory of poverty and sour luck. The past few years had aged him and bestowed hard lessons: like how quickly the world will reduce a man who turns against his raising, or how much a fool was willing to risk for a little glitter and influence. To remain steady was the greatest virtue, and his father, through boom and bust, had stayed true to himself and the people who loved him, even if he was temperamental. But even that side of Bob had mellowed, thanks to Brother Rick Jones.

Bob's mellowing had also come by way of personal loss. In March 1982, his sister Fannie passed away unexpectedly, delivering him a terrible blow. She'd left a doctor's appointment one afternoon and driven toward the cemetery to visit her son Bobby, who'd died in a car wreck in 1963. Although she tended Bobby's grave each week, her normal route seemed different all of a sudden. In fact, the city had erected detours because her sons' company was doing paving work on one of the main thoroughfares. Fannie wound up on a desolate back road, where she tried to turn the car around. But heavy rains had left the shoulders muddy, and her wheels got stuck. It wasn't until the next morning that a sheriff's deputy spotted her little Chevrolet. There she was inside, slumped against the wheel and dead from a heart attack. She'd run her car out of gas and spun the tires bald trying to get free. "Scared herself to death," Bob said. Fannie was eighty-three.

In 1979, his beloved sister Allie had died from heart disease after a long stay in the hospital. Her headaches, which had come the instant she learned her son Orville had been killed in the war, persisted until her final hour. Her husband, Tom, had maintained Orville's shrine in their home until he passed away that spring. For Bob, losing Allie had been like losing his mother for the second time, except now he was grown enough to bear the pain in full. Allie had been the one to raise him after Fannie ran off with Abe Jones to the North Texas fields. She cooked his meals, scrubbed their clothes against a washboard until her hands were raw. And it was Allie who'd guided him into the deep red water of the cattle tank and taught him how to swim.

Rheumatoid arthritis had left her bent and dependent on a cane, then a walker, and finally a wheelchair, though she still maneuvered around her tiny house trying to cook and clean for Tom. At the funeral home viewing, Bob saw not a broken old woman but the girl they once called Skinny Legs, who at one time had been so beautiful.

The year before losing Allie, his friend Davey Jones had suffered a stroke and died. His kids had buried him with a pair of dice in his hand and a five-dollar bill tucked into his pocket, hoping he could roll Saint Peter in a game and make it through the gates. Now that Davey was gone, Bob couldn't get behind the wheel of a truck and not think of him.

Around the time Bob lost Allie, Dad's sister Norma Lou suffered a series of health problems that caused Bob even greater stress. She nearly died undergoing hip replacement surgery when the doctors couldn't get her to stop bleeding. Each day Bob and Opal drove sixty miles to Odessa to see their daughter in the hospital, forgoing work and everything else.

Calamity had a way of stalking Norma. After losing her first husband, Bill Glaesman, in a truck accident, she'd married another man, James Saunders, and had a second boy, named Randy. They moved to Duncan, Oklahoma, where James worked as a salesman. One night he was moonlighting at his brother's nightclub when he tried to break up a bar fight. In the melee someone stabbed James in the chest and he died three weeks later. Norma's third marriage ended in divorce.

On top of it all, Bob's own health was failing. He'd given everyone a scare back in August when a big group of them flew to Anaheim for the General Council meeting of the Assemblies of God. The first day there, Bob complained that his shoulder hurt. His feet and legs were also starting to swell. While the others left to hear Pat Boone and Jim Bakker, Bob stayed behind at the hotel, venturing out only once — to his balcony, where he shouted to the prostitutes working the convention traffic to inquire if they offered senior discounts. "They don't," he assured Opal.

Back in Big Spring, the pain and swelling grew worse, so Bob went to see Dr. Clyde Thomas, Raymond Tollett's doctor, who diagnosed him with congestive heart failure. After a week in the hospital Bob finally went home, but with instructions not to eat salt or drink coffee. "And I don't want you working on those trucks, either," Dr. Thomas said.

Bob had no intention of quitting coffee. He drank it morning until night. But Opal really cracked down on the salt, and the result was misery. Worse, the dirt business was slow. Days passed when he didn't get a call, only adding to his depression. The two Fords sat idle in back like a pair of stabled thoroughbreds — tires checked, joints greased — just waiting to run. Whenever Bob got the itch to take them out, to feel the rumble beneath him, he took us kids for a ride.

Bob's favorite place to go was Gibson's Discount Store. The one in Big Spring was located on Scurry Street, on the site of the old tourist court where his father, John Lewis, had died. He went there nearly every day to look at tools and fishing gear, and to sniff out deals on coffee and toilet paper, which he hoarded out in the shed.

But he liked the Gibson's in Odessa much better. It was bigger, and whenever Bob visited Norma Lou, he insisted they go shopping. One Sunday in early October 1983, Bob and Opal drove to Norma's after church to have lunch. She was back on her feet from hip surgery, enough to fix her father his favorite pot roast. After eating, the three of them drove to the store. But once there, Bob took ten steps inside and said, "You know, I don't need anything," then turned and walked back to the car.

Norma called Dad that night and said, "Something's not right with Daddy. For some reason I'm really worried."

The next morning, Bob discovered a flat on one of his trucks. Defying both Opal and Dr. Thomas, he rolled out his jack, hoisted off the giant radial, and patched the tube. When he finished, he drove to the caliche pit for a load.

There was a confrontation with one of his helpers. The man had trouble maneuvering the tractor because the power steering was out. So Bob, frustrated, ran over in a huff and motioned the guy off the wheel, then wrenched the machine into submission. After loading his own truck, he pulled out of the yard in a cloud of dust.

Back at the house, Opal was waiting with his lunch, but Bob said he didn't feel well.

"If you don't mind, I'm gonna take a nap," he said, then walked back to the bedroom. Opal wrapped his food and left it on the stove. When she returned that evening, it was still there. She found Bob in bed. His body was cold.

The first thing she did was call Herman at the restaurant. "I think Bob's dead!" she shouted. It was Homer who arrived minutes later and attempted CPR, but there was no point. Papaw was gone.

The next call she made was to our house.

Mom's mother was visiting from Snyder and all of us had just sat down for dinner. I watched Mom get up and answer the telephone, then drop the receiver and run out the front door. Somebody then managed to get hold of Dad at work. They all gathered in Bob and Opal's bedroom and waited for the coroner. Bob was sixty-nine years old.

His funeral was held later that week at the funeral-home chapel, where Brother Jones led the service and wept. It was the first time I'd seen so many grown-ups cry, especially my own father, and the first I'd experienced death. (Three months later, when another heart attack finally killed Uncle Herman, I asked Mom to let me stay home from the funeral.)

That day, Brother Jones spoke about how Bob had embraced the Lord in recent years, how he'd been proud to see him living a Christian life. Opal then read out a letter the pastor had written to Bob months earlier that reflected the same sentiments. If anyone carried doubts about the fate of Bob's soul, these words put them to rest.

As for Dad, he found his own kind of solace in his father's sudden passing. On the day of Bob's death, he'd stopped by for his usual breakfast, and on the way out the door, he made sure to tell his dad that he loved him. He held tightly to that now. And he also understood that whatever he'd lost financially by coming to Big Spring and joining Grady, he'd gained in a relationship with his father. In fact, he now believed that that was the sole reason we'd come to Big Spring in the first place. That *was* the journey. The whole ordeal, like every other day on earth, was a puff of smoke, a chasing after the wind. Life was fleeting; you had to hold fast to what was true.

As for Grady and the boys, the bottom appeared closer each day. The Brass Nail, with its inflated payroll, turned out to be the extra weight that finally sank the ship. Not only was Grady's timing spectacularly wrong, but so was his audience. Now that life was returning to normal, Big Spring was again the dominion of tight-fisted cotton farmers and cattlemen who preferred chicken-fried steaks at Herman's or enchiladas at the Spanish Inn. Out-of-work roughnecks scraped by on cheap groceries and Taco Villa, and all the Yankees had driven home. Each night, Jacques and the boys watched the kitchen staff fill the dumpster behind the Brass Nail with crates of red snapper and Maine lobsters that arrived by air and went unsold. In February 1984,

Grady placed an ad in the *Herald* that read like an obituary: "Cunningham Oil Company and Grady L. Cunningham has [sic] sold the Brass Nail Restaurant. We would like to thank the public for their past Patronage and Support."

The boys hovered a bit longer, thanks to Ann's trust and shuffled oil investments. While Grady was trying to sell plates of veal Oscar to an empty dining room, the company actually drilled one of its finest producers to date, the O. O. Baker out in Haskell County, which would have paid for itself in a year.

But by then, the district attorney was calling about a list of bounced checks. One of the investors had filed a lawsuit, and a costly exploration out in Yokum County resulted in a dry hole. Through it all, Grady numbed himself with a diet of VO, diet pills, and more cocaine. One night, in a paranoid rage, he chased Hugh's car down Gregg Street, waving a .38 revolver out the window of his Lincoln.

The crash finally occurred on August 31, 1984, when Grady and Ann petitioned for Chapter 11 bankruptcy in the U.S. Federal Court in Fort Worth. According to the *Herald,* Grady's debts totaled $3,790,000, while assets were listed at just over a million. He owed money to more than twenty different creditors, including State National Bank in Big Spring, Coahoma State Bank, and First National in Midland, now under control of the federal government. The FDIC was also listed as a major creditor.

To begin satisfying Grady's debts, the banks and FDIC seized thirty-one cars, including the Excalibur, several Rolls-Royces, the Cadillacs, and a hangarful of vintage Mustangs. They also took Grady's floor-length wolf coats and the gold Rolexes. Grady was forced to sell the cotton farm back to his uncle Alvie and disband the football team. After one season of existence, the Cunningham Oilers actually finished with a record of 4–2 and went down in history as one of Grady's most successful ventures. In the end, all that Grady and Ann had left was Ike Robb's house in Highland South and one vehicle — a decidedly modest two-door Lincoln Continental Mark VII.

As for Grady's entourage, they vanished as quickly as they'd arrived, happy to walk away with a clean record and a decent story. Buddy moved to Lubbock to attend Texas Tech, but remained loyal to Grady and Ann until the end. Jacques wound up in Midland working at a hospital. For a while, Hugh joined another local firm to sell oil deals until it too went bankrupt. He later moved to Dallas and enrolled in grad school.

Somehow, amid all of this, Grady managed to pay off Dad's bank loan with State National. What he couldn't repay in interest he made up for in furniture. One afternoon, a truck pulled up the drive carrying a plush Henredon living room set that he'd salvaged from the creditors. A few of those pieces still sit in my parents' home today.

As 1984 came to a close, "Stay Alive till '85" became the new rallying cry throughout the oil patch. But as the new year dawned, prices continued to drop, and few people were hiring. As work with the FDIC began to taper off, and as retirement beckoned, Dad's boss Jeff Brown and his wife sold the real estate firm in Big Spring and moved to Central Texas. Rather than leave Dad in the lurch, Jeff arranged an interview for him with a large appraisal firm in San Antonio, where, coincidentally, Zelda and Charles had recently moved. After driving down and speaking with the firm, Dad landed the job. In April, he moved in with his sister while we finished school and Mom sold the house. The first week of June, I climbed into the cab of another U-Haul with Dad, and with Mom and the girls following behind in the station wagon, we struck out for another beginning.

The way I like to remember it, we pulled out of town just as a red sandstorm rolled off the Caprock and strangled the sun, and we did not look back. But when I checked the weather logs from that month, I wasn't surprised to find that my memory had deceived me. The day we left, the sky was clear blue with hardly any wind. We'd even gotten some rain, meaning the land was green and fragrant, the very best version of itself.

It was on this kind of day when the cattlemen used to ride out to a high place and forget all of yesterday's trials, when tomorrow men heading west would halt their fevered pace, look around, and put down roots. It's where, from the top of South Mountain, or a window in the Settles, you could look out and see the refinery sparkle like a jewel, see the land rise from an old prehistoric sea and bend north into the blue yonder. All of nature under heaven was in its proper order, and if you squinted just right, you might be able to see what it had in store for you next.

EPILOGUE

San Antonio was huge and terrifying, a separate planet altogether made of concrete and traffic. Big Spring held our family's history, but our new home was in a treeless, newly built subdivision where the houses all looked the same. No familiar landmarks or funny stories to accompany them, no hills or arroyos, and no big sky. When the sun went down in the suburbs, it just got dark.

Dad's job at the appraisal firm only paid ten dollars an hour, meaning we were broke a lot of the time, so much that Mom made my sisters' clothes until she went back to work. The good thing about our neighborhood was there were a lot of kids our age. Many of them were Hispanic, and loved to tease me about my thick West Texas accent.

"Hey, man, what's your name again?" they asked.

"Br*i*ne," I answered.

But my big-city friends introduced me to

wonderful things I'd never experienced, such as Run-D.M.C. and Metallica. That first summer, our neighborhood threw a huge block party, *puro* San Antonio, with loud *conjunto* music and lots of keg beer. My buddies blasted Grandmaster Flash from a boom box, spread out a piece of cardboard, and did backspins in the street.

In San Antonio, Dad sought to distance himself even further from his past and push closer to God. For a while we attended Aunt Zelda's church until Mom and Dad found something they liked better. The new church was located on the west side, in a building that had once been a grocery store, next door to a giant Mexican flea market.

The church was spawned from a neo-charismatic movement called the Third Wave that started in California. It wasn't rooted in doctrine, really, but in "signs and wonders" that the Holy Spirit performed through modern-day prophets and apostles. These included the head pastor, who had fronted a sixties rock band and fancied Armani suits, and his assistant, a former street fighter from England whose arms were covered in ink. It attracted people from all walks: black, white, and Hispanic, reformed Catholics and disillusioned Baptists, burned-out hippies and seekers such as

Dad. One of the weekly Bible groups was led by a former coke dealer who'd served a stint in prison.

When the preachers channeled signs and wonders, it was like the Holy Spirit on steroids. I'd grown accustomed to hearing a few people speak in tongues back in Big Spring. But here half the congregation was doing it. Once a visiting evangelist called down a "spirit of laughter" and before I could blink, people sitting around me were rolling on the floor. The preachers foretold prophecies, prayed over the sick, and even cast out demons. One evening a "spirit of benevolence" swept the congregation and people began emptying their pockets. When the pastor announced the collection would help a set of twins in our church who had leukemia, Dad stripped off his gold-plated Rolex, the last remnant of the boom, and dropped it into the plate.

Mom and Dad attended that church for ten more years, until after my sisters and I moved out of the house. Then, after a period of not going anywhere, Dad started following a local Lutheran pastor on the radio and telling Mom about his sermons. They soon joined his church, which was nothing like the Assemblies of God or the one in the grocery store. There, for the first time in his life, Dad finally felt true peace with the Lord.

We saw Grady only a handful of times after leaving Big Spring. Every so often, we got reports about him going into rehab, or his sister taking him to court for stealing money. Ann somehow managed to keep a few of their oil wells, so whenever the price of crude went up, they kicked off some royalties and gave Grady new life. He blew through town in a rented limo and took everyone for steaks, and for one long evening he was king again. But within a few months he was calling Dad late at night, asking for money. Sometimes he just called to cry. After all those years and everything that had happened, Dad remained Grady's only true friend.

One of the last times I saw him was in 1998, just before my senior year at the University of Texas. Oil prices must have been up again because Grady telephoned saying he was putting me "on scholarship." I didn't believe a word until a few days later, when a check arrived that actually cleared the bank. He also sent money for a new computer, along with airline tickets for Dad and me to join him fishing in Mexico.

In Cabo San Lucas, Grady was in old

form. He rented a six-bedroom villa over-looking the beach and employed a team of baby-faced cabana boys to deliver food and drinks throughout the day. They also brought his cocaine, which came neatly wrapped in cellophane bags and placed in a cabinet drawer. He left the taxi drivers and housecleaners fifty-dollar tips.

I moved to New York City not long after that to work in magazines, and Grady would threaten to visit. "I've got a suite at the Marriott Marquis," he told me over the phone. "All the people there, they know who I am." But he never showed, although a souvenir of Grady and my father's time in the oil business had come with me. Before moving to New York, I didn't own a proper winter coat, nor did I have any money to buy one. So one weekend while visiting my parents in San Antonio, I went into Dad's closet and found the black rabbit fur that Grady had given him for Christmas. After nearly twenty years, the coat fit me perfectly. With a few safety pins, I fastened it to the inside of a cheap brown rain jacket as a liner. That's how I survived my first East Coast winter until I found a job and could buy better clothes. The following year, in the midst of a snowstorm, I dropped the coat next to a homeless man sleeping outside the

Christopher Street subway station and never thought twice about it.

After that I lost touch with Grady and Ann. An invitation to my wedding in 2006 went unanswered, then in March of the following year, Dad called and told me Grady was dead. All the years of drinking and drugs had finally worn down his heart. He was fifty-four years old.

The funeral was held in a small Pentecostal church in Big Spring where Grady found Jesus in the months before he died. When we arrived his body lay in an open casket, his pinkish skin now waxy and gray. His favorite Elvis songs played over the PA system. Ann sat crying in the front row beside Buddy, who'd been the one to drag Grady to rehab each time, kicking and screaming, to no avail. Buddy was now an oilman and restaurateur in Lubbock, married with a houseful of kids.

In the weeks before his death, Grady had sensed what was coming and apologized to Ann, begging forgiveness for all the years of emotional anguish. Now that he was gone she was finally free, and no longer would he embarrass her in the town her father had built. Within a year she was remarried and starting over again, never happier.

■ ■ ■ ■

Four years later, in March 2011, I received another call from Dad. My grandmother Opal was dying in a nursing home. I caught a flight to San Antonio the next morning and met my parents and sisters. Together we squeezed into the car like we did when we were young and drove six hours toward the Caprock, the wooded Hill Country thinning into dry, open pasture.

My grandmother had come to the nursing home after a particularly bad year: she'd broken her hip after slipping on some ice, then suffered seizures from a botched prescription. But she'd handled it all relatively well, repeating the Bette Davis line that "gettin' old ain't for sissies." She was asleep when we arrived, her breathing deep and troubled. A hospice nurse sat by her bed, then explained to us what to expect.

A few years earlier, when Opal turned eighty-five, my wife and I had brought our one-year-old son for her birthday celebration. My wife grew up in Minnesota and had never seen Big Spring, so it was only fitting that as soon as we hit town we were met by a swirling wall of red sand. The storm darkened the sun and stalled traffic, and then it rained mud.

Inside the banquet hall that day, my grandmother was relaxed, center stage. Back surgery had left her dependent on a walker, so Zelda and Norma placed a gold-lacquered chair in the middle of the room where she received her guests. She'd come straight from the beauty shop, and her hair was freshly colored and set. She wore a long silk dress and her fingers bore the gold and emerald rings that every little girl in the family had posed with before her giant vanity mirror.

Now, in her hospice room, I saw that the ornaments were gone. She wore nothing but a simple cotton gown, and her hair flowed gray against her pillow. There was no makeup or jewelry, and the deep creases from age had relaxed in her face. Everyone agreed she looked twenty years younger.

The room was full of her children: Preston and Linda; Norma Lou and her two boys, Randy and Rodney; Zelda and Charles; Dad and Mom. Homer was there, too. Out of all the family who'd settled in Big Spring over the generations, only he and his sister Evelyn remained. Back when their father fell sick in 1983, Homer had quit his job at American Petrofina and took over the family restaurant, which he ran for nearly thirty years. Evelyn's dress shop, Miss Royale, sat across the street, its façade painted bright pink.

During those decades, not much good had come to Big Spring. The refinery remained but hadn't expanded; rust and filth covered the columns and tank batteries that Raymond Tollett had once kept spotless. Oil never experienced a comeback quite like in the beginning of the eighties, and the population hadn't recovered from losing the air base. Blight crept in. The busy downtown streets that Frances once strolled through were ghostly empty now, the shop windows caked with red dirt. Buildings crumbled onto the sidewalk or remained as burned-out husks. The Petroleum Building where Tollett had steered his empire, and where Dad and Grady chased their own, sat vacant, its top-floor suite trashed by vandals and a leaky roof. Several blocks away, the Hotel Settles, the original beacon of progress, loomed over the skyline with boarded windows and its rooms full of pigeons. There was even talk of persuading a Hollywood production company to come blow it up for a movie.

These were my impressions of Big Spring during the years when we'd visit my grandmother. But coming back now to bid her good-bye, I noticed a dramatic change. All across the Permian Basin and elsewhere, new advances in horizontal drilling and hydraulic fracturing had opened up hard-to-reach shale formations for oil and gas exploration. The Wolfcamp and Cline shales, which ran beneath the Permian Basin, were believed to hold over 20 billion barrels of unrecovered crude. The result was a boom like none Texas had ever seen. Once again the roads were crawling with service trucks and the restaurants were full of strangers. At our hotel by the interstate, we were forced to pay double rates.

The boom had breathed new oxygen into downtown. Driving up Third Street, I noticed scaffolding around the Settles. It turned out a Big Spring native named Brint Ryan had purchased the hotel and was spending $30 million to restore it to its original grandeur. Crews were rebuilding the wrought-iron staircase, reinstalling the brass elevators, even excavating the walls in order to match the 1929 paint color. Ryan had also bought the Ritz Theater and the Petroleum Building, which he planned to refurbish.

But sadly, Opal wouldn't be around to see any of that happen. With her family gathered around her bed, she opened her eyes and made one last request. She asked us to sing her home. And just as she had done for Uncle Bud back in 1936, we sang the old songs that brought her comfort: "What a Friend We Have in Jesus," "Love Lifted Me," "Leaning on the Everlasting Arms." Although those hymns hadn't crossed my lips in twenty-five years, the words came flooding back, and as our voices wove together in the family singing, I felt a sudden presence in the room, something both familiar and ancient. For two days we sang. Finally, around six in the morning after everyone except Preston and Zelda had gone to bed, Opal breathed her last. At ninety years old, she joined the big family reunion, as her father Clem used to say, while here among the living she left behind ten grandchildren, twenty-one great-grandchildren, and ten great-great-grandchildren.

The experience at her bedside left a profound impression and sent me in search of that presence I had felt. And the journey toward its roots would lead me through our history and culture and toward my own greater understanding of God.

One of my first trips was to the very hollow where my great-grandfather John Lewis was raised. Knowing nothing about our

Georgia kin, I had to search online. I discovered that his brother Daniel had had a son named Clarence Mealer who still lived outside of Atlanta. He was eighty years old, recently widowed, and eager to meet one of the "Texas Mealers."

The following summer I drove out to meet him. When he opened the door, I took one look — wide forehead, bulbous nose — and knew instantly he was one of my own. Together we headed north into the Blue Ridge, past the town of Jasper, until we reached the place where we were from, where John Lewis, Bud, Bertha, and the kids had found refuge that winter during the Great Depression. A luxury estate had swallowed the old homestead, so Clarence and I crawled under a fence and pushed our way into the brambles. After ten minutes we stopped in a place that looked no different from the surrounding thicket.

"This is where your great-great-grandfather's house once stood," Clarence said, then fell silent.

His own mind went back three-quarters of a century. He told me about playing as a child along the wagon road that ran "just there" past the poplar. About cooling jars of fresh milk in the spring that still flowed near the old barn. A pile of chimney stone was all that was left of the house. It was all trees now, choked by vine and poison oak, hardly resembling a place where family had been born and died and where an eighty-year-old man could be plunged back into the breath-less days of youth. But for that half hour or so, I saw it all.

Back home, Clarence pulled out a recent Christmas card and said, "You know, Bud still has a daughter out in Arizona." There was an address on the envelope, and within days I was on the phone with Frances, whom nobody in my family even knew was still alive. The following month I sat at her dining room table. She described the ter-rible years after her father died, trying to care for her siblings, and how they'd been left to drift. After divorcing Tommy, she'd raised her girls and stayed in politics, work-ing to get conservatives elected in order "to save America," as she described. She finally retired after Ronald Reagan won the presi-dency in 1980, her mission complete.

For eight hours we rummaged through boxes of old photos and letters. Everyone she'd grown up with was dead: her parents and siblings, Tommy and Howard Dodd, Bob Wills and the stars of the Grand Ole Opry. In fact, one of the last places Wills played fiddle before he died in 1975 was the Stampede in Big Spring. Someone else had to raise his bow and pull it across the strings.

Lately Frances had been thinking a lot about Howard, who after their divorce came home to Big Spring, became a fireman, remarried, and raised a family. One night she'd sat up in bed and realized she'd forgotten the sound of his voice, and the problem was that there wasn't anyone who could help her hear it again. Nobody left to call who could affirm the memories that surfaced at random, like how the Georgia pines seemed to stretch to heaven that summer when she was five, or the way little Jimmie held her finger when they walked to get water, his smile and the tiny chip on his tooth. Those memories were hers alone now, with all their heartache, music, and laughter.

ACKNOWLEDGMENTS

First, I'd like to thank the members of my family who have endured my many questions over the years, fed me and took me in, and have traveled with me on this journey of discovering our history. Thanks to my parents for laying it all on the line, trusting I wouldn't leave them hanging in the end, and to Ann Yanez and Frances Varner for doing the same.

Thanks to Tammy Burrow Schrecengost and Cheryl Carter Joy at the Heritage Museum of Big Spring. Thanks to Jason Blake and Ray Tollett, Pat Mares, John Currie, Myra Robinson, Verna Davis, Ben Fountain, Scott Anderson, Diana Davids Hinton, Wanda and Perry Gamble, Larry Don Shaw, Jake Silverstein, Buddy Beach, Jacques Hyatt, Hugh Porter, Hilary Redmon, Bryan Burrough, Chris Tomlinson, David Peters, Sarah Smarsh, Ray Perryman, Lonn and Dedie Taylor, and the staff at the

Hotel Settles. And thanks to Lorin McDowell for offering me a quiet place to write.

Thanks to Tumbleweed Smith for opening doors and for being a friend, and to James Johnston, whose two books on Big Spring, *Crossroads Canaan* and *Movies and Magic Houses in a West Texas Whereabouts,* fortified my research, along with our many conversations over Ruby's BBQ and Mi Madre's enchiladas.

And thanks to my readers whose comments and advice were crucial: Michelle Garcia, Lee Simmons, Kiley Lambert, and John Kenney. I owe special gratitude to John Baskin, whose wise counsel over the years, and careful attention to this manuscript, have helped to make me a better writer. And thanks to Michael Brick, who never got a chance to see these pages.

Thanks to my editor, Colin Dickerman, for urging me to find the bigger, sweeping story, and for his calm guidance throughout, and to Bob Miller for backing us up. Thanks to James Melia, Marlena Bittner, and the rest of the Flatiron team. Thanks to my agent, Heather Schroder, for helping me to continue doing this thing that I love. And finally, thanks to Ann Marie Healy, who's still by my side as we march onward into the dream.

NOTES

PART 1

Chapter 1

The Honest Man's Friend and Protector, their crimes, and fights with the law from *The Atlanta Journal Constitution,* March 6, 1890.

John Lewis helping moonshiners and descriptions of family land and death of his mother, taken from interview with Clarence Mealer, Douglasville, Georgia, July 4, 2014.

Other history of Georgia, including flora and fauna, taken from two sources: *The Annals of Upper Georgia Centered in Gilmer County* by George Gordon Ward (Thomasson Printing and Office Equipment Company, 1965) and *History of Pickens County* by Luke Tate (Walter W. Brown Publishing Company, 1935).

Migration to Texas in 1890s, population

explosion in Texas, and farming practices taken from three sources: *Lone Star: A History of Texas and the Texans* by T. R. Fehrenbach (Da Capo Press, 1968), *Gone to Texas: A History of the Lone Star State* by Randolph B. Campbell (Oxford University Press, 2003), and *The Road to Spindletop: Economic Changes in Texas, 1875–1901* by John Stricklin Spratt (Southern Methodist University Press, 1955).

Details of Newt Mealer's travels came from interview with Clarence Mealer, previously cited.

Hoboes in the 1880s and '90s taken from "In Search of the American Hobo" by Sarah White (Media project for the American Studies Program, University of Virginia, 2001).

Description of cotton market in Hillsboro, Texas, taken from *100% Cotton: A History of Cotton in Hill County* by Jack and Jane Pruitt (Jack Pruitt Books, 1993).

Bateson family history taken from *Cleburne Morning Review* and *Abilene Reporter-News,* various editions, and interview with Rosemary Bateson, via telephone, January 16, 2015.

John Lewis property in Hillsboro from deed records, Hill County, Texas.

Details of boll weevil scourge in Texas came from two sources: *The Boll Weevil Comes to Texas* by Frank Wagner (Grunwald Printing, 1980) and *Boll Weevil Blues: Cotton, Myth, and Power in the American South* by James C. Giesen (University of Chicago Press, 2011).

Details of boll weevil refugees in Roby, Texas, and crop reports from *Abilene Reporter* and *Abilene Daily Reporter,* various editions, 1908–13.

John Lewis land purchase in Eastland from deed records, Eastland County, Texas.

George Bedford shootout and death from *The Amarillo Globe,* December 27, 1927.

Chapter 2

Details of John Lewis's debt, traveling to Burnet County, interactions with Frank Day, and losing the farm were taken from John Lewis's lawsuit against Day and C. M. Murphy filed in District Court, Eastland County, Texas, April 1919.

C. M. Murphy leasing land to oil company from deed records, Eastland County, Texas.

Details of World War I, dependence on oil, and domestic fuel shortage taken from *The*

Prize: The Epic Quest for Oil, Money, and Power by Daniel Yergin (Simon & Schuster, 1991).

Texas booms in the early 1900s taken from various newspaper accounts and from two books by Walter Rundell Jr.: *Early Texas Oil: A Photographic History, 1866–1936* (Texas A&M University Press, 1977) and *Oil in West Texas and New Mexico* (Texas A&M University Press, 1982).

Details of Ranger boom taken from various newspaper accounts and from *Oil! Titan of the Southwest* by Carl Coke Rister (University of Oklahoma Press, 1949) and two books by Boyce House: *Oil Boom: The Story of Spindletop, Burkburnett, Mexia, Smackover, Desdemona, and Ranger* (Caxton Printers, 1941) and *Were You in Ranger?* (Tardy Publishing, 1935).

Frank Day and Jess Willard's relationship taken from *Were You in Ranger?* by Boyce House, previously cited.

Representative D. J. Neill statement to the Texas Legislature taken from *The Bartlett Tribune and News,* August 23, 1918.

Accounts of Spanish flu in Texas found in *Epidemic in the Southwest, 1918–1919* by Bradford Luckingham (Texas Western Press, 1984) and accounts of flu in Ranger

574

taken from *Oil Boom: The Story of Spindletop, Burkburnett, Mexia, Smackover, Desdemona, and Ranger* by Boyce House, previously cited.

Descriptions of Frank Day and the story of the farmer giving Day his land from *Were You in Ranger?* by Boyce House, previously cited. Details of Willard and Day hitting the wagon with their car from the *Hearne Democrat,* July 2, 1948.

Details of oil boom in Desdemona, Texas, taken from "Diamonds and Galoshes" by Anne Dingus, *Texas Monthly,* January 1986; *Texas Boomtowns: A History of Blood and Oil* by Bartee Haile (The History Press, 2015); *Oil! Titan of the Southwest* by Carl Coke Rister, previously cited; and the two books by Boyce House, previously cited.

Information about Goldie Mealer's death in Desdemona taken from interview with Barbara Tyler, Odessa, Texas, May 13, 2015.

Elijah Mealer's death in France from U.S. military records.

Jack Dempsey's fight with Jess Willard from various newspaper accounts, particularly *The New York Times,* July 5, 1919. Frank Day hobnobbing in Willard's suite from *Were You in Ranger?* by Boyce House, previously cited.

Kids playing with roman candles and torches from interview with Barbara Tyler, previously cited.

Details of Best, Texas, from "The Best and Worst of Times" (Associated Press, July 12, 1987), "The Discovery and Early Development of the Big Lake Oil Field" by Martin W. Schwettmann (M.A. Thesis, University of Texas, 1941), and *Oil! Titan of the Southwest* by Carl Coke Rister, previously cited.

Santa Rita well information found in *The Permian Basin: Petroleum Empire of the Southwest, Vol. 1, Era of Discovery: From the Beginning to the Depression* by Samuel D. Myres (Permian Press, 1973), and Cark Coke Rister's *Oil! Titan of the Southwest,* previously cited. Other information taken from *Santa Rita: The University of Texas Oil Discovery* by Martin W. Schwettmann (Texas State Historical Association, 1958).

James McCormick's abuse of Bertha taken from interview with Frances M. Varner, Scottsdale, Arizona, August 19, 2014. Additional interviews, via telephone and in Big Spring, Texas, continued through February 2017.

Deepest well in the world from *The Permian Basin: Petroleum Empire of the Southwest, Vol. 1, Era of Discovery* by Samuel D. Myres, previously cited.

Chapter 3

Descriptions of Permian Basin geology taken from *The Permian Basin: Petroleum Empire of the Southwest, Vol. 1, Era of Discovery* by Samuel D. Myres, previously cited, and *Gettin' Started: Howard County's First 25 Years* by Joe Pickle (Howard County Heritage Museum, 1980).

Descriptions of the Llano Estacado and Comanche presence found in *Gettin' Started* by Joe Pickle, previously cited, *The Great Plains* by Walter Prescott Webb (Grosset & Dunlap, 1931), and *Empire of the Summer Moon: Quanah Parker and the Rise and Fall of the Comanches, the Most Powerful Indian Tribe in American History* by S. C. Gwynne (Scribner, 2010).

California gold rush and details about Captain Randolph Marcy found in *Empire of the Summer Moon* by S. C. Gwynne and *Gettin' Started* by Joe Pickle, previously cited.

Buffalo hunting and bone scavenging taken

from several sources: *Howard County . . . In the Making* by John R. Hutto (self-published, 1938), *Fort Concho and the Texas Frontier* by J. Evetts Haley (San Angelo Standard-Times, 1952), *Fort Griffin on the Texas Frontier* by Carl Coke Rister (University of Oklahoma Press, 1956), and "The West Texas Bone Business" by Ralph A. Smith (West Texas Historical Association Yearbook, Vol. 55, 1979).

Railroad workers, Chinese and Irish, and the cowboy neighbors taken from *Gettin' Started* by Joe Pickle, previously cited.

Details about early Big Spring ranches taken from Joe Pickle and John R. Hutto, previously cited, and *History of Howard County: 1882–1982* (Big Spring Chamber of Commerce, 1982).

Details about drought and blizzard on the McDowell Ranch taken from interview with Lorin S. McDowell III, Big Spring, Texas, May 15, 2016.

Early farming and sod buster details from *Gettin' Started* by Joe Pickle, previously cited, and *Big Spring: The Casual Biography of a Prairie Town* by Shine Philips (Prentice-Hall, 1942).

S. E. J. Cox and his time in Big Spring from *Easy Money: Oil Promoters and Investors in the Jazz Age* by Roger M. Olien and Diana Davids Olien (University of North Carolina Press, 2009); the *Big Spring Herald,* July 4, 1975; and Cox's official "Program Souvenir: Opening and Dedication of the General Oil Company's New West Texas Field" (August 1919).

The life and times of Joshua S. Cosden from the *Big Spring Herald,* various editions; living in a tent in Bigheart from *El Paso Herald-Post,* May 11, 1932; contruction of the refinery in Tulsa from the *Muskogee Times Democrat,* December 1, 1913; capacity of Tulsa refinery and what it produced, from the *Morning Tulsa Daily World,* August 12, 1915; million-dollar life insurance policy from the *Morning Tulsa Daily World,* February 24, 1916; details of mansion in Palm Beach, Cosden's reputation in the press, and his admiration for *The Great Gatsby* from the *Palm Beach Post,* January 24, 1982; Nellie Neves divorce details, quote from Cosden's first wife ("He'll make a comeback . . .") and quotes from society writers ("Men and women whose surnames had been in Blue Book and Social Registers . . .") from the *Palm Beach Post,* January 5, 1930; millionaires buying up Cosden's racehorses from the *Brooklyn Daily Eagle,* October 13, 1927. Other information came from the books *Then Came Oil: The Story of the Last Frontier* by C. B. Glasscock (Bobbs-Merrill Company, 1938) and *The Permian Basin: Petroleum Empire of the Southwest, Vol. 1, Era of Discovery* by Samuel D. Myres, previously cited.

Details of Big Spring during the oil boom of 1927–29 and Hotel Settles construction and opening, from the *Big Spring Herald,* various editions.

Dora Roberts information from *History of Howard County: 1882–1982,* previously cited.

Cosden in receivership from the *Big Spring Herald,* November 14, 1930.

Damage to small towns during Great Depression from *The Permian Basin: Petroleum Empire of the Southwest, Vol. 1, Era of Discovery* by Samuel D. Myres, previously cited.

Trouble in the cotton market, low prices, from the *Big Spring Herald,* December 13, 1929.

The most cotton grown in United States in 1926 from "Cotton Production Remains a Multibillion Dollar Asset," *The Oklahoman,* February 23, 1986.

Chapter 4

The family's return to Georgia and details about John Lewis's parents from interviews with Frances M. Varner and Clarence Mealer, previously cited.

The family riding trains to California, railroad bulls, and rumors of a murder from interviews with Frances M. Varner, previously cited, in addition to interviews with Katie Jones Cathey, Big Spring, Texas, May 5, 2014; Rodney Lee, Odessa, Texas, December 2, 2012; and Bill White, Dickinson, Texas, April 29, 2014.

Details about kids riding trains during Great Depression from *Riding the Rails: Teenagers on the Move During the Great Depression* by Errol Lincoln Uys (Routledge, 2003).

Life on the Jones farm in Knott, Texas, came from interviews with Frances M. Varner, previously cited, and unpublished essays by Earl Jones, Odessa, Texas, (dates unknown).

Cotton boom in Big Spring from the *Big Spring Herald,* August 24, 1932, and various other editions.

The Dust Bowl's impact on Big Spring from the *Big Spring Herald,* various editions.

The federal government buying cattle under

the Agriculture Adjustment Act from the *Big Spring Herald,* June 29, 1934.

Story of government killing the Joneses' cattle from Earl Jones (essays, previously cited) and interview with John Currie, Big Spring, Texas, February 20, 2015.

Oil market rebound during the Depression from *Big Spring Herald,* various editions, and *The Permian Basin: Petroleum Empire of the Southwest, Vol. 1, Era of Discovery* by Samuel D. Myres, previously cited.

Bob finding work with oil companies from interviews with Bill White, previously cited, and Wayne Jones, via telephone, May 23, 2014.

Chapter 5

Bud working for Shell, family's new house, life in Forsan taken from interviews with Frances M. Varner, previously cited.

Dust storms in Big Spring and Forsan from the *Big Spring Herald,* various editions, May 1933 through April 1936.

Drilling deeper holes in the Permian Basin from *The Permian Basin: Petroleum Empire of the Southwest, Vol. 1, Era of Discovery* by Samuel D. Myres, previously cited.

Lee Pruitt beating Allie from interview with Barbara Tyler, previously cited.

Bud's death from interviews with Frances M. Varner, previously cited.

Bob and Opal's meeting, courtship, and wedding from interviews with Frances M. Varner, previously cited, and Zelda Odom, San Antonio, Texas, February 26, 2013.

Chapter 6

Hard times for Bud and Bertha's children after Bud's death from interviews with Frances M. Varner, previously cited.

John Lewis's death from official death certificate and the *Big Spring Herald*, June 13, 1938. Severe storm damage in Big Spring from the *Big Spring Herald*, same edition.

Reverend W. A. Nicholas details from the *Abilene Reporter-News*, various editions.

Bob Wills details from *San Antonio Rose: The Life and Music of Bob Wills* by Charles R. Townsend (University of Illinois Press, 1976) and interviews with Frances M. Varner, previously cited.

PART 2
Chapter 1

Wink, Texas, details and life during the Depression from interviews with Zelda

Odom, previously cited, *Oil Booms: Social Change in Five Texas Towns* by Roger M. Olien and Diana Davids Olien (University of Nebraska Press, 1982), and *Life in the Oil Fields* by Roger M. Olien and Diana Davids Olien (Texas Monthly Press, 1986).

Clem Wilkerson details, salvation on the street, from multiple interviews with Homer Wilkerson, Big Spring, Texas, from October 2012 through February 2017; plus interview with Bill and Veda White, Dickinson, Texas, June 27, 2012.

Mary Lou Wilkerson death from the *Big Spring Herald,* November 19, 1937, and interviews with Homer Wilkerson, previously cited.

New London School explosion from Texas State Historical Association website. Arthur Hahn beating and abandoning his family from interview with Doris Haynes, Oklahoma City, June 20, 2014.

Chapter 2

Clem as prayer warrior, Mr. Smith healing, and kids playing church taken from interviews with Norma Lee, Big Spring, Texas, October 3, 2012; Evelyn Bender, Big Spring, Texas, May 30, 2014; Doris

Haynes, Veda White, Zelda Odom, and Homer Wilkerson, previously cited.

Frank Mack bio from interview with Runelda Mack Hunsicker, via telephone, April 3, 2014.

Chapter 3

Joshua Cosden in hospital from *The New York Times,* August 9, 1930. Cosden on the veranda, buying back his company from the *Big Spring Herald,* October 2, 1979. Cosden on his deathbed, his promise to doctors, and his death from the *Big Spring Herald,* July 18, 1954.

Cosden cremation details from the *San Antonio Light,* November 19, 1940.

Cosden interview with C. B. Glasscock from *Then Came Oil,* previously cited.

Raymond Tollett claiming Cosden's body in El Paso from reporter Bob Lewis's interview with Paul Meek (deceased), Fredericksburg, Texas, September 13, 2003.

Life of Raymond Tollett from the *Big Spring Herald,* various editions; the *Cosden Copper,* various editions, 1954–67; Federal Bureau of Investigation records, obtained through the Freedom of Information Act; multiple interviews with Ann Yanez, Big Spring, Texas, from February 2012 through June 2016; multiple interviews with Jason Blake Tollett, Austin, Texas, from February 2013 through January 2017; and interview with Raymond Tollett Jr., Dallas, Texas, December 7, 2015.

Tollett running the Big Spring refinery and accomplishments from the *Big Spring Herald,* various editions; interviews with Dan Krausse, via telephone, October 15, 2015; Bobby Fuller, via telephone, October 22, 2015; Frank Parker, via telephone, December 10, 2015; Katie Grimes, via telephone, December 5, 2015; Rene Brown, via telephone, December 16, 2015.

Chapter 4

Details on Kelly-Snyder field and Spraberry Trend from *The Permian Basin: Petroleum Empire of the Southwest; Vol. 2, Era of Advancement: From the Depression to the Present* by Samuel D. Myres (Permian Press, 1977) and *Wildcatters: Texas Independent Oilmen* by Roger M. Olien and Diana Davids Hinton (Texas A&M University Press, 2007).

Details on the 1950s drought from the *Big Spring Herald,* various editions, 1950–57; "When the Skies Ran Dry" by John Burnett, *Texas Monthly,* July 2012; interviews with Jimmie Taylor, Big Spring, Texas, May 1, 2014; John Currie, Big Spring, Texas, May 13, 2015; Wade Choate, Midland, Texas, April 10, 2014; Homer Wilkerson, previously cited; Carroll and Joyce Choate, Big Spring, Texas, October 29, 2015; Lorin S. McDowell III, previously cited.

Rowan Oil Company's lawsuit taken from the Texas Railroad Commission's website archives.

Major oil companies going overseas from *Wildcatters* by Roger M. Olien and Diana Davids Hinton, previously cited.

Davey Jones family history from interviews with Katie Jones Cathey and Wayne Jones, previously cited.

Bob Mealer getting into dirt business from interviews with Preston Mealer, Pearland, Texas, August 2, 2015; Bill White, Rodney Lee, Homer Wilkerson, Zelda Odom, Norma Lee, and Wayne Jones, all previously cited.

Herman and Homer Wilkerson's cattle business, Homer driving trucks to California from interviews with Charles Miller, Big Spring, Texas, May 30, 2014; and Homer Wilkerson, previously cited.

Chapter 5

Cosden tapping into the petrochemical boom, plus Raymond Tollett's role as president of the refinery and a civic leader from interviews with Dan Krausse, Rene Brown, Frank Parker, and Bobby Fuller, all previously cited. Other information from the *Big Spring Herald,* various editions, plus Raymond Tollett's columns in the *Cosden Copper,* various editions.

Granville Hahn's rise at Cosden from family interviews and from reporter Bob Lewis's interview with Granville, Big Spring, Texas, (date unknown).

Raymond Tollett's alcoholism from interview with Iris's niece Judi Godowns, via telephone, January 20, 2016, in addition to extensive interviews with Ray and Jason Blake Tollett and Ann Yanez. Other information from John Currie, Katie Grimes, and former employees Dan Krausse, Rene Brown, Frank Parker, and Bobby Fuller, all previously cited.

Chapter 6

Most information in this chapter from extensive interviews with Frances M. Varner. Other information about Bob and

Billy Jack Wills and the various incarnations of their bands from *San Antonio Rose* by Charles R. Townsend, previously cited.

PART 3
Chapter 1

Rain returning to Big Spring from the *Big Spring Herald,* various editions.

Decline of beef production in U.S. from "Beef, Veal Imports Growing as Production in U.S. Declines," UPI, September 14, 1958.

Herman's illness and new career as restaurateur from interviews with Homer Wilkerson, Evelyn Bender, and Charles Miller, all previously cited.

Bobby and Preston's childhoods taken from interviews with Bobby Mealer, San Antonio, Texas, October 12, 2012 through January of 2017; and interviews with Preston Mealer, Norma Lee, and Zelda Odom, all previously cited.

Clem Wilkerson's death from various family interviews.

Grady Cunningham's childhood and stories about his parents taken from interviews with Selena Gould, New Braunfels, Texas, October 2, 2013; multiple interviews with Buddy Beach, Lubbock, Texas, February

23, 2013 and others via telephone through May 2016; plus interviews with Bobby Mealer, Ann Yanez, and Roe Fulgham, all previously cited.

The family's time in Odessa, plus the death of Jack, taken from interviews with Bobby and Preston Mealer, Zelda Odom, Norma Lee, and Barbara Tyler, all previously cited.

Chapter 2

Expansion of Cosden refinery and its sale to W. R. Grace & Company and American Petrofina from the *Big Spring Herald,* various editions, and Tollett's columns in the *Cosden Copper,* various editions.

Tollett's philanthropy from interviews with Homer Wilkerson and former employees, preciously cited; the *Big Spring Herald,* various editions, and the *Cosden Copper,* various editions.

Tollett's alcoholism and its damage, see notes to part 2, chapter 5.

Tollett checking into rehab facility from interviews with Jason Blake and Ray Tollett and Ann Yanez, previously cited.

Tollett's death from the *Big Spring Herald,* October 28, 1969, and interviews with his children, previously cited.

Chapter 3

Summer 1967 details taken from the *Big Spring Herald,* various editions.

Details about Ronnie, plus Bobby and Marie's relationship, from interviews with "Marie" (her name has been changed), via telephone, September 18, 2015; Merlee Dennis, Austin, Texas, September 25, 2015; and Bobby Mealer, previously cited.

Preston being drafted from interviews with Preston and Linda Mealer, previously cited.

Chapter 4

Most information from extensive interviews with Frances M. Varner, previously cited.

Barry Goldwater background from *Before the Storm: Barry Goldwater and the Unmaking of the American Consensus* by Rick Perlstein (Nation Books, 2009).

Chapter 5

Bobby and Marie dating from interviews with "Marie" and Merlee Dennis, previously cited.

Opal's involvement with *Midnight Cowboy* from interviews with Charles and Zelda

Odom, previously cited.

Midnight Cowboy filming in Big Spring from a variety of sources: the *Big Spring Herald,* various editions during August 1968; transcripts of reporter Bob Lewis's series of interviews with cast and crew, KBST radio, August 1968; and *Movies and Magic Houses in a West Texas Whereabouts* by James Johnston (James Johnston, 2015).

John Schlesinger's panic attack from interview with Jon Voight, *Esquire,* August 2013.

Grady as a teenager from interviews with Bobby Mealer, Ann Yanez, and Selena Gould, previously cited.

Ike Robb family background from the *Big Spring Herald,* various editions; *Movies and Magic Houses in a West Texas Whereabouts* by James Johnston, previously cited; and interviews with Ann Yanez, previously cited.

Bobby's drug use from extensive interviews with Bobby Mealer, previously cited.

Chapter 6

Bobby and Sharon's courtship and marriage, birth of Bryan, from Bobby's letters to Sharon and extensive interviews with Sharon Mealer, San Antonio, Texas, De-

cember 2015 to January 2017, and Bobby
Mealer, previously cited.

Chapter 7

Grady's relations with men and details
about Big Spring's gay culture from interview with unnamed source, via telephone,
September 18, 2015.

Ann's experience aboard cruise ships from
interviews with Ann Yanez, previously
cited.

Grady and Ann courtship and marriage
from interviews with Ann Yanez, Selena
Gould, and Bobby Mealer, all previously
cited.

Chapter 8

Details of energy crisis from various newspaper accounts and the *Big Spring Herald,*
various editions; *Survive & Conquer: Texas
in the '80s: Money-Power-Tragedy . . . Hope*
by M. Ray Perryman (Taylor Publishing
Company, 1990); and *The Prize* by Daniel
Yergin, previously cited.

Texas cities running out of fuel from *Survive
& Conquer* by M. Ray Perryman, previously cited.

Information about Big Spring refinery buying oil from Iraq from Joe Nocera's interview with Sam Hunnicutt, *Texas Monthly,* January 1986.

Bobby's DWI from interviews with Bobby Mealer, previously cited.

Chapter 9

Houston as boomtown, office space in demand, and stats about the "Texas miracle" from *Survive & Conquer* by M. Ray Perryman; other information from *Gone to Texas* by Randolph B. Campbell; both previously cited.

Rise in crime, addiction, and suicides in northern cities after recession from *Journey to Nowhere: The Saga of the New Underclass* by Dale Maharidge and Michael Williamson (Dial Press, 1985).

Southfork Ranch overshadowing grassy knoll from the *Big Spring Herald,* October 25, 1981.

Kenny "the Snake" Stabler details from "When Ken Stabler was a Country-Music Lyric Come to Life," by Pete Axthelm, Deadspin.com, July 10, 2015.

Mark Powell murdering mother-in-law and shooting rampage from the *Big Spring Herald,* June 9, 1980.

Chapter 10

Grady's job title and history at the refinery from Cosden/American Petrofina personnel records kept at the Heritage Museum of Big Spring, and from interviews with Ann Yanez, Roe Fulgham, and Selena Gould, all previously cited.

Iris Tollett's death from interviews with Ann Yanez, Judi Godowns, Jason Blake and Ray Tollett, all previously cited.

Cabot Corporation announces closure from the *Big Spring Herald,* August 8, 1979.

Cosden Oil and Chemical Company moves executives to Dallas from the *Big Spring Herald,* May 14, 1979.

One thousand empty houses in Big Spring following base closure from the *Big Spring Herald,* January 31, 1982.

Hotel Settles closure and details from the *Big Spring Herald,* various editions.

Bureau of Land Management oil and gas lottery information from "Good Ol' Uncle Sam Runs Energy Lottery," by Tom Tiede, Newspaper Enterprise Association, January 26, 1979; "Clear Pattern Seen in Oil-Lease Fraud," Associated Press, April 6, 1980; and "Long-Shot Oil Lease Lottery a Bonus in Wyoming," *The New York Times,* September 30, 1981.

Details of Grady winning the lottery from interview with Hugh Porter, Dallas, Texas, July 29, 2015; plus interviews with Ann Yanez, Buddy Beach, and Selena Gould, previously cited.

Bob Wilson and Orville Phelps information from interview with Bob Wilson, Midland, Texas, April 9, 2013.

Details of 1979 global oil panic from *The Prize* by Daniel Yergin, previously cited. Information on "stripper wells" from interview with Hugh Porter, previously cited, and *Survive & Conquer* by M. Ray Perryman, previously cited.

Midland banks reaping a billion in deposits from the *Big Spring Herald,* January 6, 1980.

Ike Robb death from the *Big Spring Herald,* April 10, 1980.

Grady calling Bobby about job opportunity from interviews with Bobby Mealer, previously cited.

Chapter 11

Details about Grady's "boys" from interviews with Jacques Hyatt, Hugh Porter, Ann Yanez, Bobby Mealer, and Buddy Beach, all previously cited.

Torstein the pilot information, getting hired

by Grady, Grady buying airplanes, and European vacation from interview with Torstein Faaberg, via telephone, October 28, 2015; and Linda Stegmeyer, via telephone, October 26, 2015. European vacation details also from interview with Robert Miller, Big Spring, Texas, May 29, 2014.

Ann visiting bookstore and the details of conversations from interview with Linda Stegmeyer, previously cited. Ann wanting kids from interviews with Ann Yanez, previously cited.

Bobby wrecking Sharon's Lincoln, encountering Grady and his lover, from interviews with Bobby Mealer, previously cited.

Chapter 12

Details about the Guinn No. 1 well and all others drilled by Cunningham Oil Corporation from Texas Railroad Commission records, plus interviews with Hugh Porter, Buddy Beach, Jacques Hyatt, Bobby Mealer, and Bob Wilson, all previously cited.

Bachendorf's information from company website and *Dallas Morning News,* July 13, 2010, plus interviews with Ann Yanez, Bobby and Sharon Mealer, Hugh Porter, and Buddy Beach, all previously cited.

Market analysts predicting eighty-five-dollar oil ("Eighty-five in '85") from *Wildcatters* by Roger M. Olien and Diana Davids Hinton and *Going to Texas* by Randolph B. Campbell, both previously cited.

Drug use in Big Spring and drug prices from the *Big Spring Herald,* December 20, 1981; "Elephant Hunt" parties and bowls of cocaine from the *Big Spring Herald,* June 15, 1984.

Grady's escalating cocaine habit, scoring drugs at 2 a.m., Buddy driving, from interviews with Buddy Beach, previously cited.

Chapter 13

Details on President Reagan's Economic Recovery Tax Act and its impact from PBS.org, "Reagan: The 1982 Recession."

People using less gasoline, glut of oil from the *Big Spring Herald,* February 24, 1982.

Rig count falls to half from *Survive & Conquer* by M. Ray Perryman, previously cited.

Increase in homelessness in U.S. from "Homeless Crisscross U.S. Until Their Cars and Their Dreams Break Down," by Iver Peterson, *The New York Times,* December 18, 1982.

Lines at Houston soup kitchens, police pulling over cars with northern plates, people riding the trains, from *Journey to Nowhere* by Dale Maharidge and Michael Williamson, previously cited.

Hoboes in Fort Worth from "More Hobos Riding Rails," Associated Press, September 21, 1981.

Loans up in Big Spring banks but deposits down from the *Big Spring Herald,* July 19, 1982. Unemployment in Big Spring highest since base closure from the *Big Spring Herald,* July 22, 1982.

Grady's spending and purchases, relationship with Bill Read and Coahoma State Bank, from interviews with Selena Gould, Buddy Beach, Hugh Porter, Ann Yanez, and Bobby Mealer, all previously cited.

Cunningham Oilers information from the *Big Spring Herald,* various editions.

Grady's hot check to football referee from interview with John Ferguson, Big Spring, Texas, November 1, 2012.

Grady and Ann's lavish spending in Florida and Bahamas from interviews with Charles and Zelda Odom and Bobby Mealer, previously cited.

Grady's wedding gift to Hugh Porter from

interview with Hugh Porter, previously cited.

Chapter 14

Details of Penn Square Bank lending practices, failure, and impact on industry from *Belly Up: The Collapse of Penn Square Bank* by Phillip L. Zweig (Crown Publishers, 1985); "Penn Square's Failed Concept," *The New York Times,* August 16, 1982; "Key Figure in Penn Square Collapse Keeps Long Silence," *The Daily Oklahoman,* July 12, 1992.

Continental Illinois bailout from FDIC.gov, "History of the Eighties."

Metro Bank and National Bank of Odessa details from "A Texas Bank's Ties to Oil," *The New York Times,* October 14, 1983.

Collapse of First National Bank of Midland, causes and details of closure from "RepublicBank of Dallas Gets Midland," *The New York Times,* October 15, 1983; "Bullish on the Bust," *Texas Monthly,* November 1984.

Details about FNB's community outreach from *Midland Reporter-Telegram,* October 15, 1983.

Thomas Procopio and FDIC's role in the bust from "U.S. Helps Texans Survive

Death of Bank," *The New York Times,* October 14, 1984.

Jeff Brown details from *Been There Done That: As I Remember!* by Jeff L. Brown (Nortex Press, 2000).

Brass Nail throwing out lobsters from interview with Jacques Hyatt, previously cited.

Grady waving a revolver on Gregg Street from interview with Hugh Porter, previously cited.

Grady's bankruptcy details from the *Big Spring Herald,* September 6, 1984; also from interviews with Ann Yanez and Buddy Beach, previously cited.

ABOUT THE AUTHOR

Bryan Mealer is the author of *Muck City* and the *New York Times* bestseller *The Boy Who Harnessed the Wind* — written with William Kamkwamba — which has been translated into more than a dozen languages and will soon be released as a major motion picture. He's also the author of *All Things Must Fight to Live*, which chronicled his time covering the war in the Democratic Republic of Congo for the Associated Press and *Harper's*. His other work has appeared in *Texas Monthly*, *Esquire*, the *Guardian*, and the *New York Times*. Mealer and his family live in Austin.

GREAT DISCOVERIES

DAN HOFSTADTER

The Earth Moves

Galileo and the Roman Inquisition

ATLAS & CO.

W. W. NORTON & COMPANY

NEW YORK · LONDON

For information about permission to reproduce selections from
this book, write to Permissions, W. W. Norton & Company, Inc.,
500 Fifth Avenue, New York, NY 10110

For information about special discounts for bulk purchases, please contact
W. W. Norton Special Sales at specialsales@wwnorton.com or 800-233-4830

Manufacturing by RR Donnelley, Bloomsburg
Book design by Chris Welch
Production manager: Julia Druskin

Library of Congress Cataloging-in-Publication Data

Hofstadter, Dan.
The Earth moves : Galileo and the Roman Inquisition /
Dan Hofstadter. — 1st ed.
p. cm. — (Great discoveries)
Includes bibliographical references and index.
ISBN 978-0-393-06650-0 (hardcover)
1. Galilei, Galileo, 1564-1642   Trials, litigation, etc. 2. Inquisition—Italy—Rome.
3. Astronomy—Religious aspects—Christianity—History of doctrines—17th
century. 4. Science, Renaissance. 5. Catholic Church—Doctrines—History—17th
century. 6. Catholic Church—Italy—History—17th century. I. Title.
QB36.G2H64 2009
509.4'09032—dc22

2009004325

W. W. Norton & Company, Inc.
500 Fifth Avenue, New York, N.Y. 10110
www.wwnorton.com

W. W. Norton & Company Ltd.
Castle House, 75/76 Wells Street, London W1T 3QT

1 2 3 4 5 6 7 8 9 0

For Bette

Contents

The Earth Moves

There are many books about Galileo, so the reader is entitled to ask how this one differs from any others that he or she might come across. It is, as the title suggests, an attempt to recount how the great physicist and astronomer Galileo Galilei was tried and convicted by the Roman Inquisition in 1633 for championing the Copernican hypothesis about the solar system, against which the Vatican had issued an edict. I have tried to tell this story succinctly for the general reader, and without inserting any but a few paragraphs of the simplest mathematics into the text.

In addition, I have chosen a very specific path through a thicket of information. Galileo had several reasons for embracing Copernicanism. Deductions in both mechanics and mathematical astronomy contributed to his growing conviction that the earth revolved around the sun, and I have touched upon these; I have devoted much more attention, however, to his sudden and vast improvement of the telescope in the autumn of 1609, which enabled him to actually see the

surface of the moon, the phases of Venus, and the moons of Jupiter, and so to behold celestial mechanics at work. In time these telescopic sightings would put him at greater variance with the edicts of the Roman Catholic Church than his far greater discoveries in physics ever did. This was the first great clash of religion and science, and it still has much to teach us.

What Galileo did with the telescope is in itself an exciting story. But I have a special reason for retelling it. I am primarily interested in the arts, and Galileo loved music, literature, and painting. A musician's son, he played the lute well, wrote poetry and literary criticism, taught perspective, and drew with some verve; he corresponded at length with at least one major painter. He favored the classics, yet he belonged to a world no longer classical but Baroque in orientation. A system of thought may be described as Baroque when its parts cannot be understood or enjoyed unless they are constantly related to some larger, dynamic whole, and in that sense Galileo was a prime representative of the Baroque era, which began around 1600. Geometry, trigonometry, and perspective were then still the common property of mathematicians, painters, and archi-tects, and the budding science of optics interested them all: indeed we find painters among the most enthusiastic support-ers of Galileo's discoveries with the telescope. Understandably, Galileo's position within the general context of Baroque civi-lization has not much concerned historians of science, and I have tried to offer a brief picture of it here.

Most writing about Galileo and the Inquisition has to do with philosophy, and that is as it should be. But there are already a number of excellent books on this subject, and I found myself more drawn to the psychological complexities of the 1633 trial. I have always found it hard to control my bewil-

derment over the fact that Galileo's persecutor, Pope Urban VIII, is known in art history as Bernini's and Borromini's great patron, indeed as the most devoted supporter that the Baroque style ever found in Italy; during the 1620's he also befriended and encouraged Galileo. In this book I have tried to accord him more sympathy than he usually receives.

The man who really caught my attention, however, was the all-but-forgotten Tuscan ambassador to the Vatican, Francesco Niccolini, who is the virtual hero of many of the pages that follow. Since Galileo was mathematician to the court of the Grand Duke of Tuscany, both the duke and his emissary Niccolini wished to avert a trial by the Inquisition, or, failing that, to curtail an eventual trial by means of an extra-judicial solution—what we would call a plea-bargain. Unhappily Galileo, as one of Europe's first professional intellectuals, did not fit into this scheme. Unlike earlier Italians who had taught largely by personal example—unlike, say, Saint Francis, who kissed the leper, or Giordano Bruno, who chose to die rather than think as he was told to—Galileo had faith in the transcendent value of a good argument, and he wanted to argue with the Vatican. He even wanted to argue with the Inquisition. Ambassador Niccolini, with his insight into the papal court, foresaw the danger of such a project and tried to induce Galileo to abandon it. Reading Niccolini's correspondence, I found myself touched by the abrasion between two friends representing such different ethical approaches to life.

Galileo has often been described as an inordinately suspicious man. There is some truth to this, but his friends' letters suggest that he was also subjected to a great deal of envy and hostility. The literary background to the concept of envy, or *invidia*, was raised some years ago by Miles Chappell in rela-

tion to two graphic works by Ludovico Cigoli, one of them dedicated to his friend Galileo. I have returned to this idea, stressing its background in Dante's *Divine Comedy*.

While on the subject of my debt to Galilean scholars, I wish to express how much I owe to my reading of Stillman Drake and William R. Shea (who has particularly influenced me). Much of the material on the telescope has benefited from my reading of Albert Van Helden's papers on the subject. I confess to an almost hypnotic fascination with the philosophical dimension of the trial, and here I must gratefully mention Richard J. Blackwell, Annibale Fantoli, Rivka Feldhay, Maurice A. Finocchiaro, Ernan McMullin, and Guido Morpurgo-Tagliabue, among others. Certain views expressed in this book accept Eileen Reeves's conclusions about Ludovico Cigoli in her *Painting the Heavens* (although I think that morphologically Galileo's thought is much more closely related to Bernini's). Precise references to all these writers' works can be found in the bibliography at the end of this book.

The literature on Galileo is vast, and I have surely overlooked many books and essays. I should like to make particular mention of two works that came to my attention when I had already turned in this manuscript to the editors. One is Horst Bredekamp's *Galileo Der Künstler: Der Mond. Die Sonne. Die Hand*, which contains some fascinating revelations concerning Galileo's manual involvement in fashioning pictures of the heavens. The other is Dr. Giorgio Strano's *Il Telescopio di Galileo: lo strumento che ha cambiato il mondo*, which I have so far been unable to procure.

Where I have cited translations of various texts, the translators are credited in the notes. The translation from Ariosto's *Orlando furioso* is by David Slavitt. The translations from

Dante, and the English version of the letters in colloquial Italian relative to the 1633 trial, are by myself.

FOR HELP ALONG the way, my heartfelt thanks go to Barbara Dudley, for enabling me to achieve a secure internet connection at a time when this was difficult in our rather secluded village; to David Slavitt, for his delightful, as-yet-unpublished Ariosto translation; to Nino Mendolia, for his work on the illustrations; to Prof. Norman Derby, for his clarification of several problems in astronomy; to Jim Mosher and Tim Pope, for their kind permission to use their excellent ray diagrams, and to Jim Mosher in particular for his explanations of various points relative to Galilean optics; to Dr. Giorgio Strano, of the Istituto e Museo della Storia delle Scienze, in Florence, for consenting to be interviewed at the institute about his research into Galileo's telescope during the summer of 2007, before the publication of his research; to Prof. Ricardo Nirenberg, that extraordinary polymath, for our many conversations, and for carefully reading the manuscript and notifying me about a number of errors; to Prof. Glen Van Brummelen, also for reading the manuscript from the standpoint of the history of mathematics; to Prof. William R. Shea, of the University of Padua; and to Oceana Wilson, of the Bennington College Library, Alessandra Lenzi, of the Biblioteca dell'Istituto e del Museo della Storia delle Scienze, and Mary DiAngelo, of the Schow Science Library of Williams College, for their assistance in procuring me many books and papers.

The Summons

G alileo Galilei was living in a modest house in Arcetri, in the hills south of Florence, when he learned, on October 1, 1632, that he had been summoned to Rome to be examined by the Inquisition. Arcetri now lies in the heart of what you might call *la Firenze bene*—upper-class suburban Florence. By chance I spent two years in the same area and would often walk my baby son through the village in his stroller, thinking sometimes about Galileo, who wrote a great work about physics here, or about the poet Eugenio Montale, whose ashes are preserved in a cemetery nearby. The southern Florentine countryside is a soothing spectacle, and never more so than at this season—harvest time. Every hilltop affords a vista of rolling horizon, with an occasional disclosure of the dome of the Cathedral of Santa Maria del Fiore, and from among the tender gray-green of the olive orchards and the richer green of the vineyards rise the bell towers of ancient monastic establishments, such as the Charterhouse of Galluzzo and the Convent of the Stigmatine Sisters (whose

playground my little son particularly favored). In another Franciscan convent, San Matteo, only minutes away by foot, Galileo's elder daughter, Suor Maria Celeste, then thirty-two, served as a nun. It is hard to stroll through this landscape remembering the Inquisition's summons without being struck by the contrast between the tranquil loveliness of the place and the extreme anguish that the news provoked in Galileo, his family, and his friends.

The immediate cause of the summons was the failure of the grand duke of Tuscany to get the matter dismissed or at least transferred to the office of the Roman Inquisition in Florence. But the underlying cause was a book that Galileo had published in May of 1632, the *Dialogue Concerning the Two Chief World Systems*, in which subtle arguments in favor of the geocentric, or earth-centered, and the heliocentric, or sun-centered, cosmologies were played against each other. Organized somewhat like a Platonic dialogue, it featured lucid explanations, crisp Italian prose, and entertaining arguments between three well-defined characters. The Roman Catholic Church at this time had no dogma regarding the structure of the universe, but of course it had a theology, and the Council of Trent, which ended in 1563, had confirmed that the Bible, which stated that the sun rose in the east and set in the west, was not to be freely interpreted by laymen. Galileo's book had been approved by the Vatican censor, but a number of clerics, especially among the Jesuits, still felt that it inclined too heavily toward Copernicus, the great heliocentric astronomer. In the summer of 1632, a commission was empanelled to study the book, and by late September the work was provisionally suppressed. Pope Urban VIII Barberini was an old friend and admirer of Galileo's, but he was nonetheless known to be

incensed by the Tuscan scientist's alleged infringement of restrictions as to what might be legitimately argued, as well as by other audacious features of the *Dialogue*. It was beneath the pope to summon Galileo himself, and he apparently had trouble finding other churchmen willing to do so. In the end, it fell to his brother Antonio, a morbidly timid Capuchin monk whom Urban had all but forced to become a cardinal, to notify Galileo that his presence was expected in Rome.

Galileo was then sixty-nine years old. The turning point in his life had come in late 1609, when he was forty-five, a respected mathematician with a sideline in engineering. That autumn, he had rapidly improved upon a crude spyglass circulating throughout Europe, and within months, from atop his house in Padua, had seen the rugged surface of the moon, the phases of Venus, Jupiter's satellites, and myriads of shimmering stars previously unknown to humanity. For years he had mulled over his research in mechanics and astronomy and leaned toward the Copernican theory of the heavens. He had questioned the validity of the two competing hypotheses, the Ptolemaic, or geocentric, and the Tychonic, or geoheliocentric (which seemed so preposterous, so illogical, that he never took it very seriously). But after his telescopic observations he made up his mind—there was no doubt that the sun stood at the center of the cosmos. He did not publicly proclaim this fact, however. Only here and there, in his private correspondence or in certain published works, such as his *Letters on Sunspots* of 1613 or *The Assayer* of 1623, did he frankly state that Copernicus was right about the sun and his opponents wrong.

In 1616, the Congregation of the Holy Office, otherwise known as the Roman Inquisition, had banned any advocacy

or teaching of the Copernican position, on the grounds that it conflicted with Scripture. Copernicus's exposition of his system, *On the Revolutions of the Heavenly Spheres*, of 1543, was to be "corrected" by the Vatican Index, that is, emended to indicate its purely hypothetical status. This notion of heliocentrism as a sort of counterfactual, mathematical fiction was evolving dramatically then, and as time went on Galileo interpreted it according to his own lights. Following the publication of *The Assayer*, and encouraged by the election of a new, friendly, humanist pope—Urban VIII Barberini—Galileo decided to write a magnum opus weighing the mutually opposing arguments for the geocentric and heliocentric theories. He was giddily spurred on in this task by a former student named Giovanni Ciampoli, now the pope's correspondence secretary and an influential backroom schemer, who had surprisingly little sense of the danger of the undertaking. Galileo resolved to keep his *Dialogue* on the plane of mere supposition, but, considering his deep scientific engagement on behalf of Copernicanism, this was scarcely possible.

In an orderly, logical world, Galileo, on discovering the truth about the heavens in 1609–12, would have published a tract defending Copernicanism, and this tract would have either won the Church over or provoked a scientific and theological rebuttal from the Collegio Romano, the great Jesuit scholarly institution in Rome. But as it turned out, the Collegio Romano enthusiastically confirmed almost all of Galileo's telescopic discoveries. In 1616, Cardinal Robert Bellarmine, the most gifted Catholic theologian and polemicist of the age and the director of the Roman Inquisition, merely warned Galileo not to go drawing any perilous conclusions from his

observations. Bellarmine was a very polite man, and the admonition may have been rather too gentlemanly to produce the desired effect. And though Galileo was investigated by the Inquisition at that time, he was never deposed or indicted.* Thus it was not until the publication of the *Dialogue*, in 1632, that the scientist fell fatally afoul of Rome.

If one has not read the *Dialogue*, one might suppose it consists of a grand summary of everything that Galileo had *seen* and *proved* about the heavens, mostly thanks to his improved telescope. That is not the case. The *Dialogue* is a fascinating work, but less an empirical than a theoretical, and in fact speculative, defense of Copernicanism. It is based largely on physics and celestial mechanics, and one of its most important arguments—Galileo's belief that the sun's gravitational field causes the tides—is wrong. There are other, valid explanations in the book, relating to the mathematical description of falling bodies, the confutation of popular anti-Copernican arguments, and why a ball tossed in the air as the earth rotates does not descend some distance to the west. To anyone ignorant of such matters—and almost everyone was in 1632—this material makes lucid and thrilling reading. But anyone looking for a step-by-step statement of the logical connection between what Galileo had seen through the telescope and the setup of the solar system would have been disappointed.

* It is sometimes said that Galileo faced two trials, the first in 1615–16 and the second in 1633. This is true in Italian but not English, because the word *processo* is not exactly cognate with the English "trial." A *processo* may begin, not from the moment a tribunal is convened, but as soon as a prosecutorial investigation of a suspect is initiated.

What one must bear in mind, then, is that in 1632–33 the Inquisition was not out to disprove Galileo's science or even to pinpoint all the passages of the *Dialogue* that conflicted with Scripture. The question was actually quite different. Christians had been warned not to teach Copernicanism: had Galileo heeded that warning or defied it? All the rest was beside the point.

There is thus an enormous distinction between the proximate cause of the Galileo affair, the great religion-science clash of 1633 that in some form has persisted into our time, and what really brought it about. The proximate cause was the publication of a work that appeared to flout a prohibition against Copernicanism. What brought it about was the evolution of a new science, an entire attitude toward experience that emphasized the evidence of the senses, especially vision, as opposed to metaphysics and the study of venerated texts. Over and over in his writings, Galileo expresses an almost petulant amazement that people will not *see* what nature so beautifully displays, nor search for the transcendent forms governing that beauty. In *The Assayer* this plea takes poetic flight. Perhaps because his father was a musician, and as a child he was trained as a lutenist, Galileo tells a parable about a lover of birds and birdsong who desires to look into all the ways that musical pitches are produced, whether by birds, people, or insects. This man travels the world collecting flutes, violins, even squeaky hinges and crystal goblets that produce a tone when filled with water. He examines wasps, mosquitoes, trumpets, organs, fifes, jews' harps, until at length he captures a cicada. Galileo tells us that the man, scrutinizing the cicada,

failed to diminish its strident noise either by closing its mouth or stopping its wings, yet he could not see it move the scales that covered its body, or any other thing. At last he lifted up the armor of its chest and there he saw some thin hard ligaments beneath; thinking the sound might come from their vibration, he decided to break them in order to silence it. But nothing happened until his needle drove too deep, and transfixing the creature he took away its life with its voice, so that he was still unable to determine whether the song had originated in those ligaments. . . . I could illustrate with many more examples Nature's bounty in producing her effects, as she employs means we could never think of without our senses and our experiences to teach them to us—and sometimes even these are insufficient to remedy our lack of understanding. . . . The difficulty of comprehending how the cicada forms its song even when we have it right in our hands ought to be more than enough to excuse us for not knowing how comets are formed at . . . immense distances.

If there is anything that Galileo championed it was curiosity— immense, boundless curiosity.

There are many valid ways of presenting Galileo's thought. In a short book like this, it seems to me most useful to offer, as a background to the trial of 1633, a concise account of how he brought optics into alignment with astronomy in order to see and know the heavens. Because his telescopic observations made him a convert to Copernicanism, they formed the substantive reason for his divergence from the thought patterns prescribed by the Church, and it was this divergence, between

1609 and 1632, that caused the clash between religion and science. I also devote more attention to Galileo's relations with painters and his pictorial representation of astronomical facts, and to the Church's use of the frescoed dome as an alternative celestial vision, than might ordinarily be expected. In part this reflects a personal interest; in part it embodies my belief that the sense of vision, of *seeing* as opposed to the *refusal to see*, was enormously important to Galileo. I mean this more in the figurative than the literal sense, yet Galileo is reported to have drawn well, and he was friendly with a number of artists, especially Ludovico Cigoli, who ardently supported him and with whom he corresponded regularly. Most of those who have written about his relations with artists, such as Erwin Panofsky and Eileen Reeves, have dwelt on this connection with Cigoli, which was certainly paramount from a biographical standpoint. I do not think it should obscure the fact that Galileo's conception of space has much more in common with that of Bernini, whom I do not believe he ever met.

Galileo had a good grasp of Latin and some Greek and reportedly knew by heart many long passages penned by Virgil, Ovid, Dante, Petrarch, and Ariosto. He wrote poetry, a play, voluminous letters, and numerous essays and books on physics and astronomy. When he was upset, his syntax became turgid. Otherwise he wrote a straightforward, cogent Italian, by turns playfully humorous and bitingly ironic. His prose is so free of Latinisms and ornaments, so similar to the best of modern science writing, that it has been included in textbook surveys of the history of prose composition for Italian high schools. Galileo sometimes tackles with ordinary language scientific problems that contemporary physicists would approach with algebraic symbols: he is one of the last scien-

tists who also seems something of an artist (Milton called him "the Tuscan artist"). As such, he makes frequent use of metaphor, both consciously and unconsciously. One suspects that this use of metaphor got him into trouble.

Of course, science doesn't need metaphor, and though metaphors may be embedded in scientific terms, such as "wheel" in *kúklos* (Greek for "circle"), or "overshooting" in *huperbolé*, or "comparison" in *parabolé*, the geometry would likely have been the same had they had other names. But often, as in the case of Galileo, Darwin, and Freud, scientists or their supporters enlist a set of durable metaphors that, however dispensable for science itself, somehow fascinate the public and inflame its religious segment. Such has been the case with "plurality of worlds," "laws of nature," "descent of man," "survival of the fittest," and "Oedipus complex," heuristic terms that have caught on, perhaps lamentably, as catchwords. Often it is nascent sciences, or those in periods of theory shift, that rely for a while on metaphor. (Nowadays researchers in cognitive psychology may speak of information being "encoded" or "indexed" in a subject's memory.) Sometimes Galileo deliberately enlists metaphors, such as "the book of nature," or the "repugnance" of a body to being pulled in a certain direction. At other times he violates unspoken metaphorical conventions, and this can happen almost by accident. When one denies, as he did, that the moon is a crystalline sphere, and asserts that it is a spiky, pitted ball, one appears to be making a factual statement, but that is not all one is doing: one is also offending the Marian poetic vision that associates the Madonna with the unsullied moon. I have tried to make the metaphorical abrasion between Galileo and his clerical opponents clearer than it might appear in a strictly scientific context.

All the books on Galileo's trial that I have read present his chief antagonist, the pope, in his role as prelate, theologian, political leader, and string-puller of the Inquisition. The portrait is never flattering, even when penned by conservative Catholics. But the annals of art present a quite different picture of Urban VIII Barberini, and I have tried to merge it with the one offered by the history of science. It will come as no surprise to anyone with an interest in the Baroque style that Maffeo Barberini was a great patron of Italian sculpture and architecture. A highly cultivated man, he was also a keen follower of developments in science, and that he should have been the one to silence Galileo in the trial of 1633 was a calamity for him as well as for the Church.

1

Galileo Galilei and
Maffeo Barberini

alileo Galilei was born in Pisa in 1564, to the family of
a professional musician who later found employment
in Florence. According to his first biographer, Vincenzo Viviani, he was fond of drawing and originally wished to
become a painter. His father directed him toward a career in
medicine, but after a few months at the University of Pisa the
young man abandoned his medical studies and enrolled in the
faculty of mathematics. This field then included a hodgepodge
of subjects related to number and magnitude, such as optics
and fortification, but for Galileo, mathematics meant Euclid.
The preference was prophetic. Euclidean geometry, more than
any other subject, teaches one how to think—to think logically,
deductively: it is the essence, the bread of thought. At the same
time, in the absence of algebra, with which Galileo was unac-
quainted, Euclidean geometry provided the readiest tool with
which to grasp and conceive of spacial relationships. Galileo
soon proved himself an agile geometer, and was granted the
mathematics chair first at Pisa, from 1589 to 1592, and then at

Padua, from 1592 to 1610. And it was geometry that he began to apply, as dexterously as anyone could, to such subjects as the velocity of falling and rolling bodies.

While teaching at Pisa, Galileo began to study Archimedes, many of whose surviving writings had been republished in Basel in 1544. What he learned from the ancient Syracusan was a way of conceptualizing the world in terms of intellectual machines or models, such as the lever and the balance. This way of thinking accorded well with his love of geometry and his essentially Platonic cast of mind. He saw the world in terms of Euclidean forms that not only held universally true but also were quantifiable, in the sense that extensions or areas may be proved greater or lesser than others. In Viviani's biography one reads of an experiment in which Galileo drops unequal weights off the Tower of Pisa in order to prove the equality of their velocity of fall, but the story is almost certainly untrue: while at Pisa he wrote (but did not publish) a manuscript on physics, titled *De motu* (*On Motion*), in which he argued something altogether different about falling bodies, which suggests that he had not made the experiment. Whether during this period he even conducted experiments, and how numerous they may have been, is still a matter of scholarly conjecture. Early on, he worked on what became his law of fall, which states that in a vacuum all bodies are uniformly accelerated, and that the distance fallen is proportional to the square of the elapsed time; but measuring fall was impossible then, as— leaving aside the impossibility of creating a vacuum—bodies fall too fast, and no means of retarding them by counter-weights had yet been contrived. Galileo devised geometrical thought-experiments extrapolated from actual experiments, which he carried out with metal balls on inclined planes.

It is hard for us today to imagine two oppressive conditions under which the young Galileo worked. One was the dead hand of Aristotelian dynamics, universally taught in the Italian universities of the late sixteenth century, which had no conception of inertia, force, or velocity. This sort of premodern science, which was not based on the formation of verifiable hypotheses, offended Galileo's sense of how to acquire a valid understanding of nature. The other condition was the belief that human knowledge was a fixed rather than an expanding sum, most truths of natural philosophy having been ascertained by the ancients. According to this settled view, the wise researcher could do no better in solving a scientific question than to consult authoritative texts. As early as 1589, Galileo began to inveigh against Aristotelian dynamics, as he would later do at Florence against Aristotelian astronomy. His opponents thought him a seeker after specious novelties, a flashy self-advertiser. But he was a natural polemicist, with a taste for blood, and had he not been so, he would not have survived intellectually.

Until the telescopic discoveries that wholly changed his life, Galileo subsisted on a small professor's salary, yet his family expenses were onerous. He tutored students in mathematics and designed instruments, with the help of artisans and instrument-makers. At age twenty-two, he proposed an improved version of the hydrostatic or "Archimedean" balance for weighing precious metals in air and water to determine their specific gravity. After moving to Padua, he designed, at the behest of the Venetian Senate, a one-horse pump, which has never physically been recovered but was probably helical like Archimedes' famous propeller, its blades bearing the water upward. Later, in 1598, he invented a sector, a sort of

compass used primarily in gunnery, and for a fee he instructed purchasers in its use. About five years later, he devised a water thermometer whereby a flask the size of an egg, with a reed-like neck marked off in arbitrary degrees, was filled with water, and depending on the local heat the water would rise to a specific degree.

Galileo's telescope is his most important invention, but his optical investigations did not end with its large-scale production. He also constructed a microscope, which led to some entomological illustration though never to extensive research, and his papers contain a "theory of the concave spherical mirror," complete with ray diagram, though he did not get around to constructing a reflecting telescope. He also designed a device to enable mariners to use the moons of Jupiter to determine longitude, and long after the Inquisition had silenced him on the subject of astronomy, when he was virtually blind, he bequeathed to humanity his second-greatest invention, the pendulum clock. It was based on his discovery of the isochronism of the pendulum—the fact (true within certain limitations) that even as the amplitude of the arc decreases over time, the period of each swing remains the same.

Galileo's writings reveal a novel ability to conjoin mathematical with aesthetic insights. Conceptually brilliant yet unacademic, he blends as only a humanist could the *esprit de géométrie* with the *esprit de finesse*. His own poetry is of the "occasional" and dilettantish variety, more mannered and piebald than what he professed to admire, but composing it surely helped to develop his fluency at prose composition. With respect to the great Italians, he regarded Ariosto as unequalled among Renaissance poets, delivered two erudite lectures on Dante's *Inferno* to the Florentine Academy, and

wrote an unsparingly critical essay on Tasso's *Jerusalem Deliv-ered*. He was alert to tropes in all their manifestations, and if he showed little patience with Tasso, this was in part because of the intrusion of allegory in Tasso's epic narrative, which he felt destroyed the unity of treatment. "Poetic fiction and fables should be taken allegorically only when no shadow of strain can be discerned in such an interpretation," he wrote. "Other-wise . . . it is like a work of art in which the perspective is forced and which, if seen from the wrong viewpoint, will appear absurd and distorted." The optical effect to which Galileo refers is called anamorphosis, and it is a historical irony, to which I will return, that this sort of distortion came to bedevil the very form of art, the frescoed dome, that the Church would increasingly favor with the aim (among others) of countering the spread of the Galilean world picture. Galileo himself is thought to have taught perspective for a while, and doubtless had complete command of the elaborate treatises on the subject that had recently been published; it was presum-ably on the strength of these quasi-mathematical skills that he was elected in 1613 to the Florentine Academy of Design, a glorified studio-school linked to the needs of the Medici fam-ily. His writings are studded with confident judgments on art, and he was associated with a circle of Tuscan painters, among them the aforementioned Cigoli, who were reacting against the excesses of Mannerism, which Galileo also disliked. One feels that deep down Galileo had an almost Platonic faith in graphic representation if handled with the requisite straight-forwardness and homogeneity of execution: it could probe toward general, underlying structures.

It has been noted that Galileo's critical meditations on poetry, painting, and sculpture pertain to his views on the sci-

entific method: it might even be said that they are a sort of
highly colored expression of it. To a degree, this unity reflects
not only his own temperament but also the mutual proximity
of the arts and sciences at the end of the Renaissance. Most
interestingly, Galileo did not hesitate to ventilate his fierce
contempt for all composite imagery. This included pictorial
intarsia, or the inlaying of wood veneers to create a picture;
the Aztec featherwork then common in cabinets of curiosities;
and any paintings in which some parts were not in keeping
with others. He loathed Giuseppe Arcimboldo, the painter of
fanciful heads assembled from vegetables, books, and other
oddments. In his literary criticism, he vehemently objected to
Tasso's interruption of the ongoing sweep of a lyrical tale in
order to engage in some dispensable flight of fancy. What
Galileo looks for in art is breadth: some form of treatment that
can encompass an entire picture or story or statue with one
generalizing vision and manner. The same is true of his
approach to nature itself. There is no point in so-called occult
or particularistic explanations of phenomena. If you have
found that something is true under repeatable conditions,
there is no reason not to assume that everything analogous
will behave in the same way. If you have a choice, moreover,
between an elaborate explanation of something and a simple
one, choose the latter: not only is it more economical, but perhaps it can become the initial outline of a general principle.

    There are two underlying assumptions here, so essential
that we nowadays take them for granted, but they were novel
at the time and exhilarating. The first is that nature is uniform, not subject to different sets of laws at different times or
places. The second, which is a kind of corollary, is that a few
well-chosen experiments suffice to prove a law of nature—a

scientist does not have to proceed indefinitely. Sometimes Galileo appears to work on a train of repetitive experiments, but really he is refining the same research design. If one reads the work of contemporary Galilean specialists discussing and analyzing his experiments in detail, one can follow him reformulating his law of fall or his theory of hydrostatics. As William R. Shea has pointed out, it was crucially in his experiments with floating bodies that he grew aware of the need to create conditions that could be held equal while he altered one variable at a time (a concept one could almost miss in his famous parable about the cicada's song).

Given this mindset, it was inevitable that Galileo's investigations would eventually collide with the Aristotelian conception of the universe, for the following reason if no other: Aristotle's cosmos did not recognize the uniformity of nature, positing one set of laws for the heavens and another for the sublunary sphere—that is, everything beneath the moon. The Aristotelian system is often described as a sort of onion, with the earth at the center surrounded by the three celestial spheres of the moon, Venus, and Mercury, followed by the sphere of the sun; beyond the sphere of the sun lay those of Mars, Jupiter, and Saturn, the whole being enclosed by the sphere of the fixed stars. These spheres are conceived of as circular orbits, and the heavenly orbs as perfect and incorruptible, resembling an unearthly sort of crystal. Heavenly motion is originated by divine love in the outermost sphere, and is circular because it is perfect; terrestrial motion, by contrast, is naturally rectilinear, since on earth bodies fall along a plumbline and fire rises straight upward. In point of fact, this system was emended over the centuries to account for a number of celestial observations that flagrantly contradicted it; most importantly, around

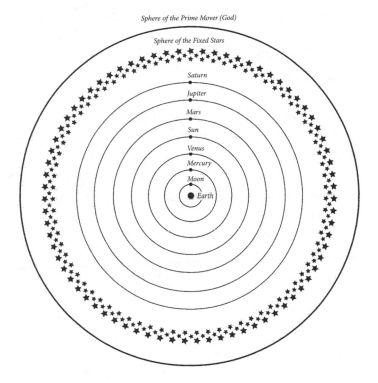

Sphere of the Prime Mover (God)

Sphere of the Fixed Stars

Saturn
Jupiter
Mars
Sun
Venus
Mercury
Moon
Earth

Simplified Diagram of the Aristotelian World System

AD 150 Ptolemy reformed Aristotle's cosmos in the vast work known as the *Almagest.* In his own writing Galileo tends casually to conflate Aristotle's and Ptolemy's systems, but since he accepts neither, he is much less concerned with demolishing their fine points than with defending the Copernican setup. Copernicus's sun-centered cosmos had already acquired many devoted followers by the 1590s, when Galileo began to embrace it, but he felt no more obliged to defend its many and elaborate computations than he did to attack Ptolemy's subtleties. What did strike him as a major scientific problem was

the fact that bodies appeared to be moving in a circular fashion in the heavens, whereas they fell straight down on earth. Galileo knew that this dichotomy was merely apparent, but how could he prove it?

As noted earlier, Galileo's first observations with the telescope, conducted between 1609 and 1612, convinced him of the overwhelming likelihood that the Copernican theory of the cosmos was correct. He championed this theory explicitly if fleetingly in the *Letters on Sunspots*, of 1613, and again in *The Assayer*, of 1623. Thus it must be remembered that there is a direct logical connection between the telescopic observations, to which the central portion of this book is devoted, and the trial of 1633. He was not content, however, with three or four key observations that tended to demonstrate the validity of the heliocentric cosmos: he wanted at least the outline of a theory of celestial mechanics decisively able to refute the old Aristotelian duality and to explain planetary motion, and this theory he put into the *Dialogue*. Using the telescope, he perceived with his eyes that the vast machine we call our solar system did not function as most people had imagined for about two thousand years; in his theoretical writing, he would try to understand with mathematics more precisely how it actually did function. Yet such an attempt did not, by itself, offend the pope or the Holy Office; indeed, the pope favored the endeavor. Galileo's sin lay in his inability or implicit refusal to explain the cosmos suppositionally, as if his research were merely a fascinating mathematical exercise; because he was not toying with Copernicanism, he was passionately supporting it, deploying empirical evidence underpinned by an all-embracing geometrical vision.

It has sometimes been claimed that Galileo, though an

excellent mathematician, had no interest in advancing pure mathematics in the sense that Descartes and Fermat did. This argument may be true, in that the ability to work out an economical demonstration in Euclidean geometry held an abiding prestige, even a sort of glamour, for Galileo and his associates, and he did not move significantly beyond it. But one must also remember that Galileo did not know about algebra, though it had been used in the West for some decades, and the mathematical culture may not have been ripe for a great leap forward. He came close to making remarkable contributions, however. Among his discoveries was a treatment in the *Dialogue Concerning the Two Chief World Systems* of the acceleration of falling bodies that outlines, in prose and with a geometrical figure, the mathematical concept of integration. Another was a discussion in the *Dialogues Concerning Two New Sciences*, the great work on mechanics that he published in 1638, after the Church had silenced him on the subject of astronomy, of a paradox that many students confront in a somewhat different form in college today. This is the fact, in arithmetic, that there are as many squares as there are integers, since the totality of both squares and integers is infinite. But since squares are themselves integers, it would seem that the part, contrary to Euclid's fifth "common notion," may be equal to the whole, in the sense of having the same number of elements, or, conversely, that one infinity may be greater than another. Galileo declined to investigate this paradox, suggesting only that terms like "longer," "smaller," and "equal" had no place (that is, required mathematical redefinition) when applied to infinite quantities. Given his interest in infinity, he has been reproached by some historians for not pursuing the question further, as if he lived in the nineteenth century and

had recourse to set theory and some notion of the classification of sets. But it would have been hard to solve nineteenth-century problems with seventeenth-century tools.

Residing in Padua as an impecunious professor of mathematics, a subject then held in low esteem, Galileo grew attached to a young courtesan named Marina Gamba. Between 1600 and 1606 she bore him three children, Virginia, Livia Antonia, and Vincenzio, though his name does not appear on any of their baptismal records. (Marina's low station in life elicited resentment on the part of Giulia Galilei, Galileo's mother, who lived with him after his father died in 1591.) When Galileo left for Florence in 1610, he took Virginia with him, leaving the other two children in Marina's care, and in 1613 she married another man, maintaining reasonably cordial relations with Galileo. For want of dowries, both girls eventually became nuns at the convent of San Matteo; the fond letters that Virginia wrote to her father, under the name of Suor Maria Celeste, were later published and translated. She died suddenly of a fever at age thirty-three, causing him great sorrow.

Galileo's money problems were compounded by illness, which began to assail him in early middle age. He was beset by a host of complaints, of which hernias, arthritis, and bad vision appear to have been the worst, and which he complains about frequently in his correspondence. In later life he became partially, then completely, blind. His medical condition must have contributed to his often dyspeptic frame of mind. It is noteworthy that at age fifty-five he made a pilgrimage to the Santa Casa, at Loreto. The Santa Casa is the Blessed Virgin's house, which, we are told, had been borne through the air from Palestine by angels in order to escape demolition by the Turks. The usual reason to visit the Santa Casa was to pray for

a miraculous medical cure, much as people do today at Lourdes. Galileo's religious faith was never in question.

It is a fact of intellectual life that people of high mathematical ability tend to feel quietly superior to everyone else. This does not necessarily mean that they are conceited or snobbish, but if challenged or obstructed by others of lesser capacity, they may abruptly show exasperation or scorn. Galileo had just such a superbly gifted mind, and as a social being he has often been put down as arrogant, irascible, and belligerent. There is ample evidence for such a view, as the list of his public quarrels is a long one. Today this may surprise us: though all these disputes are of biographical interest, only three—those with Christopher Scheiner, Orazio Grassi, and Francesco Ingoli—still prompt our intellectual curiosity, and we may fail to understand the reason for his feeling threatened or vexed by his many petty antagonists. Galileo has been said to suffer from a persecution complex; he has been called paranoid (surely a misuse of a medical term); and judging by his correspondence, he did seem to lead his life with a standing air of grievance. But here a note of caution is in order. There was no reliable intellectual property protection at this time, and a number of unscrupulous people tried to steal Galileo's ideas or to impugn his parentage of them. Many others, simply foolish or unconvinced, attacked his most profound theories or, worse, tried to get them banned. If these had been trivial notions, it would have been one thing; but they were the ideas on which modern civilization rests. In their letters, Galileo's friends, none of whom seems choleric or paranoid, often express outrage at his antagonists' envy, and they repeatedly warn him of the machinations directed against him. His life at times had a genuinely *noir*ish cast. With intermittent though

mounting intensity after the Inquisition's anti-Copernican edict of 1616, Galileo's research was subject to a widely prejudicial assessment. Rarely evaluated on its scientific merits, it was more often repudiated out of hand. Just as criminals are sometimes framed by the police, or hypochondriacs fall terribly ill, so Galileo represents that not uncommon phenomenon, the temperamentally aggrieved person who actually finds himself persecuted by the authorities because he sees what they cannot or will not see.

Finally, one must take the man's social environment into account. This is an epoch in which papal legates ransack entire cities, in which the nobility bribe and blackmail one another, in which painters poison their rivals, in which cardinals commission the theft of coveted works of art. A man is ranked at least as much by the standing of his enemies as by that of his friends, and everyone has somebody he loves to insult. To insult with flair, with humor, is a fine art, and it is one at which Galileo frequently excells.

GALILEO'S GREAT ANTAGONIST at the trial of 1633, his former friend and admirer, was Urban VIII Barberini: poet, humanist, patron of the arts, and supporter of scientific research. But Maffeo Barberini was also a warrior, a fortifier of the papal domains, and staunch upholder of the Tridentine decrees on which the Counter-Reformation was based. We see him in Bernini's marble portrait of 1637–38, the face narrow, the eyes deep-set, the nose and mouth like delicate sensory instruments, as befit an aesthete. His gaze is adrift. He seems caught in the midst of conversation, listening to someone, formulating some deft reply. It is a handsome, finely articulated visage, intelligent, a trifle devious, but not

especially commanding. An intellectual, not a charismatic presence.

Maffeo Barberini's pontificate lasted from 1623 until his death in 1644, and during this period he refashioned parts of Rome in the manner of a Renaissance pope. He also tried to make the political and military weight of the Papal States felt on the European stage, a task to which he proved unequal, and by the early 1630s he appeared embattled, inflexible, and given to impetuous decisions; but he was not always so. Scion of a great house of Florentine merchant princes, quick-witted, vain, cunning, and intemperate, he began his career as a scholar-priest, a master of Latin and Greek versification, a lover of nature, a serious follower of developments in art and architecture and of research at the Jesuit Collegio Romano. If he figures in the Galileo biographies mostly as a sort of ogre, a persecutor of the Tuscan scientist, his ecclesiastical biographers devote long passages to a different quarrel, and a tempestuous one at that.

When his predecessor, Paul V Borghese, opted for Carlo Maderno's redesign of Saint Peter's, which proposed to elongate the nave at the expense of Michelangelo's façade, Maffeo Barberini, then a cardinal, bitterly opposed the desecration. It was about as close to fisticuffs as the college of cardinals ever came. To his face, Maffeo told Paul V that should he eliminate Michelangelo's design, the next pope would tear down the new structure and put everything back to rights, at which Paul V shot back that it would be built to last forever. The next pope was Gregory XV Ludovisi, who was old and ailing and lasted only a few years; then came Maffeo himself, and though he never managed to replace Maderno's rather tame façade with Michelangelo's original and vigorous one, much else in

"Michelangelo presents his model of St. Peter's to Pope Julius II,"
by Domenico Cresti da Passignano

Passignano, a colleague of Cigoli, is credited with being the first to observe the sunspots through a telescope; Cigoli forwarded their discoveries to Galileo. In this picture Passignano implicitly criticizes the papacy for its failure to realize Michelangelo's design, a decision of Pope Paul V's that also infuriated Urban VIII. Urban also reportedly disagreed with Paul's promulgation of the 1616 edict against the teaching of Copernicanism, yet he later persecuted Galileo for infringing it.

his early career confirmed his desire to make the Church a sponsor of the finest in the arts and sciences.

Maffeo Barberini wrote books of poetry in both Italian and Latin, which together were reprinted more than twenty times during his pontificate. (Bernini illustrated an edition of 1631.) Maffeo's poetry, like much conventional verse of the period, evokes the moods of nature, the passage of time, the lives of the saints, and the intimations of mortality, but it is also bedeviled by a Baroque sense of theatrical illusion so intense as to become a metaphor for universal vanity. We are all understudies for ourselves, he seems to suggest; to become one with our mortal roles would be to show insufferable pride. Maffeo nursed paradoxical emotions toward creativity, his own included, as though his admiration for human achievement were adulterated by an almost toxic sense of futility: a certain forlornness lurked behind that mask of robust ambition. One of his pet projects was his own tomb, faced in many-colored marbles, which Bernini built slowly during his lifetime in Saint Peter's.

Urban VIII Barberini is generally regarded as one of the most shamelessly nepotistic of pontiffs. No sooner had he been elected to the office than he began to place his relations in positions of power. Within six weeks his nephew Francesco, an intelligent twenty-six-year-old who later enjoyed Galileo's friendship, became a cardinal. The position of "pope's nephew" was a formal title, universally honored; for ages every pope had had a "nephew," sometimes not genuine kin, who served as a virtual secretary of state. Yet temperamentally Francesco was poles apart from Maffeo. He had an apartment in the Vatican and handled much of the pope's diplomatic agenda, but he did so with unusual circumspection, as though submitting with resignation to inclement weather. The pope's

brother Antonio and a nephew, Taddeo, aged only nineteen, soon followed Francesco into the curia.

Maffeo also exploited his position as a source of lucre. Ecclesiastical posts not granted to family members were simply auctioned off; the distinction between the pope's personal purse and the Vatican treasury grew progressively fictive; and the Barberinis exerted their influence to acquire the palaces or landed estates of several great Roman families who found themselves short of funds. It is necessary, however, as with Galileo, to see such behavior in context. Early modern Italy (like much of southern Italy today) was dominated by an ethic whereby family ties trumped ordinary ethical considerations. No man could possibly rise to a position of power without indebtedness to a number of people in his clan, and rewarding them became a paramount obligation. The Barberinis were hardly alone in acting as they did; had they not been ruthless, like the Orsinis and the Borgheses and the Colonnas and the other great families of the period, they would have met the fate of the Gonzagas, in Mantua, who let their guard down and saw their city sacked and their palace looted a few years later. The ceiling of one of the chambers of the duke of Gonzaga bore a gilded design of a labyrinth with the phrase *Forse che sí, forse che no* running repeatedly in all directions. It meant "Maybe yes, maybe no," like a memento mori, and as Maffeo Barberini knew, the only measure likely to gain one a positive fate in this cloak-and-dagger world was surrounding oneself with blood kin and building a mighty war-chest. Not one to take chances, Maffeo made a point of fortifying Rome and the entire papal principality. Yet even by the standards of their day the Barberinis were regarded as excessively greedy. As a popular pasquinade had it, "What the barbarians didn't steal, the Barberinis did."

Urban VIII's art and architectural patronage served multiple functions. It satisfied his excellent taste, glorified himself and his dynasty, and maintained the prestige and grandeur of Rome in the eyes of foreign envoys, Catholic prelates, and pilgrims, who brought in cash. Both as a bishop and a cardinal, Maffeo acquired splendid pictures by Raphael, Andrea del Sarto, Caravaggio, and many others. As a pope he also undertook the decoration of the interior of Saint Peter's, the construction of a family chapel in Sant'Andrea della Valle, and the rebuilding of the church of Santa Bibiana, a matter of little account had it not been entrusted to Bernini.

Gian Lorenzo Bernini, who lived from 1598 to 1680, was recognized from childhood as a genius and largely defined the style we now call the Baroque. Maffeo had long yearned to gain Bernini's services for himself. At first he lacked the means to do so, but as pontiff he virtually monopolized Bernini and so became the greatest patron of sculpture Rome had ever known. Within a year of his election to the papacy, Urban VIII placed Bernini in charge of his entire artistic program, enabling the sculptor-architect, then still in his twenties, to begin to refashion the visual impact of the city. Bernini also became his intimate friend, closer to him than any councilor or member of the Barberini clan. The story is told, and there is no reason to doubt it, that when Bernini was sculpting his *David*, for whose face he used his own likeness, Maffeo would hold the mirror for him, an act of sublime (if not almost servile) admiration. Often in the evenings Maffeo would welcome Bernini into his apartment in the Vatican, where the pair would discuss their vision of Rome until the pope drifted into slumber. Then Bernini would quietly close the windows, draw the curtains, and depart.

We must look closely at Bernini because his early work is the concrete embodiment of Maffeo's own sensibility. Yet when we examine this connection, we discover a bewildering irony—that Maffeo would probably have been delighted to sponsor not only Bernini but also Galileo. Of course, such a development was impossible—Galileo was attached, at a high price, to the court of Tuscany—but Maffeo was extraordinarily well disposed toward Galileo and kept close track of his work. As early as October in 1611, Maffeo, then a cardinal, expressed support for Galileo during a scientific debate over the nature of flotation at the Pitti Palace, in Florence. In May of the following year, upon receiving Galileo's treatise on the subject, he wrote him a letter praising his "rare intellect" and suggesting that their minds vibrated in harmony. In 1620, Maffeo wrote a poem praising Galileo—inevitably, it mentioned the moon— and in 1624, as newly elected pope, he granted Galileo's son, Vincenzio, an annuity of sixty crowns and urged Galileo to write the definitive treatise on Copernicanism, a work that later became the *Dialogue*. We will see exactly how the pope came to prosecute Galileo for writing the very book that he himself had proposed, but Galileo was by no means the only adventurous thinker whom Maffeo, a late child of the Renaissance, encouraged and supported. In 1626, he rescued Tommaso Campanella from the dungeons of the Inquisition and welcomed him into his intellectual circle: Campanella, one of the most audacious philosophers of the period, had defended Galileo and posited an infinity of worlds.

I do not wish to suggest that Bernini and Galileo were in some sense conceptual counterparts, but rather that Galileo appealed to Maffeo Barberini in much the same way that Bernini did, though mostly from a tantalizing distance. Dur-

ing the Baroque period, artists and scientists did not seem as different as they do today. Not only was Galileo himself an accomplished musician, an excellent prose stylist at his best, a sometime poet and literary critic, and a friend of some brilliant artists, but he also sought a certain elegance in his geometrical proofs and in his increasingly public pedagogy. Bernini had engineering as well as artistic skills, and there was a speculative component to his art. For the entire Baroque enterprise, representing space and creating shapes meant acquiring greater knowledge about statics and dynamics, light, cast shadows, and the science of perspective: it meant reaching out and grasping space. Just as Galileo took a fixed system of the cosmos and tried to show how the earthbound observer actually revolved within it, Baroque artists like Bernini took a frontal, cinquecento vision of order and twisted it into curving and spiraling planes.

Bernini's vision was too vast and complex to be easily summed up. If one looks, however, at a sculpture like *Saint Longinus* in Saint Peter's, which he worked on for Urban VIII before and after the years of the Galileo trial (1629–38), or at the draperies on the angels on the Sant'Angelo Bridge spanning the Tiber, which came decades later but represent an elaboration of the same idea, one notes right away an original conception of gravity—and gravity of course was one of Galileo's chief concerns. In the paintings and sculpture of the High Renaissance, which Bernini wished not to subvert but to extend by novel means, the effect of gravity upon fabric is generally conveyed with simple decorum. For instance, in Raphael's Tapestry Cartoons, the great studies he and his assistants executed in 1515–16 as designs for the Sistine Chapel wall hangings, the robes of the biblical characters appear by and large to be woven

of plain, undifferentiated wool and to fall toward the ground in broad folds, without frequent hooking or complicated tubular effects. Curiously enough, if one thinks about it from the astronomical point of view, in one way or another these draperies fall toward a point at the center of the earth. They are subject to gravity, which is, after all, part of our mortal dispensation—unlike the angels, we cannot fly. But they are subject to gravity in a particularly graceful way.

With Bernini's drapery something else happens. Even though his *Longinus* stands still, and no wind billows his robe, he seems to be swept by tremendous energy. His garment—fluttering, tangled, and wavy—does not accentuate the saint's gesture but enacts a continuous independent occurrence, or a captured moment within such an occurrence, for in fact no fabric ever hangs this way—it is an antigravity effect. Bernini sees grace, in the sense of something freely, divinely given, as a medium resistant to gravity, permitting a form of existence alien to our own—potentially, many of Bernini's figures could lift off, like helium balloons—and as one looks up into some of his domes one sometimes does see *putti* playfully levitating over the cornices. In a way, this medium of grace is the super-oxygen of the Counter-Reformation, in which street procession and theater and miracle can breathe. And it is further developed in Bernini's architecture and that of his colleagues.

The Baroque style does many things, but one of them is to take the rather static classical architectural vocabulary, subject it to curvature, and rotate it at will through at least thirty degrees in plane and in elevation. If there is no strict correspondence between Bernini and Galileo, who worked in very different fields, they do share a certain mental attitude and a dynamic grasp of space. They have a common willingness to contem-

plate what might at first seem implausible solutions, a fascination with the time factor—with the combination of direction
and velocity—a keen interest in curvature and curved planes,
and a constant perception of experiential relativity. The
Baroque style attempts to put High Renaissance plastic values
into motion, to capture the semblance of an ever-shifting world.
Stage sets implicitly rotate the audience off-center, so its members observe the action from a variety of angles. Architectural
elements are enjoyable or interesting much less in themselves
than as parts of a whole perceived by a circulating observer. The
sense of reality becomes dependent on viewpoint and lighting.
In a painting, for example, if a shadow causes the limbs of two
figures to dissolve optically into each other, no contour will be
provided to establish their independent identities. Though
Bernini's art was an expressive, not a scientific, enterprise, and
certainly in no sense Copernican, it resembles the heliocentric
model in its emphasis on relativism and on the observer's position as opposed to the fixed contemplation of forms.

Bernini's baldachin, an ornamental canopy in Saint Peter's
which looks different from every angle, or his Scala Regia in the
Vatican, a long staircase with ever-changing and eye-deceiving
widths, not to mention his fictive seating areas and imaginary
audiences, all depend for their impact on the spectator's motion
through space. Bernini did not invent the Baroque style, but his
work expanded the optical Rome that had begun under Sixtus V
with Michelangelo's Campidoglio and continued radically
throughout the 1590s. The old Rome had been a hodgepodge of
evocative structures in which the pilgrim struggled to find an
axis of symmetry whereby he could orient himself; basically, he
always had to ask his way around. The new, optical Rome
offered landmarks: it was characterized by straight thorough-

fares with an obelisk centered at the focal point, on the axis of symmetry of an important façade, like a church. The visual channel of the avenue thus worked analogously to a telescope or gun sight (though without the element of magnification).* When one entered the church, one looked up into a dome like a spiritual planetarium, a model of heaven—the upper half of what astronomers called the "celestial sphere"—with, at the zenith, a glazed lantern as the focus. (In some cases, like Sant'-Andrea della Valle, the dome's interior surface bore an actual picture of heaven.)

This was the Rome that Maffeo Barberini strove mightily to renew and to expand. He couldn't help being fascinated by both Bernini and Galileo because he was so captivated by the Baroque idea of space—by his contemporaries' power to create a semblance, whether in stone or in mathematical theory, of a world that wasn't static but that changed unceasingly, like the one we inhabit. It was natural for the pope who backed Bernini to encourage Galileo, and his disappointment was natural later on when Galileo overstepped (as Maffeo saw it) the Church's clearly stated guidelines for the formation of scientific hypotheses. Unfortunately, Maffeo was more than disappointed, he was outraged, and his outrage caused so much damage that the Church has not wholly repaired it to this day.

THE GALILEO TRIAL has been interpreted numerous times, from various ideological standpoints. It is clearly the prime

---

* Anticipating the invention of the telescope, Michelangelo's Campidoglio, in Rome, begun in 1538, was also designed so as to appear magnified to the approaching observer, by means of scenographic perspectives. Its construction was delayed until after Galileo's death.

drama in the history of the conflict between science and religion, more central than the evolution controversy of the 1860s, which never formally pitted Darwin against the Church of England, and certainly more consequential than the rather jocular Scopes trial of 1925. In Galileo and Urban VIII's day, Rome was the capital of a sovereign theocratic power, the Papal States, which in 1600 had had Giordano Bruno burned at the stake for refusing to abjure his heterodox philosophy, and which reserved the right to torture (and, ultimately, to execute) Galileo, should he appear to withhold evidence. The trial also marks a turning point in the evolution of freedom of thought, since the great Tuscan physicist had been ordered not to pursue his research but had done so anyway, and had published the results. As if this were not enough, still more was at stake, for Galileo's accusers were alarmed not only by what he said but also by how he said it—by his reinvention of certain components of the scientific method, neglected in the West (though not in the Islamic world) since antiquity. Clerical scrutiny of his writings indicated that they appeared to assign logical priority to empirical observation over Scripture, which was contrary to Catholic doctrine.

The truth is that Galileo had been running a terrible risk since 1616. Did he know this? Perhaps not; perhaps he didn't much care, until the men in black vestments began threatening his freedom. At the time, the coming science-religion clash was not foreseeable, and it remains puzzling today for reasons partly unexplored, though clearly the leaders of the Catholic Reform saw the scientific revolution as forming an analogy with the rise of Protestantism, a terrible mistake. To a degree easily forgotten, dynastic rivalries and personal emotions also came into play. Historians of science have generally discussed the debate

between Galileo and the Church as though it were *cosa mentale*, the way painting was for Leonardo da Vinci. But the trial was not a mental thing, like a modern scientific colloquium, but a series of frightening interrogatories held before a tribunal of the Inquisition, and each side's perception of the other was critically defective. The Roman cardinals respected Galileo but had their own peculiar anxieties, exacerbated by the spread of heresy, by dynastic and geopolitical rivalries, and by internal, ecclesiastical rough-housing; nor did Galileo's argumentative personal style make his cause any the easier.

From the modern Anglo-Saxon standpoint, Galileo's trial bears little juridical resemblance to anything we would call by that name. It is far more like the informal hearings held by government, corporate, or religious bodies that do not require the observance of strict rules of evidence or the protection of the rights of the accused but might result all the same in his or her professional disgrace. A huge literature exists about this "trial," but almost none of it attends to the ragged formlessness of the proceedings, except perhaps to note their deviations from canon law. As a rule, such writing explores the conflict between Galileo's ideas and those of the Church, and this is understandable, because no abrasion is more central to our age than that between science and religion. Intellectually, this is where the interest lies—or ought to lie. In fact, however, the Church scarcely contested Galileo's science, but only enjoined him against pursuing it freely, with the consequence that the "debate" between Galileo and the Church is merely a mental construct, a dichotomy posited to satisfy a hunger for symmetry. It is not as if Galileo and Pope Urban VIII ever went, as we say, "nose to nose." In fact, it was the Aristotelian intellectual establishment that vociferously opposed Galileo; the Church's

substantive arguments against his methodology were assembled later, as an afterthought, when he had already been
silenced. If there is, then, a kind of intellectual weightlessness
to this nondialogue, Galileo's emotional predicament acquires
thematic tension between September of 1632 and June of
1633, as a result of the encounter between two distinct modes
of reasoning, feeling, and wanting. The scientist thought in
one way, the Church in another; they operated on different
psychic planes, a condition most clearly perceived by the Tuscan ambassador to the Holy See, Francesco Niccolini. The
question we may wish to ask ourselves, not now but at the end
of our story, is whether this great trial, so pregnant with consequences, does not reveal certain constants in the debate
between science and religion.

2

The Telescope; or, Seeing

Until 1609, when Galileo trained his telescope on the sky, the proponents of the Copernican system had in a sense been navigating blind. They had a map that was essentially correct, but no visual grasp of the world in which they were traveling. In a flash the telescope changed that. Now Galileo and his colleagues could see. They knew they were on the right track. But many others could not see, or refused to see. For this sort of outmoded savant, Johannes Kepler, writing to Galileo on March 28, 1611, had some interesting words in Latin, the scientific lingua franca of the day. Such a person was stuck in a *"mundo chartaceo,"* a world of paper, Kepler said. *"Negatque solem lucere"*—it was as if he wouldn't admit the light of the sun. *"Caecus"*—he was blind. Not by force of circumstance, but of his own foolish will.

STILLMAN DRAKE HAS claimed that Galileo had concluded as early as 1595, the date of the publication of *De motu*, a treatise on mechanics he wrote while still at the University of Pisa,

that as a geometrical construct the Copernican hypothesis came closer to explaining the motions of the celestial bodies than the alternative, Ptolemaic or geocentric system. This may be so. It is certainly true that Galileo, in a letter of 1597 to Jacopo Mazzoni, a friend and ex-colleague at Pisa, offhandedly assumed the earth's motion in a proof concerning the visibility of the stellar sphere. Indubitably his evolving conception of mechanics clashed with the geocentric cosmology, which he saw as dependent on outdated Aristotelian concepts of gravity, force, and motion.

At this period, astronomy was regarded as an exploration of mathematical suppositions, in part for complicated metaphysical reasons and in part because mankind had very scant empirical knowledge of the heavens. Observers going at least as far back as Hipparchus had made clever calculations and predictions, but they had no way of scrutinizing celestial bodies. This notion of reasoning about the heavens "suppositionally," *ex suppositione*, mind-bending to us today, is rooted in a certain premodern conception of a hypothesis not as a theory likely to be empirically confirmed, or already largely confirmed, but as kind of a logical holding pattern. Most astronomy before the invention of the telescope fell into this category. Data would be gathered and a mathematical system would be mapped into it, which approximately fit the facts ("saved the appearances," in the language of astronomers). These planetary models were required to fit the constraints of Aristotelian physics, but could not, beyond that point, be empirically verified.

In 1597, Galileo wrote Kepler that he had already accepted, on the basis of mathematical physics, both the rotation of the earth and its revolution about the sun. However, he was

unsure of his ideas and loath to publish a paper on the subject. But the appearance in the autumn of 1604 of a nova, an explosive variable star whose luminosity increased by many magnitudes during a period of hours, prompted him to give three lectures at the University of Padua, each attended by more than a thousand people, in which he noted the absence of parallax for the nova, that is, its apparent motionlessness with respect to other celestial bodies. He correctly placed it among the stars, precisely where Aristotelian doctrine asserted that no change ever took place. Though he was not overtly espousing Copernicanism, he was bluntly attacking the alternative theory, and the audience was left to draw its own conclusions.

So far Galileo's Copernicanism had been a mathematical hypothesis. What transformed it into something resembling a modern scientific theory was his radical improvement and use of a gadget recently devised in northern Europe. This was a tube with a weak lens at each end, which Galileo almost at once converted into a precision instrument. When we consider this extraordinary invention, we may wonder what a person living in 1609 would have needed to perform such a feat, and the answer is rather simple. He would have needed to know, or figure out, something about geometrical optics. He would have needed to possess a finely tuned lathe. And he would have needed to have access to superior glass and to abrasives for lens-polishing.

Geometrical optics is the science of how light behaves in an optical system, such as a telescope, where its wavelength might be regarded as negligible compared to the dimensions of the lenses or other components. We now learn in grade school that light is a wave phenomenon moving at a speed of 3×10^{10} meters per second in empty space, and that when light enters a medium such as glass its speed depends on the den-

sity of the medium and the wavelength, or color, of the light. It is also refracted, or bent, because when a light beam progresses from a less dense into a denser medium it bends toward "the normal," the plane of the medium it encounters. Light enters the medium at an angle to the normal of the surface, called the angle of incidence, creating the beam's new angle within the glass, called the angle of refraction. Snell's law of refraction is a simple trigonometric formula that tells us the index of refraction of optical glass—how beams of light bend as they pass through lenses and thus how any lens creates the image that it does. But this formula was worked out only in 1621 by Snell (Willebrord Snellius), who did not publish it, and again, independently, by Descartes, who did publish it in 1637, long after Galileo constructed his telescope. So though Galileo knew something about refraction, which had been studied since antiquity, he could not identify the exact optical properties of the lenses he was starting to make.

Glass itself is manufactured by bringing silica plus other components to liquidity and then cooling them. Venetian glass was made of quartz pebbles composed mostly of silica, which was milled into sand and combined with soda ash imported from the Levant. Optical glass was presumably discovered by accident—perhaps because it occurred in bull's-eye windowpanes or the bottoms of bottles—sometime in the thirteenth century, and by about 1260 spectacles of the sort that we call reading glasses could be bought in Florentine shops. Yet strangely, no optical instruments, no microscopes or telescopes, were produced for the purpose of getting a closer look at nature, and about eighty years ago an Italian historian of optics named Vasco Ronchi suggested a reason for this. A sort of conspiracy of silence, he said, had been

mounted against the research potential of the lens. The pre-dominant medieval theory of vision, derived from an ancient Greek idea, held that invisible films, called *species*, issued from the eyes and assembled optical data, so that anything impeding their progress, such as mirrors, prisms, or lenses, deformed and corrupted these images. Sight as such seemed an unreli-able sense, making objects appear smaller and grayer as they receded toward the horizon, so to alter sight still further with lenses, all of which had intrinsic distorting properties, would scarcely have improved matters. Scant documentation exists to support Ronchi's thesis, and it was vigorously disputed in 1972 by David C. Lindberg, a highly knowledgeable American his-torian of optics, who pointed out the scope and value of the medieval work in this field. It can be argued that no prejudice against lenticular devices as a research tool existed, but that it simply occurred to no one to devise such things; or, alterna-tively, that artisans may have tried to concoct a spyglass and failed, and being illiterate, left no record of the attempt. In any event, we do know how the first one was invented.

A primitive telescopic effect is attained when two lenses of suitable focal length are aligned on the optic axis between the object to be viewed and the retina of the viewer and a clear, enlarged image is perceived. Since the first people to jerry-rig telescopes probably did so to amuse themselves, pulling apart spectacles and lining up lenses, one wonders what corrective purpose the original lenses served and what they looked like. Most people buying spectacles, then as now, were probably constant readers over age forty—clerics, scribes, scholars, lawyers. The overwhelming popular demand would have been to compensate for far-sightedness, or presbyopia (the eyes' progressive inability to accommodate to near points) and only

later for near-sightedness, or myopia, which is less common. The earliest lenses, of whatever sort, had a faulty spherical curvature—that is, the glass piece was a sector of a sphere— and a focal length of 12 to 20 inches, the focal length being the distance at which an object viewed appears in focus in the lens. (In the measurements of today's drugstore reading glasses, that would be about 2 to 3 diopters, a diopter being the refracting power of a lens expressed in meters and in reciprocals: thus a lens of focal length 20 centimeters has a refracting power of 1/0.2 meters, which equals 5 diopters). They were made of glass blanks, which were ground on a lathe attached to a bench. Convex lenses helped correct presbyopia, concave lenses myopia, and one might assume that the concave came later, being harder to grind and harder to adjust to the customer, who might need to be fitted on a trial-and-error basis.

When did the first primitive spyglasses begin to appear? In 1578, William Bourne, in a book titled *Inventions or Devises*, mentions that "to see any small thing a great distance from you, it requireth the ayde of two glasses," a puzzling phrase that might refer to a telescope designed by Thomas Digges, an Elizabethan astronomer whose observations of the supernova of 1572 were consulted by the great Danish astronomer Tycho Brahe. Bourne, however, noted that the tiny field of view covered by this gadget was a great "impediment" to its use (though this would also be true of Galileo's telescope). The instrument, on the assumption that it actually existed, might have been made with very impure glass.

The first telescope that indisputably worked was offered in September of 1608 by a spectacle-maker named Hans Lipperhey to the States-General of the United Provinces, in The Hague. Lipperhey was a Westphalian who had moved to Mid-

delburg in Holland, and he wanted a patent. Instead he was requested to make a pair of binoculars, presumably for military purposes, and to use quartz instead of glass; within a few months, he gave the Dutch government three binoculars and was handsomely rewarded but received no patent. The reason, as Albert Van Helden has discovered, was simple. A Zeeland government document shows that a "young man" had meanwhile come forward "with a similar instrument," and "we believe that there are others as well, and that the art cannot remain secret." In fact there seems to have been considerable experimentation with such optical instruments in the glassmaking town of Middelburg, where several Italians worked, presumably introducing additional technical information from their homeland.

By the summer of 1609, the Dutch telescope in one form or another had been around for about nine months. It had been sold to the king of France and his prime minister, and to the archduke of Austria, who governed the portion of the Low Countries under Spanish rule; it had been shown at the Frankfurt autumn fair of 1608; and it had been obtained by the papal nuncio to Brussels, who sent it to Cardinal Scipione Borghese in Rome, an unscrupulous prelate of excellent taste and wide influence, from whose hands it had proceeded to the pope's. That July an English mathematician-astronomer, Thomas Harriot, working out of Syon House, the seat of the earl of Northumberland, used a telescope with the power of x6.3 to view the sunspots and Jupiter's moons. In so doing he preceded Galileo by about five months but with—crucially— about five times less magnification. Spyglasses arrived in Milan by May, in Rome and Naples by July, and in Venice and in Padua by early August.

Later recalling his dramatic improvement of the instrument, Galileo wrote that

> a report reached my ears that a spyglass had been made by a certain Fleming by means of which visible objects, although far removed from the eye of the observer, were distinctly observed as though nearby. About this truly wonderful effect some accounts were spread abroad, to which some gave credence while others denied them. A few days later the rumor was confirmed to me in a letter from the noble Frenchman in Paris Jacques Badovere, which caused me to apply myself totally to investigating the means by which I might arrive at the invention of a similar instrument. This I did shortly afterward on the basis of the science of refraction. And first I prepared a lead tube in whose ends I fitted two glasses, both plane on one side while the other side of one was spherically convex and of the other concave. Then, placing my eye near the concave glass, I saw objects satisfactorily large and near, for they appeared three times closer and nine times larger than when observed with the naked eye alone. Next I constructed another one, more accurate, which represented objects enlarged more than sixty times [that is, what we would call x8+]. Finally, sparing neither labor or expense, I succeeded in constructing for myself so excellent an instrument that objects seen by means of it appeared about a thousand times larger and more than thirty times closer [that is, x30+] than when regarded with our natural vision.

The truth was that in early August a Dutchman had arrived in Galileo's own town of Padua and was showing his spyglass to potential purchasers. So before Galileo had much of a

chance to improve his first telescopes—though he probably acquired better glass than whatever his rivals were using and was deepening the sphericity of the eyepiece—he decided that he had to design a superior instrument and stake out a market. Eight years later, Galileo recalled this key moment a little differently than in the foregoing description, hinting obliquely at the ferocity of the competition and mentioning a Dutchman active in Germany, though not his rival in the Veneto. "News came," he wrote in this alternate account of his discovery, "that a Hollander had presented to Count Maurice [of Nassau] a glass by means of which distant things might be seen as perfectly as if they were quite close. That was all. Upon hearing this news, I returned to Padua, where I then resided, and set myself to thinking about the problem. The first night after my return, I solved it, and the following day I constructed the instrument and . . . afterwards I applied myself to the construction of a better one, which I took to Venice six days later." This version, which omits to mention the letter from Badovere, stresses the rapidity of Galileo's improvement of the spyglass, and indeed he got to work with extraordinary speed. He needed to, for there was competition not only from the itinerant Dutchmen but from very knowledgeable astronomers. In England, as mentioned earlier, Thomas Harriot had already mapped the moon in August, using his weak telescope, and on the Continent four mathematicians, including Father Christopher Scheiner, were already in possession of telescopes.

Galileo was thinking fast. "My reasoning," he wrote, "was this. The device needs either a single glass or more than one. It cannot consist of one alone, because the shape of that one would have to be convex . . . or concave . . . But the concave

diminishes [visible objects]; and the convex, while it does indeed increase them, shows them very indistinctly and confusedly. Therefore, a single glass is not sufficient to produce the effect. . . . Hence I was restricted to trying to discover what would be done by a combination of the convex and the concave. You see how this discovery gave me what I sought."

Joining a bitter race against the Dutchman who was nearing Venice, Galileo wrote on August 4 to his friend the savant Fra Paolo Sarpi, imploring his help. Obligingly, Sarpi, who had great influence with the Venetian Signoria, asked to inspect the Dutchman's spyglass and advised the Signoria against its purchase. On August 21, a breathless Galileo arrived in Venice with a more powerful instrument than any then in use—its magnification was about x9—which he proudly proceeded to demonstrate to a group of Venetian senators and naval officers from atop the campanile in the Piazza San Marco. To their delight, they were able to sweep a radius of some fifty miles with the instrument, even beholding the faithful entering the church of San Giacomo on the glassmaking isle of Murano. On August 24, 1609, Galileo wrote to the doge, Leonardo Donato:

> Most Serene Prince, Galileo Galilei, most humble servant of Your Serenity, assiduously laboring by night and with unwavering spirit to satisfy not only the duties of lecturer in mathematics in the University of Padua, but with the hope of finding some useful and notable discovery to bring extraordinary benefit to Your Serenity, now appears before the aforesaid with a new device, a hollow spyglass of the most unusual perspective specifications, in that it conducts visible objects so near to the eye and represents them so large and distinct, that something which is distant, let us

say, by nine miles, appears to us as if only one mile away: an invaluable thing for any maritime or terrestrial enterprise, being able at sea, at a much greater distance than usual, to descry the masts and sails of the enemy.

In turning to the Venetian Republic, his employer and a great naval power, Galileo followed the inventor's time-honored tradition of first approaching the military to raise money. The plan worked: the Venetian navy bought his telescope, the Senate doubled his salary at Padua and renewed his position for life, and he spent the fall of 1609 trying to improve the instrument's magnification. The alacrity with which the already forty-five-year-old Galileo seized on this optical idea, the speed with which he perfected it with his own hands, the confidence with which he presented it to the Republic, and the fortitude with which he confronted his rivals—all distinguish him as a man of action as well as an intellectual.

Almost at once, Galileo's detractors began to belittle his role in the development of the telescope. But Galileo liked to argue that his knowledge of earlier prototypes made his task as inventor not easier but more difficult. The original inventor, he claimed, had merely stumbled on the device: he himself, on learning of the concept, had to see his way clearly forward to something of scientific utility. He did not admit to having handled any of the playthings then popular in certain cities, and perhaps he had not done so by the time he set to work on his first version: it scarcely mattered, since a drawing or verbal description would have sufficed to acquaint him with the basic idea. What is certain is that he had greatly outdistanced the available prototypes within a matter of weeks.

Now, despite what Galileo said about his knowledge of the

"science of refraction," it appears that only Kepler had a good enough mathematical understanding of geometrical optics at this time to trace satisfactory ray diagrams. Galileo's advantage lay rather in his deep faith in the scientific relevance of seeing, of vision. Though he worked from the start with two lenses, let us pretend, in order to better understand Galilean optics, that he first examined one lens only, affixed to a tube. Holding a plano-convex lens—a lens spherically convex on the side facing the object to be viewed and flat on the other—up to an object, let us say the moon, he could theoretically focus his eye on a tiny, inverted image of the moon suspended upside down in space, somewhat magnified, in the plane called the prime focus of the lens. In effect, the lens had converted the arriving rays of light into cones whose apexes fell approximately in this plane. Brief study of the adjoining ray diagram will disclose why the image is inverted. (If the image is captured on a piece of paper, as in the rudimentary camera obscura later devised by Galileo's disciple Benedetto Castelli and also by his friend the painter-architect Ludovico Cigoli with the help of Domenico Passignano, using two lenses positioned to throw the focal plane well to the left of the telescope, greater image stability and a wider angle may be achieved, with some loss of brilliance.)

We know, however, that Galileo was working with two lenses from the outset. We may imagine him experimenting with various pairs. And it is indeed possible, if you have a selection of weak lenses, to run through the effects produced by every possible combination of one concave and one convex lens with respect to magnification and image sharpness. You hold them in front of your eye, one close, like a loupe, the second one behind it, and gradually extend the second one out to see what happens. The exercise is tedious, and Galileo didn't

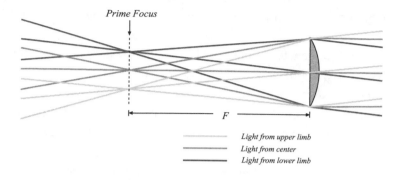

Ray Diagram of a Telescope without an Eyepiece

The focus is on a remote light source (like the moon) to the right. The location of the apexes of the bundles of light on the plane of the prime focus shows that this (aerial) image is inverted.

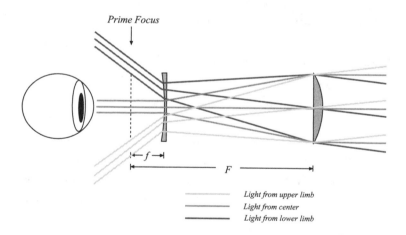

Ray Diagram of a Telescope with a Negative (Concave) Eyepiece

The eyepiece both magnifies the image and converts the bundles of light transmitted by the objective into parallel bundles more easily processed by the eye. As noted in the text, the angle at which the rays enter the eyepiece and that at which they leave it diverge, and this factor is represented by the ratio *F/f.*

necessarily perform it, but he must certainly have satisfied himself as to the effect of several lens combinations in addition to the "bagatelle"-like spyglasses (the word is Galileo's) he had heard about or got his hands on, partly out of physical curiosity and partly because the laws of refraction were unknown to him. If you were to run through such an experiment at home, you would eventually discover that you get the telescopic effect only when you hold a weak lens, preferably concave, close to the eye and move a strong convex lens gradually away, until the focal lengths of the two coincide at the point of their greatest magnification and clarity. So, having presumably satisfied himself regarding the effect produced by a single lens, Galileo began to rig up test spyglasses with two lenses. The concave eyepiece converts the converging bundles of light rays emerging from the objective into parallel bundles that the human retina can focus on more comfortably, and also permits magnification: Galileo could not have known these facts, but right away he grasped them empirically. He also saw that concave eyepieces conveniently reinverted the image. However, with the eyeglass lenses then routinely available, Galileo probably achieved in this way an initial magnification of somewhere between x2 and x3, which hardly could have satisfied him. How could he increase the magnification?

The magnifying power of a telescope is the ratio of the apparent angular diameter of a given object when viewed through the instrument to its apparent angular diameter when viewed by the naked eye. In other words, if you can find a way of measuring the angle subtended by the diameter of the object—the moon, say—when seen through a telescope and compare it to the angle subtended by the moon when seen by the unaided eye, then you know the power of your telescope.

This could be confusing back then, though nowadays, by combining Snell's law and an equation called the thin-lens formula, which allows us to regard the glass in question as a medium of negligible thickness, you can easily determine the power of a telescope like Galileo's: its magnifying power equals the ratio of the focal length of the objective to the focal length of the eyepiece. In other words, by dividing the former into the latter, you can learn the power of your telescope. In experimenting with the focal length of his objectives, Galileo must soon have had some rough notion of the equation $M = f_o/f_e$, where M is power, f_o is the focal length of the objective, and f_e the focal length of the eyepiece, because working in his own shop he soon ground objective lenses of ever-longer focal lengths.

It seems that so far very few others had been able to intuit the relation between magnification and the focal lengths of the two lenses, the eyepiece and the objective. Why did Galileo advance so rapidly where others failed to progress?

Most of Galileo's early competitors were lens-grinders. And magnification is a simple function—inverse proportion—that might elude a lens-grinder but be rapidly divined by a mathematician. Yet since there is no documentary evidence that Galileo figured out the formula for magnification, it is more likely that he came to understand the primacy of objectives with long focal lengths through trial and error. After all, he was good with his hands and interested in machines of any sort. The problem of constructing a telescope had at once fascinated him, as had that of the helical pump and the gunnery sector.

When Galileo was a very old man, in June of 1640, he wrote a letter from Arcetri to Fortunio Liceti, a professor at the University of Bologna, in which he lamented "having always been in darkness concerning the essence of light." In August of the

same year, he again wrote Liceti that "I have always considered myself unable to understand what *lumen* [light] was, so much so that I would readily have agreed to spend the rest of my life in prison with only bread and water if only I could have been sure of reaching the understanding that seems so hopeless to me." This confession, in which we hear the accent of the true scientist, has been tied to the brief time that Galileo spent in the study of geometrical optics, especially considering that Kepler's breakthrough treatise on the subject, *Ad Vitellionem Paralipomena*, had already appeared, in 1604. But actually it seems to reflect Galileo's despair of understanding not geometrical optics but something deeper, the physical nature of light itself, which he puzzled over throughout his career.

Galileo discussed magnification very warily, never offering his readers more than a simpleminded way of checking the power of a telescope they were lucky enough to possess. Such helpful-sounding but actually quite useless directions tend to support Mario Biagioli's contention that Galileo worried about others being able to replicate his telescope too readily and so deprive him of professional credit for its invention. His competitors outside of Italy have already been noted. But a mathematics professor in Rome itself, Antonio Santini, produced a telescope and saw Jupiter's satellites before the end of 1610, as did the Jesuit mathematicians at the Collegio Romano, with help from independent instrument-makers. Galileo's original, hapless intention was to retain the manufacture of the telescope as a kind of trade secret, and his exposed position partly explains the somewhat embattled attitude with which he guarded that secret and defended his recent observations. In turn, what Biagioli calls his "uncooperative stance" may help explain the ferocity of the criticism he sustained from opponents and skeptics.

One of the earliest descriptions of a Galilean telescope is contained in a letter written by Giovanni Battista Della Porta to Marchese (later Prince) Federico Cesi, the founder and principal financial backer of the Academy of Lynxes, an enlightened Roman scientific society critical of the Jesuits. (It derived its name from the lynx's capacity to see in the dark.) Della Porta was the celebrated author of *Magia naturalis*, a volume of technical games and tricks, and *De refractione*, a work on optics (which, while erroneous in itself, dispelled some earlier misconceptions). He told Cesi that he had investigated the "secret of the spyglass" and found it a "*coglionaria*," a vulgar expression for something idiotic; Galileo's invention, he claimed, was "purloined from the ninth book" of his own treatise on refraction, a wholly unfounded assertion. "And I shall describe it," he went on,

so that if you want to make it Your Excellency will at least have fun with it. It is a small tube of silvered tin, one *palmo*, *ad*, long and three inches in diameter, which has in the front a convex glass in the end *a*; there is another canal of the same, four fingers long, which fits into the first one, and in the end *b* it has a concave [glass], which is soldered like the first. If observed with that first one alone, far things are

Della Porta's Sketch of Galileo's Telescope (1609)

seen near, but because the vision does not occur along the perpendicular, they appear obscure and indistinct. When the other, concave tube is put in, which gives the contrary effect, things will be seen clear and upright; and it goes in and out, like a trombone, so that it adjusts to the visions of the observers, which all are different.

Aside from the *coglionaria* remark, the description chimes with Galileo's own. And it makes sense. A biconvex objective would have inverted the image: by refracting light rays along a slope, it would send them to the side of the optic axis opposite to the one from which the rays would have struck the retina if traveling through air alone. And this Galileo corrected by reinverting the image with another spherical lens, the eyepiece.

What Galileo (and, increasingly, other people) had by now discovered was that the principal element in a powerful telescope was a high-quality objective lens. But it was not an easy item to manufacture. During this period, the technology of lens-grinding necessitated the acquisition of glass blanks, cut from blown globes relatively free of impurities, from which the lenses would be derived; a pedal-operated lathe; forms or grinder heads, convex or concave, which would be used to directly grind the lens; and some kind of abrasive. Since Padua was so close to Venice, a major glassmaking city, one might assume that the apparatus in Galileo's atelier was derived from prototypes used on the Venetian island of Murano, where a glass manufacturer initially supplied Galileo with lenses. But lenses were ground in many cities, and Galileo went at least as far as Florence and perhaps farther in his search for lenses and blanks. No records or diagrams have survived concerning the

composition of Galileo's glass, the type of lathe he used, his forms, or his abrasives. Since he was running a small business facing many potential competitors, he kept his own counsel on such matters. It is known that after he mastered lens-grinding he increasingly delegated the task to an artisan named Ippolito Francini.

The composition of Galileo's lenses has been analyzed at the Institute and Museum of the History of Science in Florence, but the results have not yet been published. It has been determined that the greenish or roseate tint of most of these lenses resulted from impurities in the type of sand used in the manufacture of the glass. The lenses were ground on a lathe more or less as one would turn a glass ashtray to hollow out a cavity, except that certain of their properties baffle researchers. Their shape, though always a sector of a sphere, is usually somewhat irregular, especially toward the circumference, suggesting an element of unpredictability in the interface of form and glass—perhaps, for some reason, the artisan's hand was obliged to intervene and rotate the form or the lens. One also wonders what abrasives were used. Glass is a very hard substance, which basically can be ground either with glass itself, that is, with glass particles, or with an even harder abrasive, such as diamond particles in a slurry. Abrasives can be quite expensive and are often under some form of patent or proprietary lock, such as a secret formula, so Galileo had to find one that was both accessible and economical.

Galileo had access to deposits of sand, possibly along the Adriatic coast, containing substantial fossilized sediments of the microorganisms known as forams (foraminifera, or "hole-bearers"). These tiny amoeboid creatures, today much studied though mostly invisible to the naked eye, usually generate a

sort of porous shell, known as a "test," and tend to live in marine environments. There are over 275,000 species of forams, with an extraordinary diversity of tests, many quite elaborate and generally composed of calcium carbonate. Sand containing a high density of fossilized forams can be used as an abrasive, and though no one in the Veneto of 1609 yet knew this (Galileo assembled his first microscopes in 1624), certain people were obviously aware that the sand from specific zones could easily scratch glass. Typically, Galileo found out about this and put it to best advantage.

The Museum of the History of Science owns a telescope made by Galileo sometime between 1610 and 1630. It is a wooden tube almost a meter long, bearing cylindrical housings at either end for the two lenses and covered with brown leather, slightly torn where the instrument was mounted on stanchions. The optical system consists of a plano-convex objective (by far the more difficult lens to produce) and a biconcave eyepiece, with the two lenses arranged so that the focal point of the objective can easily be caused to coincide with the back focal point of the eyepiece. Precisely as described by Della Porta, the housing of the eyepiece serves as a draw-tube that can be pulled in and out for focusing. The objective is 3.7 centimeters in diameter and is positioned against a wooden ring with an aperture of 1.6 centimeters, which in turn is held by a smaller cardboard ring with an aperture of 1.5 centimeters—less wide than a thumbnail—and this is finally sealed by a leather-covered ring with a much wider aperture.

Galileo soon discovered this "stopping-down," or reducing of the diameter of the aperture (which corresponds to the diaphragm of a single-lens reflex camera). Initially, he might

have sought to compensate for his own vision deficiencies. He might also have wanted to eliminate the flaws around the circumference of the lenses caused by irregular or hand grinding, as well as inherent spherical aberrations—unexplained at the time but noticeable as blurring—caused by the fact that rays striking spherical glass far from the optic axis, the so-called paraxial region, at the center, bend more than those in the narrow bundle concentrated on this region. He must have noticed at once (perhaps from his own squinting) that such stopping-down increases the sharpness of the image, a huge observational advantage. Yet the more he stopped down his lenses in the interests of image crispness, the smaller his apertures became, and these tiny apertures, together with the increasing focal length of his objectives, eventually resulted in the sort of skinny, reed-like telescope that is preserved in the museum in Florence, which can magnify x21 but has a field of view of only 15 arc minutes. All this constituted a marvelous achievement, and some of Galileo's telescopes apparently afforded a magnification of as much as x30. But if one looks through such a telescope (there is an exact replica in the offices of the Institute), one discovers a field of view so small that it cannot encompass the moon—it is not an easy instrument to handle. More importantly, the invention in this form did not permit much technological improvement, because any increase in power would have further reduced its field of view. In fact, Galileo personally made few telescopes after the initial years of discovery, devoting his time to other interests, and eventually the field passed into the hands of much younger men, such as his disciple Evangelista Torricelli; the Neapolitan Francesco Fontana, whom he despised; Eustachio Divini; and Giuseppe Campani. The Galilean telescope was a historical

cul-de-sac, for lenses were soon combined in very different ways, mirrors were employed, and, much later, spherical aberration was corrected by the use of aspherical shapes.

Galileo conceived of the perforated plates used for stopping-down as measuring devices, graduated by size, "subtending more or fewer minutes of arc." Indeed the telescope could see a host of new stars, but technical problems in measuring angular distance were posed by the variability of the viewers' eyesight and the minuscule field of view, restrictions not shared by the ordinary quadrant, which could also measure the intervals between stars. Galileo does not seem to have actually used his aperture stops as a measuring instrument.

SOMETIME IN THE late autumn of 1609, Galileo produced a telescope with a power of x20 and undertook, with great excitement, study of the closest of the celestial bodies. The first clear night of such observation initiated a glorious period for Padua and the world, a fitting companion to the period, some three centuries earlier, when Giotto had decorated the nearby Scrovegni Chapel, reinventing the art of painting in the West. Between December 1 and 18, from atop his house, Galileo observed the phases of the moon, also creating seven wash drawings in sepia ink as a record. In his scrutiny of our satellite (or, rather, of the 59 percent of the lunar surface observable before the era of space exploration, the additional 9 percent being revealed by librations, which are very slow oscillations relative to the earth), Galileo was immediately struck by its ruggedness, a plain contradiction of Aristotle, who had insisted on the uniformity and incorruptibility of all celestial bodies. Galileo was also interested in the extremely clear contrast between light and dark surface blotches. The

very large dusky patches he called "ancient spots," because they had been seen since the beginning of history by anyone who cared to look—this was an obvious dig at Aristotle; the smaller spots had never before been detected. "The surface of the Moon is not smooth, uniform, and precisely spherical, as a great number of philosophers believe it (and other heavenly bodies) to be," he wrote soon afterward, "but, uneven, rough, and full of cavities and prominences, being not unlike the earth, relieved by chains of mountains and deep valleys." He clearly hoped this observation would put a speedy end to the recurrent theory that the crystalline moon had reflective properties and that its dark patches were only mirror images of the earth's geographic features.

Why mountains? Why valleys? He had studied the terminator, the dividing line between sunlight and shade on the moon's surface. In modern photographs the terminator shows not precisely as a line but as a transition, a zone of half-tone progressively darkening the surface of a sphere, as one might expect. But Galileo, seeing a less magnified moon with a fairly sharp terminator, had noticed a number of bright points just within its shadow side and dark points just within its bright side, and he had sensibly deduced that they were eminences and depressions. On the fourth or fifth day after the new moon, he further noticed that the small bright spots in the dark part increased in size and brightness and, after a few hours, coalesced with the bright part, which had gradually grown larger. On January 7, 1610, he wrote from Padua to Antonio de' Medici in Florence, an illegitimate but influential son of Francesco, the late grand duke. Galileo excitedly announced his improvement of the telescope, disclosing that "one can see most clearly that the moon is not of an equal,

smooth, and limpid surface, as she and the other heavenly bodies are believed to be by the great multitude of people, but seen from a closer vantage is actually rugged and unequal, and in sum the moon cannot be concluded in any sane discourse to be other than covered by eminences and cavities, similar to but rather larger than the mountains and valleys distributed around the terrestrial surface."

Galileo's letter to Antonio de' Medici—the first record we have of his observations of the moon—was accompanied by further descriptions, some small sketches, and a quick note concerning some curious "stars" in the neighborhood of Jupiter. All this he would develop further in a pamphlet, written in Latin, still the lingua franca of scientists, whose title, *Sidereus Nuncius*, may be translated either *Starry Messenger* or *Starry Message*, the former having found acceptance. It would appear soon afterward and fascinate the European scientific community. "Now on Earth, before sunrise," he wrote, further explaining his conviction that the moon's surface was craggy, "aren't the peaks of the highest mountains illuminated by the Sun's rays while shadows cover the plain? Doesn't light grow, after a little while, until the middle and larger parts of the same mountains are illuminated, and finally, when the Sun has risen, aren't the illuminations of plains and hills joined together? These differences between prominences and depressions in the Moon, however, seem to exceed the terrestrial roughness greatly." This shows Galileo's likely familiarity with the perspective treatises, often containing illustrations of the lighting of imaginary spiked objects (derived from a type of hat with an interior wicker brace known as a *mazzocchio*) that were proliferating during this period. He also addressed another question. If the moon was as mountainous as he

Sirigatti's mazzocchio

Illustration of a highly stylized *mazzocchio*, a headdress constructed on a wicker arma-
ture. From Lorenzo Sirigatti, *La pratica di prospettiva del cavaliere*, 1596. The understanding of the geometry of the *mazzocchio* came to serve as a demanding exercise for
architects, mathematicians, and students of perspective; its relevance to Galileo's conception of the moon was noted by Samuel Edgerton in *The Heritage of Giotto's Geometry: Art and Science on the Eve of the Scientific Revolution* (Ithaca and London, Cornell
University Press, 1991).

claimed—he did not speak of craters—why do we see its outline as smooth and not serrated? He explained that it was
almost uniformly jagged so that the rows upon rows of serra-
tions would, when observed from afar, obstruct one another
and appear smooth. But Galileo was not content to assert that
the moon had mountains; he wanted to measure their height,
which would strike a fatal blow at the whole Aristotelian
cosmology.

It was an audacious, invigorating idea. For one thing,
nobody in Europe (or outside of the Muslim East) had suc-

ceeded in consistently determining the heights of mountains
on the earth. With a theodolite, you could measure the angle
formed by your sightline to the peak of the mountain and the
plane you were standing on; you knew that an imaginary
plumb line dropped from the peak to the base formed a right
angle; and you could reasonably estimate the distance in
miles from your position to that right angle inside the base of
the mountain; all of which gave you enough to calculate
trigonometrically the desired result—in theory. The problem
was that unless you were looking at a mountain from the sea,
such as Mount Tenerife in the Canary Islands, a prized land-
mark for seamen, it was almost impossible to determine how
far the foot of the mountain lay above sea level, and that
queered your calculations. In 1644, two of Galileo's disciples,
Evangelista Torricelli and Vincenzo Viviani (spurred perhaps
by Descartes), had the idea of measuring heights barometri-
cally, since the air pressure in a glass tube containing mercury, if taken to higher elevations, will progressively fall—but
this method required climbing the mountain in question.
Practical methods for determining mountain heights were
not achieved in the West until the nineteenth century, and
they entailed lengthy surveying.

It was hard enough, then, to measure the mountains on the
earth. Galileo had of course never set foot on the heavenly
body whose eminences he was proposing to calculate. But he
saw this as an advantage, since some things are easier to measure from afar than from nearby: it's like solving a geometry
problem. In this case, he already possessed two constants: the
diameter of the moon, computed in Ptolemy's *Almagest*, and
the fact that sea level, often so hard to determine when one is
inland on earth, simply equaled the moon's spherical surface.

In addition, the terminator of the moon at first or last quarter—the half moon—which describes half the moon's diameter, could be considered, for geometrical purposes, to meet the sun's illuminating ray at a right angle.

Using his telescope, Galileo had noticed that some of the barely lighted points beyond the terminator lay at slightly more than one-twentieth of the moon's diameter inside this area of darkness. He took them as typical mountains. "Accordingly," he wrote, converting ancient Greek units into contemporary Florentine miles and reformulating the problem in plane geometry,

> let CAF be a great circle of the lunar body, E its center, and CF a diameter, which is to the diameter of the earth as two is to seven. Since according to very precise observations the diameter of the earth is seven thousand miles, CF will be two thousand, CE one thousand, and one-twentieth of CF will be one hundred miles. Now let CF be the diameter of the great circle which divides the light part of the moon from the dark part (for because of the very great distance of the sun from the moon, this does not differ appreciably from a great circle), and let A be distant from C by one-twentieth of this. Draw the radius EA, which, when produced, cuts the tangent line GCD (representing the illuminating ray) in the point D. Then the arc CA, or rather the straight line CD, will consist of one hundred units whereof CE contains one thousand, and the sum of the squares of DC and CE will be 1,010,000. This is equal to the square of DE; hence ED will exceed 1,004, and AD will be more than four of those units of which CE contains one thousand. Therefore the altitude of AD on the moon, which represents a summit reaching up to a solar ray GCD

and standing at the distance CD from C, exceeds four miles. But on earth we have no mountains which reach to a perpendicular height of even one mile. Hence it is quite clear that the prominences on the moon are often loftier than those on the earth.

The Aristotelian postulate of the perfection of heavenly bodies had been confidently proved false at one stroke, though the actual *height* of the mountains was immaterial.

Certain remarks are in order here. First, Galileo knew by this time that the moon shows us only one face, positive evidence of the important fact that it spins. (The time it takes for the moon to revolve around the earth is equal to its period of rotation about its own axis.) This meant that he had observed, at best, only half the mountains on the moon (not counting those concealed by its librations, of which however he would remain unaware for awhile). He used a doubtful conversion ratio for the ancient Egyptian *stadia* in which the earth's diameter was calculated, and took the ratio of the moon's diameter to the earth's as 2/7 (= 0.2857), whereas our estimate is 0.2727. Adjusting Galileo's calculations for the modern ratio, but letting his math stand, C. W. Adams found that a Galilean moon-mountain height would be at least 5,409 of our miles, or about 8,704,941 meters. The far side of the moon first became observable with the voyage of the Soviet spacecraft *Luna 3*, in 1959; if we leave this out of the picture, however, we may note that the height of the highest mountain on the normally visible side of the moon, Mount Huygens, has in fact been determined to be 5,500 meters, whereas the highest mountain south of the Alps that could be approximately measured then, Mount Aetna,

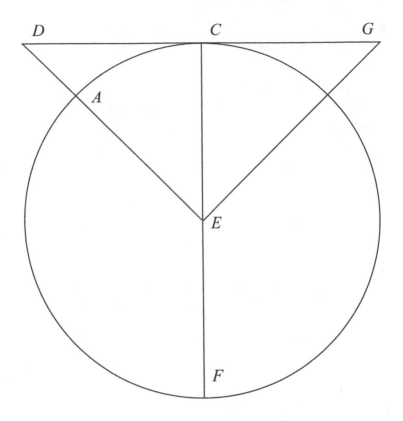

Galileo's Diagram

Galileo's proof of the minimal height of lunar mountains (from *The Starry Messenger*, 1610).

stands at 3,326 meters. (The height of Europe's highest, Mont Blanc, is 4,808 meters.)

BY LATE 1609, Galileo had managed to mount his telescope and train it widely over the skies, and in his long letter of January 7, 1610, to Antonio de' Medici he mentioned something beside the moon. "I have with respect to the stars noted this,"

he wrote. "First, that with the telescope one can see many fixed stars that otherwise cannot be discerned; and then, this very evening, I have seen Jupiter accompanied by three fixed stars, usually quite invisible in their smallness, and their configuration was in this form: ooo Nor did this form occupy more than a degree, roughly, in longitude." Three weeks later, while he was in Venice seeing to the publication of the *Starry Messenger*, he wrote his friend Belisario Vinta, the Tuscan secretary of state, that he had been struck by a wonderful circumstance: the fact that these three stars—only now there were four, since a fourth had swum into his vision on January 14—were moving in the same plane "as do Mercury and Venus and peradventure the other known planets." He asked Vinta to remember him to the ducal family, and told him that he would shortly send a copy of his new book and a good telescope to the Grand Duke Cosimo, then twenty years old, so that the duke could verify his assertions. Galileo had long yearned to return to Florence, and now he saw a golden opportunity.

In the meantime, on March 13, 1610, the *Starry Messenger* was published in Latin. Its edition of 550 copies fanned out across Europe and made Galileo an international celebrity. In it, he discussed not only the moon and various stars but also the remarkable phenomenon occurring in Jupiter's neighborhood, the one he had mentioned to Vinta. Attention has been drawn to Galileo's exceptional perspicacity in identifying this strange configuration amid the nocturnal excitement of pointing his telescope at hundreds of new stars never before seen, and then in successfully tracking, for weeks, its ever-changing appearance. What he described in the pamphlet, with diagrams, was a constant shifting of these "stars" now to

the east or west of Jupiter, but always in a straight line and always "in the line of the zodiac" (that is, the ecliptic, because the twelve constellations of the zodiac, seen from the earth, lie almost totally in the plane of the ecliptic). Since these bodies sometimes either followed or preceded Jupiter by regular distances, and since they accompanied the planet in its retrograde motion, Galileo soon concluded that they were moons. "Here we have a fine and elegant argument," he wrote, "for quieting the doubts of those who, while accepting with tranquil mind the revolutions of the planets about the sun in the Copernican system, are mightily disturbed to have the moon alone revolve about the earth and accompany it in an annual rotation about the sun." He was addressing the many skeptics who believed that if the earth revolved about the sun, it could not retain a moon, but his phrasing was patently prudent, granting a merely suppositional value to the Copernican cosmos. *If* the Polish astronomer's system is valid, Galileo seems to be saying, *then* the moon's revolutions surely track along with the motion of the earth, analogous to those of the Jovian moons.

The telescopic observations made Galileo one of the most famous men in the Western world. In a short period he manufactured many hundreds of telescopes, of which about 6 percent met his specifications—one gathers that the rest were not used—and sent the best by courier to royalty and to the aristocracy: to Duke Cosimo's cousin Marie de Médicis, wife of Henri IV of France, to the duke of Urbino, to the duke of Bavaria, and to the elector of Cologne. Cosimo himself was cleverly persuaded to use the telescope as a diplomatic gift, which turned him into a sort of ducal sales representative. The Jesuit Collegio Romano, which accepted most of Galileo's

conclusions (and would later turn against him), warmly received him in Rome. He was inducted into the Academy of Lynxes. Poets sang Galileo's praises in many languages, comparing him to Columbus, to Magellan. The *Starry Messenger* was reprinted in Frankfurt within a few months, and Kepler, the discoverer of the laws of planetary motion, wrote Galileo an enthusiastic letter.

It is never the case that a man, whatever his gifts, can get ahead on his merits alone, without securing any form of sponsorship from on high. This was all the more true in an age without intellectual-property protection, and Galileo knew it. In the past he had sought the support of the Venetian Senate, and now, at age forty-six, as the author of a revolutionary scientific pamphlet, he saw his chance to win a more powerful patron. Jupiter's moons were his to name as he pleased, and he wrote Belisario Vinta on February 13, asking which of two designations, the Cosmican Stars (*Cosmici Sydera*) or the Medicean Stars (*Medicea Sydera*), might best please Duke Cosimo. On February 20 Belisario replied that the latter appellation was—understandably—favored. Flattery got Galileo a very long way. By July of that year, after amiable negotiations with Vinta, he had been hired at an enormous salary as mathematician and philosopher to the court of the grand duchy of Tuscany, outraging the Venetians who had done so much to further his career. It was a fateful move. He would never again live in the Venetian Republic, a state that regarded with a jaundiced eye the Roman Inquisition's jurisdiction in accusations originating outside the Veneto.

Galileo's observations of the moon and Jupiter's satellites did not turn him into a full-fledged Copernican, though they certainly added arrows to his anti-Aristotelian quiver. The

turning point came with his observations of Venus. In order to appreciate their significance, however, one must first understand why the orbit of this planet was such a hot issue in 1610. Since time immemorial, observers have watched the sun describe a great eastward circle, the ecliptic, on the celestial sphere during the course of the year. The ecliptic passes visually through the twelve constellations of the zodiac, and it is also, roughly speaking, the plane of the planetary orbits. In ancient times it was noticed that whereas the stars maintain fixed positions, the planets appear to "wander," traveling at times to the east, like the stars, and at other times to the west, the latter motion being called "retrograde." Venus, for exam-

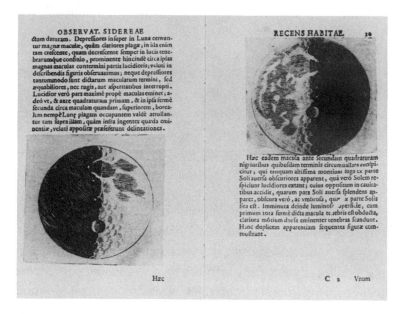

Two pages from Galileo's Sidereus nuncius (The Starry Messenger), *1610.*

These copper engravings were possibly executed in large part by Galileo himself.

ple, makes a sort of loop in the sky, which was accounted for
early on by the idea that it described a circle, called an epicycle, which when seen from earth gave it the appearance of
going in reverse.

In modern terms, this is explained by the "elongation" of
Venus. The angle formed by the line from a planet to earth
and the line from the sun to a planet along the ecliptic is called
its elongation, and a planet is said to be in "conjunction" when
the elongation is 0 degrees, that is, when it passes in front of or
behind the sun, forming a straight line. It is in "quadrature"
when the elongation is 90 degrees—when a planet forms a
right angle with earth and the sun, and in "opposition" when
the elongation is 180 degrees, when it passes behind the earth
relative to the sun. Venus is so close to the sun that its elongation never exceeds 48 degrees, and since it is an "inferior"
planet, closer to the sun than the earth, it can have neither
quadrature nor opposition. Of course, the ancient astronomers
did not see the solar system as we do now, but everybody
noticed that the remarkable thing about Venus, the "morning"
and the "evening" star, was that it kept so close to the sun. It
could not be at a wide variety of angular distances from the
sun, like the outer planets. Where, the ancients wondered, was
Venus's orbit?

Ptolemy's *Almagest*, which was still accepted in Galileo's
day and which Galileo himself had studied and taught when
younger, placed Mercury and Venus as inner or inferior planets circling around the earth between it and the sun. These
planets also described epicycles, subsidiary circles of their
own, rather like moons revolving around nothing; mathematically, the posited epicycles could be made to square more or
less with their retrograde motion as observed from the earth.

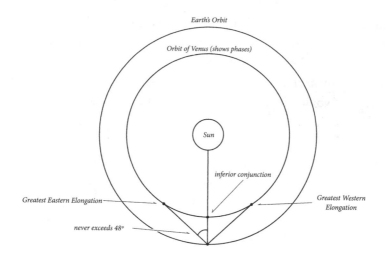

The Elongation of Venus

From this point of view, a principal objection to Copernicus's hypothesis was that Venus, being an inferior planet, should display phases yet did not. (This was also true of Mercury, a trickier case since it is closer to the sun and so harder to observe.) In other words, since in the heliocentric system Venus's orbit lay between the stationary sun and earth's orbit, we should be able to perceive varying phases of reflected sunlight on its globe—crescent, half, gibbous, and so on—rather like the moon's, depending on whether it lay in its eastward or westward elongation. Copernicus had correctly replied that the phases of Venus, though surely displayed, remained invisible to the naked eye.

Unfortunately, this also served as a rationale for the Ptolemaic cosmos, because in that setup Venus would be equally hard to see (though appearing differently from the earth, had we vision powerful enough to perceive it). In the Ptolemaic

cosmos, the planet, whirling around the earth and describing
its epicycle, would usually be illuminated, as we believe we see
it with the naked eye; but because the center of its epicycle
tracked unvaryingly along on a line between the sun and the
earth, it would never go behind the sun or assume a shape
fuller than a crescent. (It was also true that in Tycho Brahe's
system, Venus would show phases; but Galileo had never taken
the Tychonic system seriously, for reasons relating to celestial
mechanics. According to Tycho's hypothesis, all the planets but
the earth, including some much heavier than the earth,
revolved around the sun, which seemed *prima facie* to offend

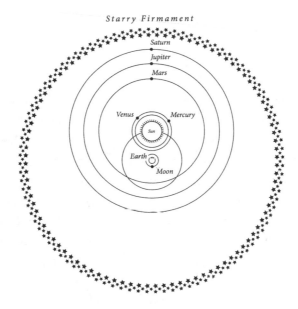

The Tychonic World System

Galileo seems never to have seriously considered Tycho Brahe's cosmology, although it
was not at variance with his observations of the phases of Venus.

the uniformity of natural law.) For most of 1610, Galileo was eager to have a look at Venus or Mercury with his telescope, largely to verify whose cosmos was the real one or had the most evidence to support it. To his great frustration, he could not do so, as both planets were usually too close to the sun.

At last, in October of 1610, Galileo began to observe the phases of Venus, and on December 11 he sent an anagram to this effect to Kepler, in Prague—such encryptions were then a common means of protecting an invention or discovery. Kepler failed to decode it, but Galileo's observations were epochal. They gave astronomy its first clear view of planetary motion and, by an intelligent inference, an idea of the revolution of the earth. In essence, Ptolemaic theory, unable to resolve the question of the earth's possible motion, had ended up regarding the planets' perceived motion—planetary motion on the observed celestial sphere—as their real motion. Denying that the earth moved, it failed to see that the celestial sphere, a visual construct, could serve as a useful tool only when astronomers bore in mind that the earth's real motion must be factored into the planets' perceived motion. Though Kepler had worked out the means for computing planetary motion about five years earlier, Galileo was now seeing that motion take place.

If his earlier observations had led Galileo to decisively reject the Aristotelian cosmos, it was his viewing of the sunspots, between 1611 and 1613, that caused him to declare outright his support of the Copernican system. The sunspots themselves had been seen during the realm of Charlemagne, if not before, and three scientists had recently preceded Galileo with similar observations: Johann Fabricius in Wittenberg, Thomas Harriot in England, and Father Christopher Scheiner, a brilliant Jesuit

professor of mathematics at the University of Ingolstadt in
Bavaria. But Galileo knew nothing of their work, and in any
case no one had proposed a valid theory concerning the nature
of this phenomenon, though Ludovico Cigoli, the painter-
architect, had corresponded with Galileo about the sunspots
and had sparked his further interest. Sunspots, as "spots," are
actually optical illusions: since the temperature in a sunspot, a
complex and impermanent electromagnetic event, ranges from
about 4300 kelvin in the umbra, the dark center, to about 5500
kelvin in the penumbra, the lighter surrounding region, the
sunspot itself, which is lower in temperature by more than a
thousand degrees than the photosphere, appears more or less
silhouetted against its background. Fascinated and perplexed
by what he could see of this phenomenon, Galileo intensively
studied the sunspots from Venice, from Rome, and from a
friend's villa in Tuscany during much of 1611.

Galileo's *Letters on the Sunspots*, of 1613, were penned as
replies to Mark Welser, a wealthy Augsburg merchant who
had sent a paper by Scheiner to Galileo. Scheiner, in publishing his speculations, had had to adopt a pseudonym,
"Apelles," for fear of compromising the Jesuits. On becoming
acquainted with Scheiner's views on the sunspots, which supposed them to be stars or planets moving at some remove
from the sun, Galileo formulated a lengthy reply, which
despite its civil tone embittered Scheiner for decades. "I confess to your Excellency that I am not yet sufficiently certain to
affirm any positive conclusion about [the sunspots'] nature,"
he wrote Welser.

> The substance of the spots might even be any of a thousand
> things unknown and unimaginable to us, while the phenom-

ena commonly observed in them—their shape, their opacity, and their movement—may lie partly or wholly outside our general knowledge. . . . Let them be vapors or exhalations then, or clouds, or fumes sent out from the sun's globe or attracted there from other places; I do not decide on this— and they may be any of a thousand other things not perceived by us. . . . I do not perceive the spots to be planets, or fixed stars, or stars of any kind, nor that they move about the sun in circles separated and distant from it. If I may give my opinion to a friend and patron, I shall say that the solar spots are produced and dissolve upon the surface of the sun and are contiguous to it, while the sun rotating upon its axis in about one lunar month, carries them along.

Galileo insisted to Welser, correctly of course, that the darkest sunspot was at least as bright as the brightest part of the moon, and only appeared murky because of simultaneous contrast. He went on to observe that though sunspots moved in a disorderly manner, they traveled as if the sun were carrying them along in a west-to-east rotation; that they were a "tropical" occurrence with respect to latitude; and that as they neared the sun's circumference and passed around out of view, they gave no sign, *pace* Scheiner, of separation or extrusion from the parent body.

Most of these speculations have been verified: the sunspots do emit great light; they are a tropical occurrence, seldom seen at heliographic latitudes greater than ±45 degrees; and they are not detached from the sun. Most importantly, the sunspots as markings could indeed allow astronomers to determine whether the sun rotates, and Galileo was the first to grasp this fact. (In 1860, Richard Christopher Carrington dis-

covered that the sun's rotation varied with respect to latitude, the speed at the poles lagging behind that at the equator by a ratio of about 0.74; the "proper" motion of the slower sunspots was so slow that they traveled from east to west with respect to the sun.)

In *Letters on the Sunspots*, Galileo often freely admits that he does not know or understand one thing or another. It could be, for example, that the sun is not revolving; that only the spots revolve; or that the spots might look different when seen from elsewhere on our globe, though this seemed implausible. In a remarkable passage, he grants that the sun probably revolves and then wonders what causes its motion, and here his earlier research in mechanics connects with his astronomical observations. Aristotle had had no conception of impetus, and thus no conception of motion corresponding to what we may see and measure. He thought that the medium through which objects travel sustains their motion. By contrast, Galileo wrote, "I seem to have observed that physical bodies have physical inclination to some motion," which he then described—lacking the mathematics for an exact characterization—by a series of "psychological" metaphors, themselves of partly Aristotelian origin: inclination, repugnance, indifference, and violence. "Inclination" meant gravity; "repugnance" indicated resistance to being pulled in a direction opposed to the force of gravity; "indifference" referred to a certain tendency of bodies to stay as they are; and "violence" referred to an external force setting something in motion or increasing its motion. Galileo's conception of the sun's motion is necessarily hesitant and ambiguous, and he was wary of flatly stating general principles. But one can perceive here the rough outline of what would become Newton's first law of

motion, or principle of inertia, namely, the tendency of bodies
to remain at rest or to continue to move with constant velocity
unless acted upon by an external force—what Galileo called
violence.

In response to another treatise by Apelles, the pseudony-
mous Father Scheiner, Galileo again wrote Welser, elucidating
his views on the sunspots and on the elongation of Venus. He
asserted among other things that the axis of the sun's tilt was
perpendicular to the ecliptic, an error he would soon rectify.
More importantly, disputing certain of Scheiner's claims about
Venus, he claimed in an excess of exasperation that any
knowledgeable astronomer who had read Copernicus's *Revolutions* would realize that Venus revolved around the sun and
that his understanding would also serve "to verify the rest of
the [Copernican] system." Toward the end of the letter, in a
discussion of the elusiveness of Saturn, whose form he felt
unable to establish with certainty, he suggested in a breathtak-
ing phrase that "perhaps this planet also, no less than horned
Venus, harmonizes admirably with the great Copernican system, to the universal revelation of which . . . propitious breezes
are now seen to be directed . . . , leaving little fear of clouds or
crosswinds." Though technically such confidences to Welser
did not imply the public teaching or imparting of heliocentrism, both men belonged to the Academy of Lynxes, and its
president, Federico (now Prince) Cesi, soon published their
correspondence. At last Galileo had openly declared for
Copernicus. That did not exactly put him in the good graces
of the Holy Office.

A number of scholars who had no connection with the
Inquisition were infuriated at Galileo's discoveries, or scornful
of them. One of these was the Aristotelian Cesare Cremonini,

at Padua, who loudly declined to look through the telescope. Another was Giulio Libri, at Pisa, who marshaled textbook arguments to wish away what the instrument plainly showed. A scamp named Martin Horky wrote a letter to Kepler claiming that Galileo's telescope revealed nothing when trained on the heavens. Such examples could be multiplied. The sheer refusal of these men to accept the usefulness of the new research tool, or sometimes even to put their eye to it, reminds us of the classical literary etymology of the Italian word for envy, *invidia*, whose Latin root (-*video*, "I see," plus the privative prefix *in-*, which reverses its meaning) means "not seeing," or "refusal to see," to accord recognition. There was, for awhile, this extraordinary need on the part of many people *not to see*, mostly in the sense of acknowledgment but sometimes in the brute physical sense.

For the Galilean telescope, essentially a tube fitted with glass, the human retina was part of the optical system, and it is possible that certain novice viewers—the *profani*, as they were known—really saw nothing on occasion. Many people, especially scholars, wore spectacles; the sky could be partly obscured; and the diameter of Galileo's eyepiece was, as noted earlier, about 1.5 centimeters. In addition, the eye had to be placed in a central position, which in this instrument was very near the eyepiece, in order that the arriving bundle of light rays could strike the retina intelligibly. This position could be a little hard to find, especially if one was impatient, skeptical, or pigheaded, as some of Galileo's detractors were.

When Galileo reported the motions of the moons of Jupiter, it drove certain people to distraction. They refused to believe that he had seen such a thing. "I must write of a harsh objection leveled at me by all the astrologers and many of the

physicians," a Neapolitan philanthropist wrote to a scholar at Padua in the spring of 1610. "These people say that if so many new planets are added to the number of those known, this will of necessity ruin astrology and demolish most of medicine, in that the distribution of the houses of the zodiac, the essential dignity of the signs, the quality of the nature of the fixed stars, the records of the star-chroniclers, the government of the ages of men, the months of the gestation of the embryo, oh, a hundred and a thousand things which depend on the sevenfold order of the planets—all will be destroyed from their foundation up." The writer wasn't convinced by these objections: after all, he quaintly noted, since the *luminosity* of Galileo's newly discovered celestial objects was the same as ever, why would their astral influence increase by one iota? But one senses the ferocity of the traditionalists' resentment. What the eye could detect with this funny "eyepiece-reed," or *cannocchiale*, as it was called in Italian, destroyed "a hundred and a thousand things," reaching into every aspect of human life. They didn't want to see; the invention had to be a hoax.

So the question immediately arose as to how Galileo could secure credibility or authority for his telescopic observations. Not only did no system of experimental repeatability exist, in that initially no one else had a telescope of the requisite power to verify his discoveries, but also he did not want mathematicians (as opposed to princes) to possess such instruments. It was a real dilemma. Very soon, of course, telescopes began to proliferate, in part because Galileo's invention passed into the hands of other scientists, and in part because rivals learned how to manufacture them. But in principle the problem of confirmation remained: six months

after the publication of the *Starry Messenger*, the only witnesses to the existence of Jupiter's moons whom Galileo
could produce were the Grand Duke Cosimo de' Medici,
Giuliano de' Medici (the Tuscan ambassador to the Hapsburg
court at Prague), and himself. One solution, introduced to
Galileo by Benedetto Castelli in 1612, and also practiced by a
number of others, including Ludovico Cigoli and Domenico
Passignano, the painters in Rome, who had obtained a telescope, was to transform the instrument into a primitive camera obscura by projecting the enlarged image onto a piece of
paper. Unhappily, this system assumed strong light-
collection—the light grasp of a telescope is proportional to
the square of its aperture—which was hardly true of Galileo's
instrument. So the camera obscura idea, though useful for
studying the sunspots, failed to work with the planets and
stars. Another possibility was to offer viewing sessions to
select notables, but such occasions were so dependent on the
weather and on the eyesight and open-mindedness of the
notables in question that they acquired a doubtful, séance-like
aura. Finally, and perhaps most interestingly, there was the
very modern option of the "virtual witness," or pictorial
record that could be universally consulted. Single images
would have certified nothing and would not have satisfied the
Baroque-era demand for a sense of how things moved. But
sequential diagrams of phases, rotations, and sunspot movements had every claim to credibility and provided a picture of
developments in time. In a way, they were like the frames of a
movie. Galileo perceived this almost instantly and offered in
the *Starry Messenger* sequential, comparative engravings of
the moon, in the first and last quarter. These engravings rendered the terminator as a sharp line to prove his point about

lunar geology: one could see the same prominences and depressions lit from opposite directions. He carefully indicated the bright spots falling just inside the area of shadow, the spots he identified as mountains, and it became hard to gainsay his evidence.

(There was an alternative solution to the problem of verification, which never greatly interested Galileo. It was to organize public sessions for independent observation by persons of truly irreproachable integrity, as Kepler did in Prague in September of 1610. Over several nights, using the telescope that Galileo had sent to the elector of Saxony, Kepler and three other viewers of standing confirmed the existence of Jupiter's moons.)

Galileo's wash drawings of the moon, on which the engrav-

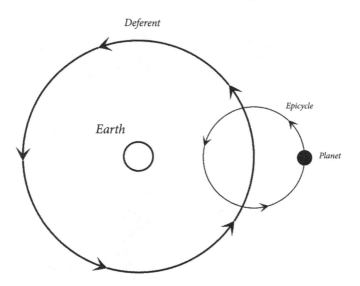

Deferents and Epicycles

Aristotelian astronomers posited deferents and epicycles to explain retrograde planetary motion.

ings were based, excite perennial fascination. Nothing of the sort had been done before. Botanical and zoological illustration abounded, but the preconditions for astronomical illustration—being able to see the subject—hadn't previously been achieved. One imagines how Galileo must have struggled with these little moon pictures, considering his tele-

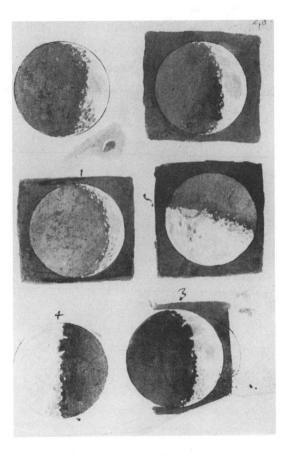

Galileo's Wash Drawings of the Moon (December 1609)

scope's minuscule field of view and the need to constantly refocus his eyes on a sheet of paper; yet this was the birth of an extremely important field: astronomical representation. Until recently, only seven studies were known to exist, all in sepia ink, six on one page and one on another. They were probably first drawn in faint chalk and later rendered in a succession of washes, the chalk traces afterward removed with *mollica*, compressed bread. As might be expected of one who had received a modicum of artistic training, the Tuscan physicist showed in these drawings, which were as delicate as anything one might hope to produce under the circumstances, an ability to capture the proportions of the moon's geological features. But the subsequent copper engravings in the *Starry Messenger*, which Galileo apparently had a large hand in, deliberately altered those proportions. In particular, several scholars have noticed the enlargement of one crater, probably Albategnius, to demonstrate the existence of lunar surface depressions analogous to terrestrial ones (Galileo had the valley of Bohemia in mind). In mid-March of 1611, he wrote to Cosimo de' Medici that he planned a deluxe version of the book with illustrations of the moon in all its phases. But this never came to pass.

Rather recently we have learned more about these drawings. On the basis of a libration of nine degrees vertically measured from Albategnius, Gugliemo Righini established in 1975 that Galileo drew the moon's first and last quarter on December 3 and 18, 1609. Librations are lunar oscillations relative to the earth that result in 18 percent of the moon's face being alternately visible and invisible, 9 percent of this surface being visible at any given time; they are caused partly by the moon's rotation being slightly out of sync with its revolution

around the earth and partly by the gravitational pull of other bodies. And another discovery was to come. The March 27, 2007, issue of *Corriere della Sera*, the Milanese daily, revealed that five such drawings, previously unknown to scholars, had been discovered in a first edition of the *Starry Messenger* that had recently come on the New York antiquarian book market. Galileo had drawn them directly onto its pages, so they might represent initial sketches for illustrations to accompany the deluxe edition for Cosimo—or maybe some other edition. These wash drawings have not been made accessible to the public, but the two photographs in the *Corriere* show versions of earlier images and not the beginning, so far as one can see, of a complete illustrated lunation.

Lunar cartography would enjoy a marvelous expansion in the seventeenth century, leaving us documents both informative and beautiful, but Galileo had no part in it. He was busy, his eyesight was poor, and the Galilean telescope's field of view could not be plausibly expanded. He limited himself to savaging other people's verbal and graphic descriptions of the moon, most of which he found contemptible. But one suspects that maps did not interest him much—they had too much distracting detail. Galileo had a deep faith in the epistemological value of images, and like the "ideal heads" of Leonardo, his moon pictures approached the world of Platonic form: they were truth made visible. Enlarging them or adding complications would only have compromised their geometrical integrity.

IT IS OFTEN asserted, in brief references to the subject and on the dust jackets of popular books, that Galileo's observations with the telescope convinced him of the truth of the

Copernican hypothesis and, in fact, proved it. This is, of course, mistaken. Nor did his observations with the telescope demonstrate the truth of the Copernican hypothesis, and he was under no illusion that they did. What they showed, momentously, was that the evidence for the heliocentric theory now greatly outweighed the evidence for the geocentric theory. Galileo was very mindful of the logical axiom known as the law of the excluded middle, which pushes you into one assertion if its contrary is demonstrated to be false, and the Aristotelian worldview (which he rather too easily conflated with the Ptolemaic) began to seem so false to him that he explored every possibility that the only alternative he considered was likely to be true. But if this intuition came closer to confirmation with each passing year, he never found a definitive proof for it. Such a proof could only have been provided by some evidence of the stars altering, even slightly, their apparent positions in the celestial sphere, which would signify that the earth was moving. As Tycho Brahe had asserted, having searched in vain for it with his vast battery of naked-eye instruments at Uraniborg, on the Danish isle of Hven, this evidence did not, or not yet, exist.

The apparent relative displacement of the stars, or stellar parallax, is not the only evidence that the earth moves.* Light from the stars is also affected by the earth's constant motion, arriving at an angle—the "aberration of light"—so that to view a star, a telescope must be tilted in the direction of the earth's revolution. Consequently, when James Bradley first measured

---

* The Doppler effect also proves the earth's motion, though this necessitated the discovery of the wave properties of light and so is not relevant to this period.

this tilt, in 1727–28, a moving earth could be inferred. But the apparent angular shift of the stars against nearby bodies as our planet revolves about the sun had since Aristarchus been regarded as decisive proof for the heliocentric theory, if anyone should be able to obtain evidence of it. Stellar parallax was proved only in 1838, by Friedrich Wilhelm Bessel, a feat that also rendered possible the trigonometric computation of the distance of the closer stars.

One sometimes reads claims to the effect that the Vatican rejected Galileo's heliocentrism because of the inadequacy of his proofs for it, but this is mistaken. Galileo never claimed to offer absolutely conclusive proofs. And the Holy Office, as stated earlier, did not rigorously dispute his science but merely warned him not to champion unorthodox ideas—at least until his condemnation in June of 1633.

GALILEO'S STUDY OF the moon helped validate his long-standing suspicion that the direct observation of nature would undermine the postulates of Aristotelian natural philosophy. But it also ran counter to religious beliefs and folkloric traditions, and one may speculate that offense to those beliefs and traditions stoked the resentment of some of his opponents. As it happened, the moon, identified in Greek antiquity with the goddess Selene and associated with female fertility, had acquired a nearly unassailable status in two branches of Roman Catholic thought. One was the theology of Saint Thomas Aquinas, in which the moon enjoyed an incorruptibility directly derived from Aristotelianism. In his "Three Articles" concerning the fourth day of creation, in Supplement III to the *Summa Theologica*, Aquinas asserts that the heavenly bodies are "living beings" which can "impart life" in

their service as agencies of an "intelligent power." Though to question Aquinas was by no means heretical—his opinions were not articles of faith—such probing might be viewed as dangerously rash. Small wonder, then, that in the spring of 1611, when the Jesuits of the Collegio Romano were studying the *Starry Messenger*, whose conclusions they approved on the whole, two things bothered them: the description of Saturn, on purely observational grounds, and the imperfection of the moon. Father Christopher Clavius, the distinguished German mathematician who headed the Jesuit Commission, absolutely insisted on the moon's smoothness, suggesting that any blotching of its surface was caused merely by a tint unequally spread throughout its body.

The other tradition was the rising doctrine of the Virgin's immaculate conception. Though still unorthodox, this enjoyed wide popularity in the early seventeenth century, especially among the Franciscans, whose patron saint, in his *Canticle of the Creatures*, had devoted a verse to "our Sister Moon . . . clear, precious, and beautiful." The fundamental idea that Mary, like Jesus and unlike the rest of us, had been born without original sin was deeply rooted in popular piety by 1609 (though it did not become Roman Catholic dogma until 1854). It stems from the perception of a paradox: that a child, as the product of sexual union, must itself be naturally sinful unless divinely exempted, and that nonetheless Mary, as the mother of God, could hardly have been born of sin. The learned doctors of the Church were divided on the question. Saint Bernard of Clairvaux attacked the idea of Mary's sinless conception as a mere superstition. Saint Thomas Aquinas held that Mary was not exempt from original sin because Christ was the savior of all men, and so Mary had need of His

redemption, though he conceded that Mary's soul had been cleansed of her sin some time before her birth. Duns Scotus, a Franciscan, subtly argued that although sinless, she would have contracted original sin had it not been for Christ's intervention, which seemed to reconcile her sinlessness with her need for salvation. For centuries, the doctrine of the immaculate conception was preached by the Franciscans and opposed by the Dominicans, with the Franciscans gradually gaining the upper hand.

That Mary's freedom from sin should be connected in later iconography with the image of the moon belongs to a strain of Catholic thinking, illustrated by much beautiful art, which teaches that shadows or intimations of what will come to pass may be discerned in antecedent revelations. This notion of recurring "types" of human behavior is essential to Christian thought: as the tree of life foreshadows the cross, for instance, so the prelapsarian or initially sinless Eve foreshadows Mary. Though it was hard to dispute the frivolous and probably pagan origin of the Virgin's symbolic association with the moon, the first two verses from Revelation 12 appeared to confirm it: "A great and wondrous sign appeared in the heaven: a woman clothed with the sun, with the moon under her feet and a crown of twelve stars on her head. She was pregnant and cried out in pain." For this reason, the moon, though accorded no cult, became associated with Mary in folklore and honored in later medieval and Renaissance symbolism and devotional painting. Luna was Mary's *impresa*, her heraldic emblem, and to impugn Luna was, in some minds, almost to impugn Mary herself.

Whatever he may have made of such notions (and he

appeared to criticize them *sotto voce* in a letter to Gallanzone Gallanzoni of July 16, 1611), Galileo's favorite work of Renaissance verse, Ludovico Ariosto's *Orlando furioso*, the long heroic poem first published in 1516, which he frequently cited and of which he possessed a heavily annotated copy, features a delightful moon voyage that presages Galileo's own debunking of the crystalline moon. One might even say that rather in the manner of a Christian "type" or prophecy, it strangely foreshadows Galileo's own lunar observations. Ariosto's passage was intended in part as a sort of reprise, in a somewhat burlesque key, of Dante's moon ascent in the *Paradiso* (Canto II, lines 31–36), a passage that Galileo, as a lover of Dante, must have known.

Dante was a serious student of medieval astronomy, and the *Paradiso* was intended to reflect the actual world-picture of Italian savants in the early fourteenth century. Here, as we behold Beatrice and Dante the protagonist ascending into the heavens, he attentively notes the moon's physical nature:

> *It seemed to me that a cloud surrounded us,*
> *brilliant, dense, solid, and unsullied; adamantine,*
> *and as though penetrated by a sunbeam.*
>
> *Into itself the eternal gem received*
> *Us, as water may receive a ray of light*
> *And yet will not divide, but stays united.*

Curiously, the moon's substance, at once solid and cloudlike, is conceived here as entirely enveloping the pair. Without fussing unduly over the five qualities the bard attributes to the moon—

brightness, density, solidity, glossiness, and adamantine hardness—we may note that it reflects standard Thomistic doctrine. The "gem" stands for the moon itself, while the metaphor of light penetrating water without decomposing it, referring to the pair's ecstasy, was also a standard trope for the fertilization of the Virgin's womb.

Now what does the jaunty Ariosto make of the traditional Aristotelian moon? The issue only arises because Orlando, the hero of his poem, has lost his sanity, as many a good knight has done before him. In the thirty-fourth canto of the *Furioso* we hear Saint John the Evangelist tell another paladin, Duke Astolfo, that the only way poor Orlando can ever recover his wits is if Astolfo undertakes a journey to the moon, traditional instigator of "lunacy." So when night falls, the evangelist hitches four horses to a chariot and they climb through a sublunary ring of fire, believed at that time to produce the comets, toward our gleaming satellite.

Having crossed that fiery sphere they arrive
at the realm of the moon, which looks like a steel plate,
entirely spotless, and about the same size, I've
been told, as the earth—that would include our great
oceans which add to our globe. After the drive,
which has not taken long, I would estimate,
Astolfo expresses his astonishment and surprise
that the moon, which looks small from earth, is of such size.

The other unexpected revelation is that
the earth is hard to discern, emitting no light
by which to perceive it. From the distance at
which Astolfo is standing the earth is quite

indistinct, its pleasant habitat
not projecting far out into the night.
With difficulty he can get some hint
of where the oceans are—but he has to squint.

On the moon there are rivers and lakes and hills and dales,
Like those we have but different. And also towns
With houses and public buildings, but on such scales
As we are not accustomed to. He frowns
In amazement and concentration, for earth pales
In comparison. There are also woods and downs
Where nymphs and fawns are hunting fierce moon beasts
And celebrating afterwards with feasts.

Duke Astolfo does not pause to explore
every feature of the moon. He is
there to transact business which is more
pressing. The Apostle is aware of this
and leads him downward to a valley floor
where all that we have lost or has gone amiss
through Time or Fortune or our own grievous fault
is collected and stored as if in a huge vault.

I do not speak only of realms and gold
That Fortune's unstable wheel gives or takes back.
But also those things beyond what she can hold
Or give—Fame, for example, which the attack
Of Time can devour before it has grown old.
Up there as well are countless prayers our slack
belief has offered up, and vows that were broken
very nearly as soon as they were spoken.

Ah, and the tears and sighs of lovers, and all the hours
That gamblers lose and ignorant men waste,
The plans that we make that are well within our powers
But require perseverance if not haste
Before they fade away. And books of ours
That we intended to study rather than taste.
Those can often be heavier losses than
Material things in the life of any man.

Among other lost or mislaid items that the apostle and Astolfo find on the moon are women's charms, amorous vows, flopped love affairs, threats, royal crowns, gifts and flattery lavished upon princes, disastrous plots, villains' schemes, charity postponed until the benefactor's death, and—finally—a large flask, full of a "thin liquid, apt to evaporate," containing Orlando's wit. So Orlando will recover his sanity. Leaving aside Ariosto's other astronomical details, such as the fact that the earth reflects so little light, what Galileo must have remembered in 1609 was the moon's initial appearance in these stanzas as a uniform "steel plate" and its subsequent revelation to be a body possessing "hills and dales, Like those we have but different." One can imagine him smiling wickedly at these verses as he calculated the minimum height of a lunar mountain. Yet, in a development at once delicious and dismaying, on reading the *Starry Messenger* Kepler wrote Galileo a letter describing in loving anthropological detail the likely attributes of a lunar civilization somewhat similar to Ariosto's: "They have, as it were, a sort of underground city. They make their homes in numerous caves." The level-headed Galileo demurred: He had seen no clouds upon the moon's face. Without water, how could there be life?

In the meantime, the Catholic Church was proceeding apace with another system of representing the heavens, one that was not scientific but imaginative and propagandistic. The Church had invested an enormous capital of intelligence in astronomy, partly for calendric and doctrinal reasons and partly because certain churchmen, often Jesuits, happened to have the mental gifts and education to pursue the subject. It was largely squandered with the injunction against Copernicanism in 1616 and with the Collegio Romano's espousal of the Tychonic, or geoheliocentric, system. Yet as the construction of domed churches proliferated and architects and designers acquired the same mathematics that astronomers used—that is, Euclidean geometry and trigonometry, including spherical trigonometry—it became possible to enlist a specialized branch of perspective to create the illusion of celestial space on the inner surface of a hemisphere or similar shape. Thus, the tradition of the *cupola dipinta*, or painted dome, came about, together with all the correlative ambiguities of the *sottinsù*, the view from below looking upward.

Both domes and planetariums had been erected since antiquity, and the parallel between the church dome and the celestial hemisphere—the apparent firmament in which we perceive the stars, with the observer's zenith shifting as he moves—was perfectly obvious. The circumference of the base of the dome was the metaphorical equivalent of the astronomer's celestial equator. Within the *spazio divino*, or consecrated space of the cupola, it fell to the commissioning prelate, in tandem with the painter and his team, to agree on some manner of representing heaven, virtually always as an ecstatic apparition. By the early 1500s, small frescoed cupolas of this type were already appearing in

northern Italy, but the first great painted dome was executed by Pordenone in 1519 for the Cathedral of Treviso. Between 1520 and 1530 Correggio in Parma created *Vision of Saint John at Patmos* for San Giovanni Evangelista and *Assumption of the Virgin* for the Duomo. The latter in particular, with its vision of heaven opening to a vast angelic flutter, set the tone of future ceiling frescos, while the expanding girth of the domes themselves reflected a growing awareness of the size of the universe.

Many early painted ceilings showed concentric images of heaven, whose processional circularity, reminiscent of a vast carousel, bore an eerie and ironic resemblance to diagrams of the Copernican solar system, especially when centered on a light-transmitting lantern. (A lantern is the glazed cylindrical or polygonal structure that usually sits atop a large dome.) I have found no evidence that the general shift during Galileo's lifetime to dome-painting characterized by a degree of asymmetrical turbulence, observable notably in Giovanni Lanfranco's *Assumption of the Virgin* in Sant'Andrea della Valle, in Rome, of 1625–27, betokened a conscious attempt to depart from this "Copernican" concentricity (and actually the concentric scheme continued in use). The painted dome, with its gravity-free angels, saints, and cherubs, lasted about three and a half centuries as a living art form, defying Newton's law of gravity, and its morphology does not follow a linear progression. However, it passed through critical moments of renewal, and a hinge figure in its evolution was Galileo's friend Ludovico Cigoli, an architect and painter of sacred history whose work, like that of Federico Barocci and the Carraccis, anticipated major features of the Baroque style.

In 1610, Cigoli was fifty-one years old, five years older than Galileo. He would die in 1613. Partly because of Galileo's

Cigoli's Dome

Ludovico Cardi Cigoli's fresco of the Madonna Immacolata in the cupola of the
Pauline Chapel of the Basilica of Santa Maria Maggiore, in Rome (completed 1612).

influence, he had become convinced of the truth of the
Copernican system. Cigoli felt that those who persisted in
upholding the doctrine of the crystalline moon not only erred
scientifically but revealed their impious envy, for *invidia*
implied not only the refusal to see what genius had discovered
but also what God had created. On March 18, 1610, Cigoli,
then living in Rome, wrote Galileo that

upon the arrival here of Sig.r Ferdinando Martelli, I am
fairly obliged to salute you with this missive; and receiving a
letter written to me from Florence by Sig.r Amadori, to
rejoice with you who have raised your telescope to such perfection that it has been able to perceive and observe marvelous things in the heavens; and I read that you have given
a lecture of some sort concerning all this [a reference to the
Starry Messenger] and have been to Venice to have it
printed. . . . Finding myself as it happened with Sig.r Cardinal del Monte [Caravaggio's principal patron, a supporter of
other artists, and an optics enthusiast] and turning to this
topic, I read him Amadori's letter, and he at once ordered
one of his envoys in Venice to procure the said telescope,
and as soon as the book should be printed to send it to him.

Between 1610 and 1612, Cigoli, corresponding with Galileo
and experimenting with perspective projections upon curved
surfaces, created the *Immacolata*, a superb fresco in the cupola
of the Pauline Chapel of Santa Maria Maggiore, the major site of
Marian devotion in Rome. During this period he and his colleague Domenico Passignano, took time off to peer through a
telescope they too had obtained, sometimes using the great
church as an observatory from which to view the monuments
of Rome and, eventually, celestial bodies, and notifying Galileo
of their sightings. (Passignano saw the sunspots before Galileo
did, using a colored filter to protect his eyes and describing
them as "lakes or caverns.") Cigoli's *cupola dipinta* is not merely
concentric in structure but dynamically rotatory in that its
shapes suggest a whirling motion, perhaps inspired by his read-
ings in astronomy. The one static element is the immaculate
Virgin, who stands upon a crescent moon clearly pitted and

Cigoli's Madonna

Line drawing of the figure of the Madonna in Cigoli's fresco in the Pauline Chapel of Santa Maria Maggiore. She stands on her emblem, a crescent moon, but it is pocked with crater-like cavities, a violation of the Aristotelian doctrine of the perfection of celestial bodies.

scarred, just as Galileo had drawn it. So Cigoli's moon, floating in a chapel in anti-Copernican Rome, is defiantly un-crystalline and un-Aristotelian, and geologically akin to Galileo's, though it does not depict any part of the moon's actual geography.

Throughout the late sixteenth and seventeenth centuries, dome after dome was erected and painted to inspire faith and awe, and to hold out the promise of salvation. But if the Church

hoped that these lofty extravaganzas would offer devotional tran-
quillity, that hope was frustrated by weird perspective problems.
Had the imperatives of religious decorum and, increasingly, a
certain theatrical seduction not been paramount, the problems
could have been readily solved. But the designers soon recognized that given the demands of visual legibility, they were very
hard to solve, for within these great temples one had to consider
mobile observers in relation to huge hemispherical or polygonal
ceilings, and in their proposed solutions a curious perceptual relativism crept in that was cousin to some of Galileo's discoveries.
Galileo argued that our perception of moving bodies was subject
to a kind of relativity, because we ourselves conducted our observations from a moving body, the earth; now a somewhat analogous relativity cropped up with respect to our perception of very
large hemispheres.

We are not comparing two mathematical models here, but
juxtaposing a mathematical model and an expressive art form;
it is, however, the optical rather than the aesthetic properties of
the painted dome that concern us. And we see at once that an
aesthetic problem, of a socio-religious nature, generated a host
of difficulties for the dome-painter. Suppose your aim as a
dome-painter is to create a ceiling that works as an illusionistic
extension of the worshipper's actual space—you want to co-opt
him, to include him in a moment of prearranged ecstasy. If so,
the entire perceived design of the room, both real and fictive,
will have to function as a continuum, no matter where on the
floor the observer stands. And this is indeed possible in a very
small room, such as Andrea Mantegna's Stanza degli Sposi in
Mantua, with its little trompe l'oeil cupola. As soon as you're
dealing with a big church dome, however, two problems arise.
First, the total area of the wraparound hemisphere is larger

than the arc subtended by the angle of human vision, so that wherever the observer happens to be, even at the center, some of the painted architecture will appear anamorphic, that is, distorted, and certain of the painted figures will almost certainly appear upside down. Second, an accurate perspectival projection of figures on the dome will leave them too small to have any psychological impact, and too spatially compressed to be comfortably legible. Your teensy saints and angels will diminish to dwarfishness as they ascend toward your vanishing point, which is heaven's zenith, and they will lack the foreshortening that includes them in the observer's world. You, the dome-painter, know that such effects will undermine both the religious decorum and the sensuous enchantment that the Counter-Reformation has come to embrace.

What could be done? Well, the theorists of the Baroque devoted a lot of thought to these problems, culminating in the invention of *quadratura*, a branch of perspective devoted to projecting geometry onto vaults and domes (perspective being a system of drafting rules derived mathematically from geometrical optics). But all such designing, whether aimed at depicting a concentric or an asymmetrical heaven, faced the same persistent problem of figures that were too small and too spatially compressed. In response, some painters intuitively concluded that the composition on the dome should reflect not the real observer's viewpoint, which sent the painted heaven too far away, but a virtual viewpoint much closer to the ceiling, which allowed for much larger and more legible figures. As such a viewpoint ruptured the illusionistic continuity of the dome's space with the observer's space, however, and left it convincing only if seen from one part of the floor, the idea arose of using balustrades, grilles, or other devices to confine

the observer to this optimal position. One of the major exponents of *quadratura*, Andrea Pozzo, inserted a marble disc in the pavement of a Jesuit church, the Sant'Ignazio di Loyola in Rome, whose ceiling he had painted, and indeed from this viewpoint the illusion is overwhelming; but as one moves away, the soaring trompe l'oeil architecture creates a dizzying effect and threatens to cave in. Pozzo was well aware of this

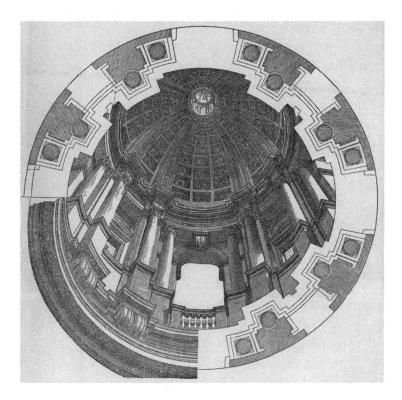

Andrea Pozzo's Design for the Cupola over the Crossing of Sant'Ignazio di Loyola, in Rome (painted ca. 1685).

Pozzo stipulated that the cupola was to be seen only from the vantage point implicitly indicated by this view, the vanishing point being equated with God's glory.

illusion, and at the end of his life wrote an influential treatise on *quadratura* in which he pleaded for a unifocal, off-center, stationary viewpoint. Harking back to what the art historian Samuel Edgerton has called "the traditional medieval belief that God spreads his grace through the universe according to the laws of geometric optics," he exhorted painters to view the perspectival vanishing point, or infinity, as an expression of God's glory. Pozzo's marble disc symbolized all too clearly and unhappily the doctrinaire standpoint of the Collegio Romano vis-à-vis mathematicians like Galileo and Kepler. Just as the

Computerized, Anamorphic Image of Pozzo's Cupola When Seen from a Peripheral Viewpoint Not Authorized by Him

Actually, the cupola looks even stranger than this when seen from beneath its circumference, but today it has darkened so much that it cannot be usefully photographed. Galileo despised the anamorphic effect, and in fact the binding of viewers to a fixed point under the dome came to serve as a metaphor for the Jesuit refusal to accept Copernican reasoning.

Jesuits insisted that the earth stood still, so they also pinned observers to a single point in the church of Sant'Ignazio: move, and everything collapses. It was the perfect visual symbol of a gigantic intellectual self-deception.

Meanwhile, some painters opted for the alternative strategy: diffuse, dissociated imagery designed to be seen successively from a moving axis. Mattia Preti seems to have decided that a multifocal image could be best enjoyed by a mobile observer whose itinerary imaginatively coincides with an array of virtual viewpoints. This approach also acquired many adherents: in Giovanni Battista Piazzetta's *Glory of Saint Dominic*, a late example in Santi Giovanni e Paolo in Venice, one can make out no fewer than five scattered vignettes on the ceiling, all bathed in the same golden soup, yet each with its private vanishing point.

There is, in this tacit admission of the need, under any vast frescoed ceiling, to circulate in order to align oneself with multiple focal axes, a Baroque parallel with Galilean relativism, though of course it concerns optics, not celestial mechanics. Yet if Galileo's astronomical contribution consisted in turning the Copernican theory into an actual model, capable of verification, the Baroque church-dome was not a rival model: if it had been, it would have resembled a gigantic armillary sphere in stone. It was, rather, a metaphor, something of a pious distraction, and as the century progressed, the decorative skills it enlisted were unabashedly derived from those used for the stage. It must be granted that many people no longer find frescoed domes very compelling: of all the excesses of the Baroque style, these neck-bending heavenly carnivals are among the least popular, and even devout believers may find them too extravagant to touch the soul, which is

what they were intended to do. But they are marvels of engineering, design, and virtuoso painting, and they represent the Church's liveliest response—if only an implicit one—to the threat of Copernicus and Galileo.

GALILEO'S DISCOVERIES AND essays had not failed to produce an effect. On February 24, 1616, alarmed by the spread of writings sympathetic to the idea of the earth's motion, the Inquisition issued a decree condemning Copernicanism as "foolish and absurd in philosophy, and formally heretical" because it contradicted Holy Scripture. On March 3 of the same year, the Congregation of the Index—the branch of the Holy Office responsible for censorship—declared that two such works, Copernicus's own *Revolutions* and Father Diego de Zuñiga's *Commentary on Job*, were to be "suspended until corrected." A third, the *Letter on the Pythagorean and Copernican Opinion of the Earth's Motion*, by the Carmelite friar Paolo Antonio Foscarini, was "prohibited and condemned." Galileo was not mentioned, presumably out of respect for the Tuscan ducal family and the prestige he enjoyed with the Jesuits. In case he should fail to get the message, however, he had been summoned a little earlier, on February 26, to the residence of Cardinal Robert Bellarmine, the head of the Inquisition and the greatest Roman Catholic theologian of the period, to receive a special injunction. The minutes of this meeting reveal that Bellarmine warned him to completely abandon his Copernicanism, and "henceforth not to hold, teach, or defend it in any way whatever, either orally or in writing; otherwise the Holy Office would start proceedings against him. The same Galileo acquiesced in this injunction and promised to obey." In the meantime, during much of 1615 and 1616, the Inquisition

investigated Galileo on suspicion of heresy but found little in the way of solid evidence, and the case was dismissed; since the charges were revived in 1633, however, we will examine them shortly.

It is a chastening task to read through the documents pertaining to Galileo's relations with the Church in search of some longed-for, blessed text in which Copernicanism is substantively discussed. Revealing, though hardly substantive, is the implicit verbal duel between Bellarmine and Galileo that took place when Galileo was being investigated. Saint Robert Bellarmine, or San Roberto Bellarmino, as the Italians call him (he was canonized in 1930), is generally regarded as the most incisive champion of the Counter-Reformation. A tiny, gentle, exquisitely well-mannered man, he had written what the Vatican regarded as the definitive refutation of the Protestant heresy, the multivolume *Disputations on the Controversies of the Christian Faith* (1581–93), whose careful argumentation had earned even its opponents' respect. He had lectured at Louvain, practically on the front lines of the wars of religion, and had spoken out eloquently against English King James I's anti-papal Oath of Allegiance. As a Jesuit cardinal he had served on the tribunal that condemned Giordano Bruno to death—the free-thinking philosopher was burned at the stake in the Campo dei Fiori in Rome in 1600—though Bellarmine later confessed to being plagued by remorse over Bruno's refusal to repent and receive last rites. Bellarmine had reached his early seventies by 1616 and was nominally bound by the Jesuits' "rules of study," which closely followed Saint Thomas Aquinas; yet his own theology hewed much closer to the literal meaning of Scripture than had that of Aquinas, a grave misfortune for Galileo.

Robert Bellarmine, in a letter of April 12 warning Friar Antonio Foscarini to abjure his heliocentrism, explicitly cited Galileo along with him. In the third numbered point of this elaborate missive, Bellarmine stated,

> I say that whenever a true demonstration would be produced that the sun stands at the center of the world and the earth in the 3rd heaven, and that the sun does not rotate around the earth, but the earth around the sun, then at that time it would be necessary to proceed with great caution in interpreting the Scriptures which seem to be contrary, and it would be better to say that we do not understand them than to say that has been demonstrated is false. But I do not believe that there is such a demonstration, for it has not been shown to me. To show that the sun in the center and the earth in the heavens can be made to match our astronomical calculations—this is not the same as to prove that in reality the sun is at the center and the earth in the heavens. The first demonstration, I believe, can be given, but I have the greatest doubts about the second. And in the case of doubt one should not abandon the Sacred Scriptures, as interpreted by the Holy Fathers.

To this argument Galileo silently responded with a ferocious memorandum, which today goes by the title "Considerations on the Copernican Opinion," in which he explained that neither Copernicus nor he regarded the heliocentric setup as suppositional at all, but in all probability as factually true. Luckily for Galileo, this piece never got farther than Prince Cesi, who apparently advised him not to send it to the head of the Inquisition.

Bellarmine's words merit close analysis. Elsewhere Bellarmine had pointed out that to deny the spoken is to deny the speaker, which in the case of the Scriptures is God; so far so good, one follows his reasoning. To this divine authority he adds that of the Holy Fathers, on the grounds that the Church as an institution is inerrant, which is already going rather far; but again one follows his thought—after all, the cardinal represents the embattled Counter-Reformation. What happens next, however, is fascinating. Bellarmine makes a perfectly correct distinction between mathematical or suppositional astronomy and physical astronomy, as though he were quite familiar with such matters, and declares himself willing to accept a purely suppositional Copernicanism. But what if somebody like Galileo were to insist that the sun, as a physical body, really occupies the center of our world? That is where the trouble starts. If a sound proof were offered him, Bellarmine says, he would have to reinterpret Scripture; but can a sound proof actually *be* offered? Of this he has the greatest doubt, and when in doubt, he relies on Scripture, as interpreted by the Fathers. These are circular sophisms, as the Italian philosopher Guido Morpurgo-Tagliabue has written, which "under the pen of Cardinal Bellarmine reveal that to command it is not necessary to reason: in fact, one loses the habit." This letter of the otherwise enormously intelligent Bellarmine, which is the closest the Church ever came to a rational argument specifically directed at Galileo, seems nothing but a gentle command.

If we continue our search, we discover that a few years later, in 1619, Galileo became embroiled in a controversy with Father Orazio Grassi, a Jesuit and later the architect of Sant'Ignazio, over the origin of the comets, which Grassi held to be a

but had declined to adopt heliocentrism because of its scientific and theological shortcomings. Perhaps Galileo never fully understood the pope's expectations, or perhaps he forgot his own job description somewhere along the line. In a way, this was understandable: after all, the Inquisition's "corrections" to Copernicus's *Revolutions*, applied in 1620, amounted to only thirteen passages that failed to specify the system as suppositional (and many extant copies from that period, including the one in the very library of the Collegio Romano, show no emendations). It was reported, moreover, that the pope had laughed out loud at dinner at Ciampoli's reading of *The Assayer*, which made open sport of the Jesuits. So Galileo slowly began work on the huge, self-incriminating manuscript that would become the *Dialogue Concerning the Two Chief World Systems*; after long, intermittent efforts, the book appeared for sale in Rome in May of 1632.

By the summer of that year, the Inquisition was once again investigating Galileo, this time on grounds that turned into the two implicit counts of an unwritten indictment. The first was that he had hoodwinked the Vatican into issuing him a license to publish the *Dialogue*, a point we will come to later on. The second was that the new book championed Copernicanism, which, as we have seen, was a formal offense since 1616. Considering, then, that Galileo did indeed write the book to suggest the likely truth of the heliocentric thesis, and that this was the charge leveled against him, the modern reader might expect it to adopt a straightforward polemical line. One might assume, for example, that it would offer a complete exposition of the heliocentric cosmos, and then rebut to objections; or that it would take the opposite tack, and demolish the geocentric cosmos, supplanting it at once

with that of Copernicus; or that it would juxtapose the two, and reveal which suited the facts better deductively and empirically—one can imagine a number of obvious outlines for the book. But Galileo followed none of them, because they would not have interested him. Nor would they have made much sense in the 1620s.

Galileo chose the dialogue form, spun out in the Italian vernacular rather than Latin, the international language of science, for three important reasons. First, in showcasing his literary skill, he hoped to reach a large audience whom he might captivate and entertain. Second, he could place his scientific arguments in the mouths of fictitious characters, and so perhaps dodge ultimate responsibility for them. Finally, by adopting a modified form of the Platonic dialogue, he could deploy a variant of the Socratic method, a form of conversational teaching that enlists the antagonist's own power of reason to convince him of the truth of an idea. Galileo apparently believed, like Plato, that the mathematical shapes of the world and those conceived by man's mind are corresponding reflections of each other. Human intelligence, properly honored, consists in a search for the simplest and most beautiful geometrical forms that will fit and explain the empirical facts, and as Socrates maintained in certain of Plato's dialogues, we human beings possess these forms somewhere in the recesses of our memory. True philosophy consists in remembering what is, at bottom, inborn knowledge. In this sense, the argument with those who believe in the earth-centered universe is really an attempt to induce them to remember what they intuitively know but have forgotten. The most poignant moment in the *Dialogue* occurs when Salviati, the champion of Copernicanism, draws out of his opponent Simplicio, the exponent

of Aristotelianism, the long-submerged, almost dreamlike awareness that the earth circles the sun.

As it happens, however, the dramatic back-and-forth of the conversation is punctuated by another of the author's aims—to eliminate certain objections he has encountered—because along with the teasing Socratic method, Galileo resorts to a fair number of pugilistic jabs. Indeed, all the quasi-journalistic sparring may at first bamboozle the contemporary reader. So it is that we find Galileo's characters addressing the question why, as the world turns, flocks of birds are not left behind in the air; the problem makes us smile but had to be disposed of promptly, as it probably troubled many perfectly sensible readers. The upshot of these overlapping complications is a large, entertaining, but very difficult volume, which it would not be opportune to explicate here, least of all in the form in which it was written. Certain of its arguments were fallacious, as a number of readers, including Fermat, already noted in the 1630s. Yet it also disclosed, by fits and starts, its author's philosophy of science.

It is as an odd truth that one sometimes discovers revelatory aspects of Galileo's methodology by noticing what he does *not* do. It seems logical, for instance, that he would use telescopic observations as the cornerstone of his argument. He doesn't, and for good reason. The most important observations, such as the sightings of the phases of Venus, are in fact reconcilable with the Tychonic system, and if presented sequentially, as they would have to be, they would run the risk of sequential dismissal. The *Dialogue* is an attempt to give its readers a whole new world picture, which entails changing their habits of mind, so Galileo declines to offer them a set of empirical facts, however startling. He wants first and foremost to con-

vince them that doing mathematics is not a petty intellectual chore, analogous at best to the measuring and computing a carpenter must perform when building a house or repairing a boat, which is what a lot of philosophers thought at the time. Repeatedly the Aristotelian interlocutor, Simplicio, is scolded for his lack of mathematical agility in approaching various problems of physics, which require geometrical expression. Of course the attempt to conceptualize problems involving mass and velocity in the terms of Euclidean geometry now strikes us as archaic, if undeniably ingenious, but that doesn't diminish the force of Galileo's insight. Galileo sees experience as a mental snare unless situated within a larger frame of reference: the horizon, after all, bisects the heavens, an experience that ought to place us at the center of things but doesn't. He knows that practically every advance in astronomy in the past hundred years has violated common sense. Certain of his most important laws, such as the law of fall, cannot in fact be verified. "For the Aristotelians," as William R. Shea has written, "this is an insurmountable barrier. For Galileo, it merely proves that the frontiers of science are not coterminous with the frontiers of experience." Only an imaginary experiment makes the law of fall possible, yet only that law can account for what happens to actual falling objects. "One can, indeed one must, go beyond sense experience, but this presupposes a philosophical conviction that the real is described by the ideal, and the physical by the mathematical." Of course we are at the threshold of a philosophical issue here—how we can legitimately claim to have proved anything—that can hardly be encompassed by this book. I might add that scholars have also discussed whether the sources of Galileo's "conviction" lay primarily in Plato; in some sort of quasi-mathematical aesthetics; in his

Christian faith; or—and this is most likely—in some delicate blend of all three.

In the early fourteenth century, Dante achieved a fusion of astronomy and theology in the *Divine Comedy*. His system consisted of nested spheres, within which the Inferno could be elegantly described as a series of conic sections, which Galileo as a very young man had explained in two lectures to the Florentine Academy. Galileo, who loved Dante, did not read him as science—we will return to this point later—but Dante's moral vision had such a firm hold on most literate people's imaginations that they couldn't quite shake it as a descriptive cosmology. And this cosmology, which owed much to that of Aristotle and Saint Thomas Aquinas, pictured the world as a sort of colossal tool for humanity's instruction and guidance. The zodiac, for instance, was relatively close to men and women and governed their fate at God's behest, like a sort of divine keyboard. Did the Copernicans situate the stars at a vast distance from Saturn, the outermost planet then known? Galileo has his Aristotelian character Simplicio recoil at the thought of the intervening void. "Now when we see this beautiful order among the planets," he asks, ". . . to what end would there then be interposed between the highest of their orbits (namely, Saturn's), and the stellar sphere, a vast space without anything in it, superfluous and vain? For the use and convenience of whom?" At such a remove, the stars would have no earthly purpose. Though the earth is a place of corruption, it remains at the center of everything, and the universe exists for the salvation of men's souls.

We have all had the experience of being unable to solve a spatial problem, in any realm of activity from map-folding to chess to auto mechanics, until we have abandoned a blocked

approach in favor of some sudden, fresh insight. Galileo had observed that the ingrained habit of picturing the earth as stationary made it difficult even for receptive listeners to follow Copernican reasoning. Etched in their brains was what Shea has termed a false "graphic representation" of the heavens, in which the observer continued to visualize his viewpoint as motionless. Philosophically, this representation came from Aristotle and Ptolemy; psychologically, it came from Dante. According to this picture, for instance, planets like Mars and Venus appeared at certain points in the year to reverse their course against the background of the stars, which the Ptolemaic system explained by assigning them epicycles like little gears: this could explain their apparent rearward motion, though what caused them to revolve around empty space remained unclear. In place of this jumble, Galileo's Copernican spokesman, Salviati, urges the reader to think in terms of a different graphic representation. If he would only forget about looking at the celestial sphere from terra firma for a moment, and situate himself at a point on the moving earth, he would see why Mars, which revolves around the sun more slowly than the earth, and Venus more rapidly, would seem at times to travel backward.

One by one, Galileo's protagonist demolishes the typical Aristotelian arguments of the day. An objection to the rotation of the earth is that if it were true, a projectile fired to the west would travel farther than one fired to the east. Salviati disagrees and suggests a thought experiment. Mount a crossbow on a carriage and shoot an arrow first in the direction of its motion and then against it: the arrow will partake of this motion or be hindered by it. Let us say that the arrow would travel 400 meters if shot in the direction of the carriage but

only 200 if shot in the opposite direction. If we grant, for the purposes of argument, that the carriage travels 100 meters during the time of the arrow's flight, the arrow will in either case land 300 meters from the carriage's new position. The same is true of any projectile fired on the surface of a planet in motion—it lands an equal distance from where it was shot.

Galileo's aim in the *Dialogue*, as he put it, was to purge the world of the Ptolemaic system with a medicine distilled of the Copernican. But he never found the panacea, the decisive proof, he sought. For a moment, a device in the church of San Petronio in Bologna seemed to provide what he needed. This conjoined a peephole in an exterior wall with a line of marble inlaid in the interior pavement running along the earth's north-south meridian line; it showed that the sun's image, passing through the hole and striking the meridian line, had slightly altered its position at certain moments in the past few years. The measuring "instrument," the church, was unreliable in that it could have shifted (as a result of subsidence, for example), but Galileo worked out a geometrical way of detecting such a possibility and compensating for it. What interested him as a Copernican was the hint of a decrease in the obliquity of the ecliptic, which would indicate some sort of movement in the polar axis relative to the background of the stars: such a decrease would inevitably push the sun's image a tiny bit farther south along the meridian line. But the measurements were inconclusive, and Galileo had not developed a notion of the "wobble" of the earth's axis (which does in fact take place). So he alluded to the meridian line only in passing, noting that he had high hopes that the data collected at San Petronio might prove more reliable in the years to come.

In the end, Galileo came to rely heavily on a solar theory of
the tides. There is evidence that he had been thinking a lot
about fluid dynamics ever since some long discussions he had
with his friends Giovanfrancesco Sagredo and Fra Paolo Sarpi
in Venice, a city filled with water, in the mid-1590s. Now, in
the *Dialogue*, he worked out a cohesive tidal theory. In
essence, it had to do with the earth's "two motions," about its
own axis and around the sun. To conceive of the action of the
tides, he devised an extraordinarily brilliant mental model,
which, however, lacked the mathematical notation (at the very
least, analytic geometry and algebra, though in the long run
calculus would have been necessary) to attack this problem. It
was as though he were using a meta-language, Italian, to refer
to a mathematical language that had not yet been invented. In
any case, fellow scientists soon noted that his model was
insufficiently analogous to the way the seas actually shift on
the terrestrial globe.

From the start, Galileo had refused to consider the idea that
the moon's differential gravitational action on the seas might
cause the tides (with, as is indeed the case, some lesser pull from
the sun). If this strikes us as eccentric, we must remember, as
Shea has observed, that "the idea of gravitation or attraction
was embedded in a philosophy which made much of sympa-
thies and antipathies, of occult forces and mysterious affinities,"
and that Galileo "had worked himself out of this universe of
discourse and was in open revolt against it." We have seen that
for many Italian Catholics during this time the moon had a
kind of magical aura, and in throwing out the magic Galileo
threw out a more plausible explanation of the tides.

It was not, however, the truth or untruth of Galileo's tidal

theory, or of any other theory in the *Dialogue*, that caused the Holy Office to summon him to Rome in the autumn of 1632. The Inquisition and the pope were not concerned with verifiability but with obedience. An edict against Copernicanism had been issued in 1616. Had Galileo defied it? Had he championed the banned cosmological hypothesis? And if so, could he be induced to mend his ways and abjure it?

3

The Trial; or, Not Seeing

At the time of the summons to Rome, Galileo was still philosopher and mathematician to the Grand Duke Ferdinand II of Tuscany. We may wonder therefore why the duke never raised a finger to save one of his highest-paid and most prestigious courtiers. Treachery? Pusillanimity? No: Ferdinand, then twenty-two, was neither perfidious nor cowardly; rather, the offhand betrayal of Galileo fell into an already established pattern of Tuscan subservience to the Vatican. Discussions of Galileo's trial seldom raise the question of his extradition, which is accepted as an inevitability to be passed over in silence. In fact, it was such an inevitability, but how it came to be so requires some explanation.

The grand duchy of Tuscany, though constrained like any small principality to pursue a cautious foreign policy, possessed the formal attributes of sovereignty. Florence was then suffering a political decline, but Ferdinand theoretically was an absolute ruler, and in the absence of a concordat between Tuscany and the Papal States, the surrender of one of Europe's

leading scientists to a tribunal of the Holy Office at first seems self-destructive on his part, an act of suicidal lèse-majesté. After all, it wasn't as if the ecclesiastical realm, whose territory extended over a portion of central Italy, enjoyed great respect in the peninsula—far from it. Sneered at by the Venetian Republic and threatened by Spain, which ruled Naples and the south, it often found its agencies thwarted or mocked. Venice imposed severe limits on the jurisdiction of the Roman Inquisition within its boundaries, assigning many cases of blasphemy and witchcraft to its own secular courts, insisting that all inquisitorial judges be citizens of the Serenissima, and tending to reject denunciations and testimony proceeding from outside the Veneto. The Roman Inquisition was also seriously hampered at Naples, whose people had rioted in 1547 against the introduction of the Spanish Inquisition, and secular rulers elsewhere meddled in its activities. If Galileo had remained at Padua, on the Venetian mainland, it is doubtful that he would ever have fallen into the hands of the Roman Inquisition, especially in that he stood accused not of heresy but of "rashness," a lesser offense.

At first, Galileo and his allies tried to secure a change of venue, to the Inquisition's office at Florence, where he might defend himself orally and in writing. He was unwell; a plague was rife in some Tuscan provinces; ever short of funds, he feared the expense of detainment in Rome; and the vague possibility of a so-called rigorous examination, or torture by the *corda*, a variant of the rack, theoretically hung over him should he be deemed to be withholding evidence. That putative secrets or intentions can be extorted through the application of pain is an ancient and universal idea. Though torture was practiced far more readily by civil courts than by the

Roman Inquisition, it was far from unknown in this religious context.

Fully aware of the peril Galileo was running, Ferdinand did nothing to protect his most famous subject. Yet the grand duchy had not always been Rome's obliging bedfellow. In the standoff between Venice and the Papal States, for instance, Tuscany long tried to remain discreetly neutral. In 1575, as the Counter-Reformation was being implemented, the arrival of three bishops dispatched to Tuscany to oversee the application of the decisions of the Council of Trent met with resistance, especially from the local clergy, and the reigning Medici grand duke dissolved a society of pro-Inquisition vigilantes who were hounding the townsfolk of Siena. Then there was the sorry dispute between the Vatican and the Este family, lords of Ferrara. A splendid walled city in the Po Valley, known for its frescoed palaces and lavish state banquets, Ferrara had distressed the Papacy by appearing likely, during the 1550s, to fall under the magnetism of a Calvinist princess, Renata di Francia. The Estensi were never forgiven for this little lapse. They had patronized great painters and poets and brilliantly redesigned the entire town, but in 1598, when their succession fell into doubt—it was the classic issue of bastardy—Pope Clement VIII elbowed them out and claimed their domain. An orgy of papal looting followed. Attempts on the part of Ferdinand I of Tuscany to rescue the Estensi, with whom he was allied by marriage, were foiled by Spain and France.

During the Counter-Reformation, no principality, and indeed few significant public figures, could ignore the wishes of Rome, Spain, and France and survive. Besides, the Papacy, if courted, was a source of political legitimacy; the Vatican's aid had been enlisted in transforming Tuscany into a grand

duchy in 1569. Some major popes had been Medicis themselves, or otherwise Florentine, or closely allied with great Florentine houses. And so a habit of deference grew in the early
seventeenth century, a sort of coziness which the eighteenth-
century Tuscan historian J. R. Galluzzi would describe as
"feeble acquiescence" to the papal court and a "pernicious tol-
erance" of its scheming. The Tuscan property held in mort-
main by the Church grew by leaps and bounds, and the
Inquisition snooped about unopposed. By the time Ferdinand
II reached majority and assumed power in 1628, he was
bound by paralyzing conventions: hearing in late 1630, for
instance, that the inquisitor of Siena had been arrested for
arming his retainers in defiance of the law, the poor duke simply let him go, at the behest of the inquisitor of Florence. The
following year, Ferdinand, although he had married the duke
of Urbino's daughter, did nothing as Urban VIII (in a move
analogous to the ouster of the Este family from Ferrara) formally annexed Urbino to the Papal States. Ferdinand was in
no position to oppose Galileo's extradition.

It fell chiefly to the lot of three other men to help the
embattled genius. One was Andrea Cioli, the grand duke's secretary of state, who tried to monitor information from his
perch in Florence. More crucial were Francesco Niccolini, the
Tuscan ambassador to the Holy See, and the pope's correspondence secretary—and Galileo's former student—Monsignor
Giovanni Ciampoli.

Niccolini was the most influential of the three. Now nearing fifty, he had served as a page in the Florentine court and
studied as a novice, abandoning his ecclesiastical ambitions to
marry Caterina Riccardi—the Riccardis were a great Florentine family—in 1618. He became Tuscan ambassador to the

Vatican in 1621, fondly hosting Galileo during his triumphal visit to Rome in 1623–24. Niccolini united in his person three prominent characteristics. He was a Tuscan, totally loyal to the duke and his homeland, and wary of the papal court; he was a diplomat, who could tell the possible from the impossible; and he was a friend, which Galileo, a difficult man in the best of circumstances, could not say of everyone.

Giovanni Ciampoli was a trickier case. It is hard to judge someone's character on the basis of scanty documents, but the records seem to show Ciampoli as a man whose ambition gradually detached itself from his initial field of endeavor to become a free-floating hunger for power. He met Galileo in 1608 and studied mathematics under him in Padua before taking holy orders, in Rome, in 1614. With his considerable talent for poetry, he attached himself to the circle of Maffeo Barberini, who, as we know, was an accomplished poet himself. Without question there is a resemblance between Ciampoli's verse and Maffeo's, especially in their common emphasis on the fleeting nature of earthly delights. Any lover who follows the fugitive forms of joy, Maffeo wrote in a famous couplet, would find such fruit bitter in his hand; Ciampoli, though he deemed the world a "theater of marvels," likewise wrote that he was "weary in my suffering"—meaning not personal suffering but the harshness of confinement in a sensory world. Politically, Ciampoli came to occupy a pivotal point between the Vatican and the Jesuits, whom he knew well but quietly loathed. After Maffeo's papal election, he managed to become not only the pope's confidential secretary but also an important member of the Academy of the Lynxes. The Lynxes were at daggers drawn with the Society of Jesus, and Ciampoli (who, among other diplomatic achievements, man-

aged to improve relations between the pope and Cardinal Richelieu) worked tirelessly to promote Galileo's standing with Urban VIII and to reduce the influence of the Jesuits in the Vatican.

It was Ciampoli who obtained the imprimatur for Galileo's *Assayer*, and who later urged the Tuscan physicist to heed the pope's encouragement to write the more comprehensive work that became the *Dialogue*. His letters to Galileo often have a saccharine, fawning tone, as if he were keen to bask in the light of the great man's friendship. In late 1623, he drafted a letter from the pope to Grand Duke Cosimo that referred to Galileo as "my beloved son." Yet while toadying to the pope, Ciampoli came to despise him, and his encouragement of Galileo despite the pope's anxieties about Copernicanism betrayed the irresponsibility of the self-regarding meddler, the manipulator whose social perceptions are defective and dangerous. Fortunately for later generations, however, Ciampoli, like Cioli and Niccolini, wrote informative letters during the Galileo affair, and through them we may follow the scientist's fate.

From the outset, Ambassador Niccolini and Secretary Cioli were convinced that Rome's concern about Galileo was the consequence of a "calumny of . . . envious and malicious persecutors," especially considering Galileo's undisputed Catholicism. It was perhaps a matter of no serious concern. By September 5, 1632, however, we find Niccolini growing very worried. He writes Cioli that he has spoken at length with Urban VIII, and

I began to think, as you so rightly say, that the world is coming apart. While His Holiness was talking about these distressing matters before the assembled Holy Office, he

flew into a rage and suddenly told me that "Our Galileo has been too eager to enter where he does not belong, in matters more grave and dangerous than should be raised at this moment." I countered that Galileo had not published without his own ministers' approval, and that I myself, to that end, had obtained and sent the book's preface here. He replied in the same apoplectic rage, saying that Galileo and Ciampoli had circumvented him, and that Ciampoli in particular had impudently told him that Signor Galileo wished to do everything that his Holiness commanded and that everything was fine, and that this was all he knew, without his having actually seen or read the work.

And so the furious pope went on to regret ever having anything to do with such people as his own correspondence secretary and the Vatican censor, who had strangely let such a book slip through their hands. None of this prevented the ambassador from humbly and bravely beseeching the pope to let Galileo defend himself, which he felt was only fair; to which the pope replied, "In such affairs of the Holy Office, one only censures, and then calls upon the accused to recant."

But Niccolini, as one Tuscan to another, did not give up. "Doesn't it seem to your Holiness," he persisted, "that Galileo ought to know in advance all the questions and criticisms and censures being leveled at his work, and just what is troubling the Holy Office?" The pope was appalled. "I tell you," he shot back in a rage, "that the Holy Office does not do such things, and not in such a way, nor does it ever give anyone such things in advance—it just isn't done! And besides, he knows very well where the difficulties lie, if that's what he wants to know,

because we ourselves have discussed them with him and he has heard them from our own mouth!"

Dispirited by this encounter, Niccolini advised Cioli that Maffeo Barberini had appeared so ill-disposed toward Galileo that they had best turn to somebody else, perhaps Maffeo's nephew, Cardinal Francesco Barberini. If one tried to oppose the pope frankly, Niccolini said, "he gets his back up and shows no respect for anyone. The likeliest course would be to win him over with time, to skillfully work on him, without causing a stir, and also to go through his ministers, according to the nature of the business at hand."

The diplomat had tried to nudge the autocrat away from his inclination to rely on the sort of procedure that we now call (without reference to the Inquisition) "inquisitorial," in which magistrates compile evidence instead of hearing the argumentation of two adversarial teams. He had suggested instead the application of what were, in fact, the norms of canon law, which would be recognized 150 years later, in a different form, as basic civil rights: the prisoner's right to know the charges against him, and his right to defend himself. In the following days, however, Niccolini learned that such rights were not likely to be honored, because the pope was particularly incensed at the *Dialogue*'s intrusion into matters of faith and was disposed to turn the matter over to the theologians of the Inquisition. Niccolini's most forceful plea, to the effect that Galileo after all was mathematician to the grand duke and universally respected as such, fell on deaf ears. The pope replied that despite his friendship with Galileo he had warned him off these topics sixteen years ago, and now Galileo had wandered into a dreadful thicket and had only himself to blame. When Niccolini quietly suggested that Galileo could

hardly be seen to be straying from the basic dogmas of the Church, and that his case had arisen only because everyone in this world has his enemies and invidious detractors, Maffeo snarled, "*Basta, basta!*" quite as if he suspected that the diplomat, who tended to veil his phrases, might be insinuating something about Maffeo himself.

Maffeo Barberini was certainly angry at Galileo, and Niccolini and Cioli and Ciampoli probably wondered how much personal animus there was in his anger. It was true that in 1615, when still a cardinal, he had told Galileo through Ciampoli and another friend, Piero Dini, a former archbishop now living in Rome, and very likely even in person, that though he admired his scientific work, he should stay clear of theology and "speak cautiously, like a professor of mathematics." Yet Maffeo and Galileo had indeed corresponded and been on the most cordial terms. When in the summer of 1623 Maffeo's nephew Francesco had received a baccalaureate and been accepted into the Academy of Lynxes, of which Galileo was a member, the scientist had sent him a letter of congratulation. Maffeo had written back, noting "my great esteem for you" and "your affection toward me and my House, . . . and my ready disposition to serve you always."

Now, in the autumn of 1632, the pope felt strongly that Galileo had taken up Copernicanism in a manner explicitly forbidden and, abetted by Ciampoli, had hoodwinked the Vatican censor. Something else irked him too, something pertaining to the *Dialogue*'s contents. This work, which the inquisitors would later closely examine, does not, as mentioned earlier, attempt to prove the truth of the Copernican system but instead plays certain arguments for it against the

geocentric, which turns out to be more vulnerable to polemics. The dialectical schema allowed Galileo to claim, as he did in his eleventh-hour preface, that the whole thing was a kind of intellectual exercise, a demonstration that the Catholic philosophers of Italy knew all the best arguments in favor of Copernicanism and had in the end rejected them, but only because of "reasons that are supplied by piety, religion, the knowledge of Divine Omnipotence, and a consciousness of the limitations of the human mind." Unhappily, this lame preface was printed in a typeface differing from that of the rest of the book, which revealed it as an afterthought, according in substance with nothing in the body of the dialogue but its equally lame conclusion. For in the last few paragraphs, out of the blue, Simplicio, the character who all along has been gamely championing the geocentric cosmology, states in abbreviated form Maffeo Barberini's own theological view on such matters, which Galileo had probably heard directly from the pope around 1624. This view held that God's power to create the world is limited in one way only, namely, that it must obey the law of contradiction—the logical axiom stating that no proposition can be simultaneously true and untrue—and that otherwise God is omnipotent. Consequently, because He can create any number of worlds, we cannot hold with absolute certainty that the earth revolves about the sun. Thus the entire *Dialogue* is bracketed by two passages that attempt, however feebly and bizarrely, to unsay its own arguments.

Much has been written about this dialogue, but almost all of it conflates or confuses three distinct issues. First, it has sometimes been claimed that Simplicio is a literary caricature of the pope, an extreme contention that does not follow from

the sense of the text and would assume that Galileo was a self-destructive madman. It has also been claimed that the pope's vanity was stung by seeing his theological opinions placed in the mouth of the callow and student-like Simplicio, and this is certainly true, and also perfectly understandable. Finally, Maffeo Barberini's theological idea is generally taken to be silly or sophistic, but in fact it is not, nor would it have seemed so to Galileo. The inquiry into how we know what we know, and how our knowledge may relate to God's omnipotence, had a long pedigree in scholastic philosophy and was in the process of being taken up by Descartes, among others. The trouble is that it is a metaphysical concern, logically anterior to scientific investigation, and to bring it to bear on celestial mechanics represents what we would now call a category error. Prevailed upon by others, Galileo at the eleventh hour used it to get himself off the hook, with the consequence that a category error became an insult to the bishop of Rome.

So it was that Maffeo Barberini, both doctrinally dismayed and personally offended, turned the Galileo affair over to the Holy Office. On October 3, Galileo learned that his request for a change of venue had been rejected. His presence was urgently requested in Rome.

ALL WAS NOT lost, however. An extraordinary letter from Galileo was delivered on October 13, in Rome to Cardinal Francesco Barberini, Maffeo's nephew, whom Galileo thought of with reason as a friend. The missive, of the favor-currying sort, complained bitterly of the hatred to which the great man had been subjected by his rivals, who had been "thrown into shadow," as he characteristically put it, by the "splendor" of his scientific writings. Galileo informed the cardinal that his ene-

mies had somehow prevailed upon the Inquisition to suppress the *Dialogue* and, more alarming still, that the dread office expected him to appear that very month before its tribunal. "This torment causes me to rue all the time I have spent in my studies . . .," he lamented. "And in making me regret having shown some of my work to the public, it may induce me to suppress and burn whatever remains in my hands, thus satisfying the yearnings of my foes, whom my ideas so greatly annoy." For all his high self-regard, persecution was turning him against himself.

Francesco, a friendly, meditative man with a pointed mustache and a little pointed beard, was entirely unlike his uncle, from whose political decisions he programmatically dissociated himself. Temperamentally an aesthete, he had become a great francophile while serving as papal legate to Avignon in the 1620s, and he owned several pictures by the young Nicolas Poussin. In 1625, he had bought what soon became Palazzo Barberini, on the Quirinal, and had assumed responsibility for its decoration. With the help of Cassiano dal Pozzo, he turned its library into one of the finest in Rome.

Galileo nursed reasonable hopes that this sensitive soul would take up his cause, or at least deftly mediate between himself and the pope. His letter does not appear to have furthered these expectations, but the cardinal gave proof of forbearance, and perhaps a dram of sympathy, in refraining (so far as we know) from disclosing the missive's contents to the Inquisition, for it went on to make some very strange and intemperate claims. Galileo, in lieu of going to Rome, offered to put all his ideas in writing, which would redound to the glory of the One True Church—which was fine in itself, but he didn't stop there. He went on to say that he had been con-

vinced of the value of his offer "when I heard a saintly and admirable pronouncement issue from the mouth of a person most eminent in doctrine and worthy of veneration for the sanctity of his life, setting forth in no more than ten words (strung together with keen-witted loveliness) as much wisdom as may be found in all the discourses of the sacred doctors . . . I will for now withhold its admirable author." Was the mathematician claiming to have visitations from the other world? Or was he referring to the pope himself? The cardinal might well have wondered. Farther along in the same wildly rambling, deeply melancholy letter, Galileo insisted that "if what I write should fail to mitigate whatever bill of indictment is brought against me, when objections are raised I shall not fail to respond as God dictates to me." *As God dictates*? For much of the previous century the phrase "dictated by the mouth of God" had been used by churchmen to refer to the Bible. So if Galileo in an unguarded moment was claiming to receive revelation in the manner of Moses or Paul, Cardinal Barberini kindly let it pass. He did, however, forward the letter to his uncle.

Niccolini must have had word of this, for soon afterward he wrote Galileo a shrewd note, warning him not to defend himself but to "portray yourself as the Cardinals of the Inquisition see you; otherwise you will have the greatest difficulty in defending your case, like many others before you. Nor, Christianly speaking," he went on, in his diplomatic doublespeak, "can you claim anything other than what they, as a supreme inerrant tribunal, may wish." And having assumed a role very like a modern public defender angling for a plea bargain, he went off to haggle with the pope.

"I tried to awaken in him some compassion for poor Signor Galileo," he told Cioli,

> now so old and loved and venerated by me, all on the assumption that his Holiness had seen the letter [Galileo] had written to his nephew the cardinal. His Holiness replied that he had seen the letter but could not waive the requirement that Galileo come to Rome. I then suggested that his Holiness ran the risk of forfeiting his case whether here or there, since Galileo's age, together with his extreme discomfort and misery, might cause him to expire on the way. He replied that he could come slowly, in a litter, with all due comforts, because he wanted to examine him personally, and may God forgive him the error of getting involved in such a tangle after his Holiness himself, when still a cardinal, had strongly cautioned him to stay out of it.

After which the pope had again proceeded to attack his *bêtes noires*, Ciampoli and Niccolò Riccardi, the Vatican censor (who in fact was Niccolini's wife's cousin), calling them flunkies who brazenly thwarted their master. So Niccolini left empty-handed. And in the end Galileo received nothing from the nephew in scarlet but a promise to shorten the quarantine at the border crossing, which happened to be under his purview.

Between them, Niccolini and Cioli now foresaw two dangerous eventualities. One was that Galileo, with his bumptious self-belief and blindness to his own provocative tendencies, might further arouse pontifical ire or stir up the Roman wasps' nest. The other, more imminent, was that he might die

on the way. He was sixty-nine years old—they thought or claimed he was seventy—afflicted by terrible arthritic pains in his midriff and lower extremities and by a bad heart, vertigo, incipient blindness, migraines, a hernia, and chronic depression. Worst of all, the *peste nera*, the Black Plague, had been rife in northern Italy since 1629 and had been creeping into Tuscany as well.

The pestilence, manifesting as buboes and *carboni negri* (which one takes to be coal-black carbuncles or patches of some sort), had incubated in conditions of war. The issue of the Mantuan succession and the pope's incompetent interference in the crisis had lured the Hapsburg Empire into Lombardy, dragging able-bodied men out of the countryside and causing a shortage of grain. Famine was compounded by drought. As fields were left fallow, Mantua and other cities were sacked by the imperial troops and many of their inhabitants slaughtered or raped. Hordes of filthy vagabonds began to besiege the intact towns and villages, and disease soon spread among the encampments of refugees. When the pestilence itself inevitably arrived, probably brought by the undisciplined German soldiers, the authorities began to isolate the sufferers, yet still the Black Plague raged throughout Lombardy, harvesting hundreds of thousands of lives. It crept south toward the papal principality, where a quarantine of forty days was imposed on those intending to cross into papal territory. At times in various parts of Tuscany, women and children, and sometimes men as well, were forbidden to leave home for long periods, and on some days all but those delivering foodstuffs were barred at the gates to Florence. The plague never ravaged Tuscany as it did the north, but only in May of 1633, when the Madonna of Impruneta, an icon of the

Virgin supposedly painted by Saint Luke, was borne in solemn procession from Impruneta into Florence, did the contagion die away.* Though on December 17 Galileo garnered three doctors' attestations to the effect that the journey to Rome posed a danger to his survival, the Inquisition insisted on his presence, and it was decided that he would travel by closed litter and undergo a quarantine reduced (thanks to Cardinal Barberini) to eighteen days. His exposure to the plague was thereby diminished, but given his fragile state, the slow journey would entail a health risk of its own, as would the long delay at the border crossing of Ponte Centino. The forced journey from Florence to Rome was, for a sick old man, a form of torture in itself, or at the least, a display of what lawyers today might call depraved indifference: there was some question even then as to whether the juridical procedure—extradition, detainment, interrogation—would not be more painful than any sentence eventually handed down, a plain violation of canon law. Was Maffeo Barberini vindictively chastising Galileo in order to salve his wounded pride? So Niccolini and Cioli might have wondered, though they couldn't do so in their letters. In the Tuscan ambassador's correspondence, however, the word that stands out whenever he recalls his conversations with the pope is *aggirato*: over and over, the pope complained that he has been *aggirato*—gone around, circumvented. He was infuriated, outraged, galled that so many people—Galileo; the Vatican censor; Ciampoli, his own scamp of a letter scribe who has made common cause with a hostile, pro-Spanish camarilla in Rome; and heaven knows

---

* This plague, one of the worst ever to affect Italy, is described in a famous passage in Chapter 39 of Alessandro Manzoni's *The Betrothed*.

who else—have "gone around" him. In Niccolini's portrait of
the pope we have a picture of power without charisma, without any graceful, natural capacity to command. As a ruthless
Tuscan politician, Maffeo readily yields to spite—one recalls
Niccolini's remark that he has "no respect for anyone"—but
his dismay also appertains to his role. As Urban VIII, leader of
the Catholic Church and chief administrator of the Counter-
Reform, he legitimately worries that doctrinal chaos is erupt-
ing on his watch, and he has no clear view of what is
happening. Though his motives are mixed, this is something
that no manager can tolerate, and only the Holy Office, under
his eye, can bring it back under control.

At this point, one thing was certain: Ciampoli, who had
been conspiring all the while on Galileo's behalf, had it coming. And late in November, another old friend of Galileo's,
Benedetto Castelli, now mathematician to the pope, told him
that Ciampoli had been disgraced. Castelli was a former student of Galileo's at the University of Padua, a Benedictine
priest, and an eminent scientist in his own right. He wrote, a
little naively, that Ciampoli, having "astonished all Rome with
his frank spirit and shrewd conduct," had left the city for an
insignificant governorship at Montalto della Marca. Though
all Ciampoli's friends were upset at this development, the former correspondence secretary was behaving like a Stoic, not
only unbowed but quite as though he had never been involved
in a struggle at all. "Fully self-possessed, he's more light-
hearted than ever, applies himself to his studies, and, best of
all, actually shows irreproachable reverence toward the
Church fathers, submitting quietly to the will of God." Visiting
the inveterate schemer in his study, Castelli found him in a
state of perfect composure, troubled only by Galileo's ordeal.

As for Galileo himself, Castelli said that since he had never failed Mother Church in any way, "your malignant persecutors wish for nothing better than your arrival in Rome, so they can raise cries from the ignorant mob and call you rebellious, insubordinate. . . . For that reason alone, you should vigorously resolve to hold your own against the weakness of age and the inclement season, and gird yourself for the journey. . . . And recommending yourself to God, sir, may you come joyfully, too, because I do believe you will surmount every obstacle."

Galileo left Florence on January 20, 1633, traveling in a litter provided by Grand Duke Ferdinand, an expense he was later asked to cover out of his own pocket. After submitting to the quarantine, he arrived in Rome on February 13, lodging at the Villa Medici with Francesco Niccolini and his wife, Caterina.

READERS WHO HAVE visited Rome will recall that the Villa Medici stands on a hill called the Pincio, which commands a magnificent view of the city. Built for a Medici cardinal about sixty years before Galileo's arrival in 1633 to face trial, the villa had soon afterward become the Tuscan embassy. In this edifice, Galileo was nobly but uncomfortably lodged, since its stairways proved trying for an elderly man with severe arthritis. Held under house arrest, he was not allowed to leave the grounds of the embassy unescorted until early March, when he received permission to stroll in the fragrant garden beside the church of Santa Trinità dei Monti. Apparently, he never entered the church itself. Sometimes, standing by the villa's fountain, which in those days was still surmounted by a stone carving of a Florentine lily, he must have gazed in frustration

across the rooftops of Rome at the dome of Saint Peter's, though with his poor eyesight it would have been something of a blur. Perhaps he knew of Maffeo Barberini's wish to preserve Michelangelo's original design for the façade and now remembered his struggle with Paul V. Surely, he recalled how he and Maffeo had dined together and talked of poetry and philosophy. He must have wondered how his old friendship with the pope had come to this.

That winter, Galileo depended entirely on Francesco Niccolini, but he often disagreed with him too. As ambassador, Niccolini had the habit of cautiously cultivating others' good graces, whereas Galileo grew ever more embattled, more entrenched: his argumentativeness was alarming. Only a few days before leaving Florence, Galileo had written a long letter to his old friend Elia Diodati, an erudite Parisian lawyer, explaining in impassioned terms why Copernicanism did not contradict Scripture: considering that the Bible is couched in terms accessible to simple folk, he asked, "Why must we begin our investigations with God's word rather with His works? . . . God himself is subject to anger, to repentance, to forgetfulness" in the Bible. To take such capriciousness literally would be pure heresy. "But I would ask whether God, to accommodate himself to the capacity and opinions of the same simple folk, has ever changed his creations, or if nature, his inexorable minister, impassive with respect to men's opinions and desires, has not always preserved and maintained the same sort of movements, geometrical figures, and dispositions in the parts of the universe, the moon being always spherical."

Galileo continued to insist upon these views during the course of the winter. And, well, maybe he was right, maybe he

was wrong, but as far as Niccolini was concerned, this wasn't the moment. Was he going to raise debating points with the Inquisition, as if this were a seminar at the University of Padua? Niccolini decided it would be wise to take Galileo to see Monsignor Alessandro Boccabella, who had just left his position as assessor at the Holy Office and was amiably disposed toward him; they also inquired after the commissary general, Fra Vincenzo Maculano da Firenzuola, who would likely have more power over Galileo at his trial than anyone but the pope, but Maculano, who would later serve as the papal principality's chief military architect, was out, unfortunately. In the meantime, Cardinal Barberini told Niccolini to keep visitors away from Galileo, and please, the less jabbering the better: nothing he would say was likely to help him. Niccolini agreed: it was best not to rock the boat. "Though the affairs of this tribunal can never be discussed clearly or with any certainty," he wrote Cioli, "it seems that no great evil is upon us yet."

Left to his own devices, Galileo tried to persuade himself that all was going well. He wrote Cioli on February 19 that his case was being handled "according to a very mild and benign principle of treatment, wholly unlike the threatened racks, chains, dungeons, etc." And he went on, in the dressy style he fell into when upset, "To have heard from many, and also to have seen, that there is no lack of persons, and powerful persons too, whose affection for me and my affairs reveals itself to be nothing if not well disposed, is a source of consolation." One notes that the reference to torture is immediately followed by tortured syntax, and also that if the threat of a *quaestio*, a "rigorous examination," had been waived in his case, as certain writers have claimed, no one had told him or Niccolini

about it. It had certainly not been waived but would reappear, amid all the apparent official benignity, in the form of a specific pontifical request for the "threat of torture" some four months later.

Over the next few days there was a curious development. A chatty Inquisition official with the title of consultor, a theological position, turned up repeatedly at Villa Medici. His name was Lodovico Serristori, and he came, Niccolini thought, "in the guise of a visitor," as if on his own account. But he always inquired about Galileo's case, prying even into its technicalities, so the ambassador concluded that he had been sent to sound out Galileo—to gauge his rhetorical skills and weigh his defense of his cause—in order to report back to his superiors. In other words, Serristori was a spy, but in this Rome of 1633, that scarcely excluded his being a sympathetic spy, even a spy with divided loyalties. And Galileo, forgetting Cardinal Barberini's advice, poured out his soul to the officious caller, whom it seemed he knew from way back. Together they reviewed Galileo's writings, which Serristori said he admired profusely. "I think Serristori has cheered up the dear old fellow," a bemused Niccolini told Cioli. "And then again it comes back to Galileo, how strange it is, this persecution. I told him, Whatever you are ordered to do, always be ready to submit and obey. Because that is the way to assuage the ferocity of those who are fired up against you, those who treat this cause as something personal."

In their correspondence, Galileo and his friends Niccolini, Cioli, Castelli, and others often refer to this cabal of personal enemies, animated (so they claim) by envy. They do not name them, since their letters, transmitted by courier, are in no way secure. But surely they have specific people in mind, and one

wonders whom they refer to. Galileo specifically names his former supporters, the Jesuits. Among them were Christopher Scheiner, the gifted astronomer who had quarreled with Galileo in 1611–13 about the sunspots, though one questions his weight with the Holy Office; Father Orazio Grassi, who bore a long-standing grudge against Galileo after the dispute over the comets; and Melchior Inchofer, a Hungarian-born polemicist who would later file a scathing report against Galileo at the behest of the inquisitors. And of course Galileo, like any important man, had a host of petty critics and belittlers. Yet the great question is whether Niccolini and company weren't really talking about the pope. Clearly Maffeo Barberini had personal reasons to be embittered against Galileo, but he had objective reasons, political and doctrinal, to be worried about him as well. Whether he would have behaved any differently had he not had an axe to grind is a question that will never be resolved.

Curiously, Grand Duke Ferdinand, anxious to intercede on Galileo's behalf, did not turn to the pope but elsewhere. He decided to write two cardinals, Desiderio Scaglia, a respected theologian, and Guido Bentivoglio, a cultivated Ferrarese whose family had for a while owned Palazzo Borghese and who had been painted by Van Dyck, urging an expedited trial for Galileo on account of "the compassion that he deserves" and "the love that I bear him." Bentivoglio politely called on Galileo; Scaglia did not. Yet despite Galileo's conviction that these princes of the Church were now in his corner, the duke's entreaties failed to achieve their desired end.

This may have been because a disconcerting development, in the form of fresh evidence, had arisen. "From what I gather," Niccolini told Cioli in late February, "[Galileo] was

given an injunction as early as 1616 against either discussing or teaching the [Copernican] opinion: but he says that the commandment was not at all in this form, but was rather that he neither hold nor defend it." This was perplexing, and Niccolini inquired further. One thing he knew for certain: that if the Inquisition could prove that Galileo had personally received a formal warning of this sort, his conscious infraction of it would trump the subsequent permission he had received to print the *Dialogue*. The Holy Office would have an open-and-shut case.

Here are the circumstances that soon came to light: Everybody knew that in March of 1616, the Congregation of the Index had issued a decree against Nicolaus Copernicus's *Revolutions*, against a letter by the Carmelite monk Antonio Foscarini, and against the *Commentary on Job* by the Spanish priest Diego de Zuñiga, the latter two being pro-Copernican works. The Copernicus and the Zuñiga publications were ordered "suspended until corrected," Foscarini's was suppressed outright. Galileo, though he had explicitly backed Copernicanism in *Letters on the Sunspots*, of 1613, was not mentioned, perhaps out of regard for his patron, the grand duke of Tuscany, but he could hardly have failed to get the message. Of course, since the summer of 1632 the Holy Office had strongly suspected that Galileo had violated this decree in publishing the *Dialogue*— that was the reason for his presence in Rome.

More recently, the Special Commission appointed to investigate the matter had unearthed a memorandum in the Inquisition's archives. It stated that on February 26, 1616, at a meeting with Cardinal Bellarmine, then the commissary general of that body, Galileo had received not only an oral but also a written injunction, similar to a "charitable admonition,"

ordering him to abandon forthwith any leanings he might have toward Copernicanism. "At the Palace," this memorandum stated, "the usual residence of . . . Cardinal Bellarmine, the said Galileo . . . was . . . admonished by the Cardinal of the error of the aforesaid opinion and that he should abandon it; and later on in the presence of myself [the notary], other witnesses, and the Lord Cardinal, who was still present, the said Commissary did enjoin on the said Galileo, there present, and did order him . . . to relinquish altogether the said opinion, namely, that the sun is the center of the universe and immobile, and that the earth moves; nor henceforth to hold, teach, or defend it in any way, either verbally or in writing. Otherwise proceedings would be taken against him by the Holy Office. The said Galileo acquiesced in this ruling and promised to obey it."

This document shows several technical irregularities, and experts have written much about it, even in the past few decades. But it was accepted as genuine in 1633 and it enraged the pope, who claimed that Galileo had concealed it from him. It put a whole new complexion on the trial.

Galileo's response to this development astonished Niccolini. Indeed it was here that Niccolini's tact and worldliness came to the fore and that the two friends' views on the trial parted absolutely, for some time to come. Strangely, a buoyant Galileo wrote to a friend in Florence that "those numerous and most grave imputations [against me] have all been reduced to one point"—he meant the 1616 written injunction from Bellarmine—"and all the others have ceased; on this basis alone I shall have no trouble securing my release." But Niccolini, confessing to Cioli that he had lost all hope for an expedited trial, suggested that Galileo, if he knew what was

good for him, should proceed with *amorevolezza*, or loving affection, a curious locution in the circumstances.

Amorevolezza: this word reveals at one stroke the enormous psychological abyss separating Niccolini's and Galileo's estimation of the perils ahead. And because Niccolini understood the papal court so much better than Galileo, it also reveals the abyss separating Galileo and the Church. The emotion of *amorevolezza* is generally reserved for members of one's own family—it is something you feel for your mother or your favorite aunt or uncle, or a grandchild who particularly cares for you. Niccolini's advice to approach the Inquisition with *amorevolezza*, which to us may sound utterly bizarre, resulted from his shrewd sense of how that body viewed Galileo's position. Ever since *Letters on the Sunspots*, Galileo had wanted to argue as a believer with the Church. He had wanted to convince the Church that nature (that is, the sun's true position in the universe) and Scripture were in harmony. One could line a bookshelf with volumes about the supposed debate between Galileo and the Church, but it was a debate that never took place. It never took place because the Church did not engage in debates with people, especially not with people who had wandered off in the direction of rash or erroneous philosophical ideas. No: if what Galileo sought was intellectual engagement, what the Church sought was discipline. The Inquisition's aim was not intellectual but disciplinary, in the sense that a parent needs to discipline a wayward child. And in this sense discipline is ultimately an exercise of love. The strategist in Niccolini grasped that the inquisitors saw Galileo as a kind of prodigal son, a famous and gifted Catholic who must at all costs be reconciled with Church doctrine. So *amorevolezza* was the best tactic to soften them up.

Meantime, to forestall disaster, the ambassador again tried to intercede with the pope. He mentioned Galileo's age, his ill health, his prestige at the Tuscan court. But the pope spoke of how ponderously the Holy Office deliberated and of Galileo's audacity in expressing his opinions. He also intimated darkly that the detested Ciampoli was behind the whole affair (which suggests, interestingly, that he somehow saw a tie-in between Galileo and a pro-Spanish camarilla in the Vatican, with whom Ciampoli had nefarious connections). Worst of all, Galileo had flagrantly disobeyed the written order given him in 1616 by Cardinal Bellarmine.

Niccolini had come up empty-handed, but at least Maffeo hadn't lost his temper—that was something. So Niccolini decided to try the nephew, who was usually so much more sympathetic. And indeed Francesco conceded that "he cared very much for Galileo and admired him as an extraordinary man; but this issue is delicate and might introduce some fantastic dogma into the world and especially into Florence, where wits were subtle and curious." And then wasn't there that little matter of his having argued so much more forcefully in the *Dialogue* for the earth's motion than for its stability? In the end, Niccolini was given to understand that he should be grateful he had obtained Galileo's detention at the Tuscan embassy, instead of the dungeons of Palazzo Pucci or the Castel Sant'Angelo.

Niccolini secured one more audience with the pope before the opening of the trial. The meeting took place on March 13. This time, the ambassador invoked Duke Ferdinand's willingness to assume redoubled indebtedness to the Holy See if the matter could be expedited, but the pope dismissed the offer. "May God forgive [Galileo] for considering these questions,"

Maffeo said, as reported by Niccolini to Cioli, ". . . and may God forgive our little Ciampoli also, since he too is fond of them and is a friend of the new philosophy. Galileo was my friend, and we have conversed and broken bread together several times at my home, and I regret having to displease him, but this bears upon the interests of the faith and religion." It occurred to Niccolini to suggest that if Galileo "might speak for himself he could easily give every satisfaction, with all proper reverence to the Holy Office; but the pope answered that he would be examined in due course, and there is an argument that they have never been able to answer, which is that God is omnipotent and can do anything; and if he is omnipotent, why do we wish to restrict him with contingencies?" This of course was Maffeo's pet thesis. Niccolini, pouncing on the opening, told Maffeo that he had clearly heard Galileo say that "he himself didn't believe in the earth's motion, but since God could make the world in any one of a thousand ways, one couldn't deny that he might have made the world in this one."

Ah, diplomacy.

But the trouble with diplomacy, as Niccolini knew, was that it seldom worked with Maffeo Barberini. As a temporal prince, Maffeo received continuous visits from foreign ambassadors, usually representing Catholic states, and often they came away stunned. The pope did not listen, cutting off his interlocutors; or he listened only to himself, holding forth for hours on end; or he harangued one envoy and then carried the dispute forward to the next audience, causing total bewilderment; or he refused petitions or demolished arguments merely to show off his reasoning powers. The Venetians, with their antipathy toward the papacy, treated him like a spiteful

child, requesting the reverse of what they desired in the conviction that he would take the opposite course.

This time, Maffeo's response to Niccolini was annoyance mounting toward ungovernable rage. "One cannot impose necessity on the blessed Lord," he barked. And Niccolini, beginning to worry lest he drift unguardedly into some heretical opinion and end up before a tribunal of the Inquisition himself, passed on to other business.

On April 9, only three days before the trial was to open, Niccolini visited Cardinal Barberini to see whether Galileo might not remain at the Tuscan embassy instead of being confined for the duration of the proceedings in the palace of the Holy Office. Niccolini assured Francesco of the esteem of the Casa Medici, lamenting that he could not hope to paint a suitably vivid picture of the "failing health of the dear old fellow, who for two nights together has cried out and complained continuously of his arthritic pains and of his wretched old age and the suffering it causes him"—all good reasons to permit Galileo to return each night to sleep at the embassy. Francesco regretted his incapacity to make such a concession, but he promised to arrange comfortable lodgings for the scientist in the chambers of the Inquisition's treasury. There Galileo would be free to come and go, to take exercise in the courtyard, mingle among the churchmen, and receive his meals from the kitchen of the Villa Medici.

Meanwhile, Galileo prepared to defend his philosophical position, much to Niccolini's dismay. Convinced as always that fortune favored him, he appeared to have some polemic or piece of evidence up his sleeve that would sink his persecutors' arguments. The ambassador pleaded with him to abandon this project and to toe whatever line the Holy Office

might lay down, especially with regard to "that particular" of the earth's motion; but he soon saw that such thoughts depressed Galileo so deeply he began to fear again for his life. The great man deserved all the best, Niccolini felt; "this whole house," he told Cioli, "which loves him exceedingly, feels an indescribable pain."

The trial of Galileo Galilei opened on April 12, 1633.

"*QUOMODO ET A quondam tempore Romae reperiatur.*"

"I arrived in Rome the first Sunday of Lent, and I came in a litter."

"*An ex se seu vocatus venerit, vel sibi iniunctum fuerit ab aliquot ut ad Urbem veniret, et a quo.*"

"At Florence the Father Inquisitor ordered me to come to Rome and present myself before the Holy Office, this being the command of the ministers of this Holy Office."

"*An sciat vel imaginetur causam ob quam sibi iniunctum fuit ad Urbem accederet.*"

"I imagine that the cause for which I was ordered to present myself before the Holy Office in Rome was to give an account of my recently printed book."

With these words the trial opened in an upstairs chamber of the magnificent Dominican convent of Santa Maria sopra Minerva, near the Pantheon. Galileo was interrogated in the third person, in Latin, by Friar Carlo Sinceri, the proctor fiscal, in declarative sentences beginning with the word *an*, "whether." The mental picture that most of us have of this notorious event is an amalgam of many impressions: the memory of old paintings and engravings; perhaps Bertolt Brecht's effective but factually inaccurate play; surely a strong sense of what was historically at stake; and, I would argue, an

unsuspected residue of Perry Mason. I say Perry Mason because we in the English-speaking world have come to see a trial, any trial, as an occasion in which two sides contend, each presenting evidence on behalf of a narrative model held to be essentially valid. So we assume that the dispute in Rome in 1633 concerned the nature of the solar system and the accuracy of assertions about it. We know that Galileo was right about that, and we know that his accusers were wrong, and this gives us our vision of the trial. A row of pinch-faced, black-clad priests in funny hats sit facing an elderly, rather stooped, but nobly defiant scientist, a good Catholic as it happens, who raises his arm in a demonstration of a physical principle. In response, the judges quote scriptural chapter and verse, until at last the old man is silenced. As he leaves the courtroom he mutters, *"Eppur si muove"*—and yet it *does* move—unable to betray the truth.

Of course, this picture (and, unhappily, Galileo's parting line, of which there is no record in the transcript) does not correspond to reality. For one thing, the idea of an adversarial trial would not have seemed natural to the cardinals of the Inquisition. For another, most of them (though not all) were products of the Renaissance, sophisticated humanists as well as post-Tridentine prelates, and God only knows what inner reservations some of them may have had about the proceedings. We cannot know their secret thoughts, or enough about this trial in general, but one thing we do know is that like all such trials it was "inquisitorial" (this time with a small "i") in its juridical form. This means that no Perry Mason–style debating took place, but rather the Commission of Inquisitors balanced the evidence for and against the accused, as is still the case in certain Continental judicial

systems. In accordance with this procedure, Galileo was deposed as needed, being formally addressed in Latin, and permitted to answer in Italian, with his statements recorded by a clerk.

The indictment was arrived at during the course of the proceedings, which strikes us now as outlandish, but it raised no eyebrows at the time. The central accusation—whether Galileo was guilty of "temerity," that is, rashness (contradicting the Church fathers' interpretation of Scripture), or out-and-out heresy, a violation of the faith—was to be decided on the basis of the evidence obtained as the trial unfolded. Galileo's astronomical ideas were pondered only once, in a short, dismissive paragraph written by a churchman without scientific qualifications, and no expert witnesses in the matter were called. However, theologians were enlisted to comb his writings for signs of a positive endorsement of the heliocentric opinion.

If the Vatican's extant transcript of the Galileo trial is complete, and it does appear to be, the proceedings show signs of a departure from standard practice. The Roman Inquisition has hardly been studied, in part because its archives were closed to scholars until 1998, and we know little of the general run of cases in Galileo's day. Yet the scrutiny of canon law, legal manuals for inquisitors' use, and a body of records preserved at Trinity College, Dublin, have given us some sense of Inquisitorial practice in early-sixteenth-century Italy.

The Roman Inquisition was founded by the papal bull *Licet ab initio* of 1542 as a response to the Protestant threat in the peninsula, and a commission of ten cardinals, later called the Congregation of the Holy Office, headed by a prosecutor called the commissary general, was set up to

administer it. The task of the Congregation, which was always understood to be under the direct authority of the pope, was primarily to pursue heresy, usually Lutheranism, and this mission must be borne in mind. The Inquisition was unquestionably a form of mind police. But because it was a Christian mind police, it aimed not only to discover what the accused was thinking but also to reform his thoughts, that is, to return him, fit for salvation, to the body of the church militant. In the business of saving souls, it was more lenient than any Italian civil authority of the same period. Our chamber-of-horrors vision of the Roman Inquisition is derived largely from tales of the earlier, medieval Inquisition, or from the Spanish Inquisition, or from pulp fiction penned in the Protestant world. Whereas civil courts concerned themselves overwhelmingly with what defendants had done, and were often ready to hang them for it, the courts of the Roman Inquisition wanted to know what they were thinking, and so were inclined simply to imprison them until they changed their mind: in fact, the ecclesiastical courts were the first in Europe to use prison terms as a form of punishment and not only as a means of pretrial detention. This fact is sometimes cited in mitigation of the Inquisition's proto-totalitarian nature, but really it cuts two ways. The civil courts after all did not care what you thought, only what you did, whereas if the Inquisition got wind of any public doubts you harbored about the Virgin birth, you would be in very hot water indeed. By the same token, the Galileo affair came about not, as one might imagine, because Galileo had one view and the Church another, but because Galileo had a certain view and the Church insisted that he change it.

Since the Inquisition cared so very much about what the

accused was thinking, he was generally permitted to read the articles of his indictment and to prepare a defense. Theoretically, he could also study the evidence against him and rebut the testimony of his accusers, call his own witnesses, and enlist the aid of a court-appointed attorney well-versed in canon law (though not in a manner resembling current American or British practice—he had, for example, no opportunity to question the legal merit of the accusation or the court's jurisdiction, no right to cross-examine witnesses, no in-court litigator, and no access to a court of appeals). If, in the end, he threw himself on the mercy of the tribunal, his petition might be favorably considered, since a change of heart was the desired outcome. Nowadays we assume, in accordance with Article Ten of the Universal Declaration of Human Rights, that a tribunal must be "independent and impartial," that the accused must know exactly who is accusing him, that a trial must be public, and that the jury must be composed of one's peers. Galileo enjoyed none of these rights. A defendant moreover is to be presumed innocent, yet in the more heavily consulted jurisprudential manuals of this period, the issue of the presumption of innocence or guilt is not unambiguously posed. To sum up, if the evidence against Galileo was thoughtfully weighed in the spring of 1633, and he was permitted to speak in his defense, we should not imagine that his hearing resembled a trial in our sense of the word.

The *quaestio*, or judicial torture, a holdover from the medieval Inquisition, was applied above all in heresy cases. The culprit, it was felt, did not have the right to withhold evidence about himself or his accomplices or to issue an insincere confession. To attain this information, which the court conversely had the right to know, a limited amount of physical

torment could be applied, and the accused, whether determined to be guilty or innocent, was to suffer no permanent
harm by the end of his ordeal. Because of these restrictions,
and because the instrument used, the *corda*,* allowed for the
incremental application of pain, there is little difference
between the Inquisition's prescription of torture and that
practiced by modern Western governments, including the
United States. The aim was and remains to discover some
secret intention by means of a measurable administration of
pain that does not—theoretically—leave any permanent
damage.

The Inquisition's use of torture was curtailed by numerous
cautionary restrictions, some of which bear directly on the
Galileo case. Ideally, both the bishop and the chief inquisitor
in question had to be present. The exercise was not to last
more than an hour at most, and clerics, the gentry, and persons of unimpeachable character were exempt, as were those
already convicted, children, the old or infirm, and those who
had eaten within six to ten hours, lest they vomit. "Torture,"
went a canon-law maxim, "is a fragile and dangerous thing,
and the truth is frequently not obtained by it. For many defendants because of their patience and strength are able to spurn
their torments, while others would rather lie than bear them,
unfairly incriminating themselves and also others." Sometimes indeed the accused bore torture without altering anything in his original testimony, and if he denied harboring

---

* The corda was a mechanical device by which the culprit's arms were
bound behind his back and he was lifted into the air by a rope looped
through a pulley. Pain was applied by dropping him repeatedly and
abruptly to within several inches of the floor.

heretical beliefs, his ordeal might be considered proof of his innocence and a prelude to absolution. In any event, confessions obtained by such means were not considered valid until ratified by the accused at least twenty-four hours later, outside the torture chamber.

The most cursory review of these canon-law guidelines reveals that they were not followed in the Galileo trial. Galileo was never given the right to have a lawyer, to prepare a thorough defense, or to call friendly witnesses. Moreover, as his letter of February 19 to Benedetto Castelli reveals, he felt himself under threat of torture, or to put it more mildly, no one had seen fit to relieve him of this perhaps unwarranted apprehension. (As we will see, the trial transcript for June 16, 1633, holds a clear pontifical order to interrogate Galileo *super intentione*, "concerning his secret intentions," *etiam comminata ei tortura*, "also under the threat of torture.") Yet at almost seventy he was too old to sustain the *corda*. He was manifestly unwell, and his character, if testy, was irreproachable. And finally, curiously enough, he was technically a cleric, having been tonsured on April 5, 1631, by Monsignor Strozzi, the archbishop of Siena, in order to receive a church pension.

One can only conclude that canon law was not properly followed during the Galileo trial or, less plausibly, that it was being generally applied at that period in a manner for which our contemporary legal historians have found no substantial precedent.

There is another matter that we ought to consider before we examine the course of the trial. Very often, when a question bearing heavily on the public interest is decided by a high or influential court, the popular press automatically assumes that the issue itself has been substantively adjudicated. When a

closer look at court documents discloses that the question has been resolved on the basis of some point of law such as jurisdiction or untimeliness, the pundits speak of "technicalities" or "loopholes." Of course they are nothing of the sort for lawyers. In the Galileo case, analogously, two substantive issues were theoretically at stake: whether the Church might accept the Copernican hypothesis, and whether it might countenance, with respect to cosmology, a layman's figurative reading of some verses of the Bible. In reality, however, the issues to be resolved were precisely what we might call technicalities: whether Galileo had a proper license to print the *Dialogue*; whether he had disobeyed Cardinal Bellarmine's 1616 injunction not to teach Copernicanism; and whether his discussion of Copernicanism was hypothetical or conclusive. The truth is that just as no churchman had been authorized by the Vatican to challenge Galileo's heliocentrism before the trial (though several did, very weakly), none seriously challenged it during the trial. The reason is obvious. It contradicted Scripture. It was a nonissue.

If there is little interest in the Galileo trial from an intellectual point of view, however, there is much interest from other points of view. One aspect, not often considered, is the psychological. For three months Francesco Niccolini, with his diplomat's insight into the papal court, had been coaching the fiercely resistant Galileo as a future deponent, trying to get him to abandon his argumentativeness and egocentrism, trying to get him to submit. Thus, when Galileo appeared in court, the language of his depositions and his replies to the magistrates sometimes seems strangely binary, as though a small, rebellious defendant were hiding inside a larger, more compliant one.

On the other side, the inquisitors were hostage to the Vatican's own equivocation, an ambivalence that went back to the decree of 1616, which only "suspended" Copernicus's *Revolutions*, pending certain emendations, and failed to censure Galileo, whom Bellarmine had merely admonished not to advocate the heliocentric opinion. The equivocation had been renewed, even cemented, with Maffeo Barberini's accession to the papacy in 1624 and his promotion of Galileo's friends Ciampoli and Riccardi to positions of authority, all of which had given Galileo hope that the 1616 ban might be revoked. Why had Prince Cesi of the Academy of Lynxes, that avowed foe of the Jesuits, found such favor with Urban VIII? Why had Galileo been encouraged to plug away for a decade at his pro-Copernican dialogue? Why had Riccardi allowed it to be printed, if the Vatican was so opposed to its contents? The answer to these questions cannot be found in the realm of ideas, but rather (if at all) in that of psychology, in the rivalry of personalities and parties. The equivocation, which arguably resulted from the papacy's desire to play two conflicting roles simultaneously—on the one hand, promoter of Tridentine orthodoxy; on the other, renewer of humanism and patron of a Tuscan client-state—was a terrible trap for Galileo, but it was a trap for the Holy Office as well, for that body now found itself in the derisory position of considering whether to suppress a work that bore its own imprimatur.

Galileo at his first deposition must have seemed remarkably optimistic. Under interrogation he began to walk very boldly on very thin ice. He stated that he had come to Rome in 1616, and again in 1624, with the express intention of "making sure of holding nothing but holy and Catholic opinions"; actually, he had hoped to induce the Roman Church to become

the sponsor of his own opinions. After conversing with several cardinals, he said, he had discovered that the Vatican regarded the Copernican system as "repugnant to the Holy Scriptures, and admissible only *ex suppositione*, in the manner in which Copernicus himself took it"*; actually, neither Copernicus nor Galileo had regarded the Copernican system as anything but a likely hypothesis in the modern sense. This sort of suppositional reasoning was exactly what Galileo had demolished with the telescope and physical astronomy. Now he was invoking it for the purposes of self-exoneration.

But Galileo wasn't through yet, for a piece of surprise evidence was about to make a fairy-queen appearance. Ever certain that fortune favored his cause, he held up a document for the court to see: one imagines the inquisitors bending forward to peer at it. So *this*, during those long, hard weeks of winter confinement, was what had mysteriously cheered him and convinced him of his impunity.

"I was notified," he began, "of the decision of the Congregation of the Index, and the one who notified me was Cardinal Bellarmine. . . . In February of 1616, Cardinal Bellarmine told me that to be of the Copernican opinion was absolutely contrary to Holy Scriptures—one could neither hold nor defend it—but suppositionally one could take it up and make use of it. In conformity with which I have here a certificate from the same Cardinal Bellarmine, executed on the 26 of May, 1616, in which he says that the opinion of Copernicus

---

* The editor of Copernicus's *Revolutions*, Andreas Oseander, had supplied an unapproved and not particularly believable preface stating that the work was suppositional, and of use primarily for mathematical computations.

may not be held nor defended, being against the Holy Scriptures, of which certificate I hereby offer a copy, and here it is."

What the court gathered was that Galileo, outraged by rumors to the effect that he had been censured by the Holy Office, had in 1616 sought further contact with Bellarmine of his own free will, and had received a written clarification and attestation. Bellarmine had died in 1621, but Galileo's copy of the document, twelve lines long, was read aloud, marked with the letter "B," and dutifully entered into evidence:

> We, Robert Cardinal Bellarmine, hearing that it has been calumniously rumored that Galileo Galilei has abjured in our hands and also has been given a salutary penance, and being requested to state the truth with regard to this, declare that this man Galileo has not abjured, either in our hands or in the hands of any other person here in Rome, or anywhere else as far as we know, any doctrine or opinion which he has held; nor has any salutary or any other kind of penance been given to him. Only the declaration made by the Holy Father and published by the Sacred Congregation of the Index has been revealed to him, which states that the doctrine of Copernicus, that the earth moves around the sun and that the sun is stationary in the center of the universe and does not move east to west, is contrary to Holy Scripture and therefore cannot be defended or held. In witness whereof we have written and signed this letter with our hand on this twenty-sixth day of May, 1616.

When pressed further by the interrogator, who must have sounded not a little surprised, Galileo described the circumstances of the meeting and the possible presence of certain

clerics. He remembered that perhaps he had also been instructed "not to teach" the heliocentric system, as well as not to hold or defend it. "I don't remember any particular *quovis modo*," he slyly added (referring to a Latin expression in Bellarmine's earlier warning to him meaning "in any manner whatsoever"), "but I didn't think about it or recall it, because several months later I received this certificate from Cardinal Bellarmine." In other words, the letter from Bellarmine, which did *not* exclude hypothetical discussions of the heliocentric position, seemed to override the more sweeping prohibitions of other injunctions, which, besides, he had not remembered.* "After receiving the above-mentioned precept," he went on, "I did not request permission to write the book in question . . . because I do not believe that in writing it I would transgress in any way against the precept made to me, not to hold nor defend nor teach the said opinion, indeed to refute it."

Everything had gone swimmingly for the defendant until the last few words: *indeed to refute it*. There he had gravely erred. As we know from detective stories, superfluous compliance can be self-incriminating: consider the witness who offers more testimony than anyone asked for. Galileo had now done something analogous. Perhaps he might have convinced the Inquisition that he had explained the Copernican system as a sort of philosophical exercise, but did he really think he could convince it that the *Dialogue* humbly obeyed an ecclesiastical command to *refute* that system? The Inquisition had been

---

* He may have received a further warning from a priest named Michelangelo Seghizzi de Lauda, but if this happened, he apparently forgot about it as well.

studying the *Dialogue* for eight months, so Galileo's preposterous claim can only have suggested that he was speaking insincerely, telling the interrogator whatever he wanted to hear. It may be that he had taken Niccolini's counsel too much to heart in the end. Unfair trials, however, have an extraordinary capacity to induce otherwise honest people to lie: the trial's unfairness and Galileo's lying must be taken as correlatives.

If, then, for Galileo, Bellarmine's certificate was a ticket to exoneration, the Inquisition must have begun to see it very differently. "Malice" and "deception" were what the tribunal most abhorred. Remember that the culprit needed to explain why he had not told Riccardi, the Vatican censor, of Bellarmine's earlier warning. The reason, he intimated, was that it had been clarified and superseded by this document. But if one listened carefully to Galileo, here he was claiming that his *Dialogue* aimed to "refute" the Copernican system, which sounded like simple perjury. Moreover, Bellarmine's certificate had itself clearly admonished him neither to defend nor to hold the Copernican "opinion." Had he, in fact, "held" it? Soon a panel of three theological consultors would issue their reports. Suppose the *Dialogue* should indeed be found to defend or to hold Copernicanism (not to mention teaching it or discussing it, both of which Bellarmine had forbidden); wouldn't that provide a perfect justification for Riccardi and the entire Holy Office? Instead of being in the doubtful position of banning a book they had licensed, they could now claim that Galileo was maliciously availing himself of Bellarmine's certificate in order to justify obtaining permission for a volume injurious to the faith.

So the interrogator logically turned to Galileo's request to publish the book. Galileo recalled his trip to Rome in 1630 to

obtain an imprimatur from Riccardi. He cited his reception of it, on Riccardi's condition that he, Riccardi, be allowed to "add, delete, and change as he saw fit" before publication, which had led to the pope's brief and clearly inadequate review of the *Dialogue* and to the insertion of its contradictory preface and conclusion (an echo of the pope's own theology, which actually contradicted the book's thesis and were voiced by the callow Aristotelian, Simplicio). But upon Galileo's return to Florence, the matter had been suspended, as Riccardi fretted over the book's contentions. Meanwhile the Black Plague broke out, rendering travel hazardous. Riccardi agreed to turn the request over to the inquisitor in the Vatican's Florence office, as long as the preface and conclusion be submitted to him for final approval. So it had come about, Galileo said, that the *Dialogue* had been printed in Florence, with no less than one imprimatur from Riccardi and another from the Florentine inquisitor (and—though he didn't mention this—the preface and conclusion in an incongruous typeface).

This account left him open to an obvious question. When he had asked Riccardi for permission to print the book, the interrogator asked, had he revealed the injunction previously given to him by the Inquisition?

"I did not," Galileo replied, ". . . because I did not judge it necessary to tell him, since with this book I had neither held nor defended the opinion of the earth's motion and the sun's stability."

This was nonsense. For he had held and defended it, if only (as he claimed) suppositionally; though, when you came right down to it, he had held and defended it conclusively and affirmatively. After one short hearing, Galileo stood on the verge of losing his freedom forever.

He also stood on the verge of winning it. The Tuscans had a certain clout in Rome, and Niccolini, for one, had been active behind the scenes. He had asked the grand duke to draft pleas to all the inquisitors not already canvassed. Soon Niccolini received a batch of missives, which he attempted to deliver to them, with scant luck in some instances. He also became aware that his fellow-Lyncean Francesco Barberini and, of all people, the chief inquisitor, Vincenzo Maculano, known as the commissary general, were now laboring to get the trial expedited, exactly as Niccolini had wished from the start. Maculano, it has been suggested, felt his case undercut by the appearance of Bellarmine's certificate. But he was not analogous to an American district attorney—his "career" was not on the line. He was charged not with "winning" the case and securing the maximum penalty, which would have meant seeing Galileo charged with heresy and imprisoned, but with securing Galileo's confession and renunciation of Copernicanism, if it should be proved that he supported it; the book could be condemned or emended as needed. One may speculate that the authority of Bellarmine's certificate might only have softened Maculano's already malleable position. On April 16, Niccolini wrote Cioli that Maculano had told Niccolini's secretary that the trial would soon be concluded; indeed, Maculano had informed Niccolini himself that Francesco Barberini had been pleading the scientist's case to his uncle, trying to "palliate" the pope's "emotions." He wrote Cioli a jubilant note on April 23, declaring that Galileo would probably be released on or right after Ascension Day, when Urban VIII returned from his country residence of Castel Gandolfo.

What had happened? The day before, in a letter discovered only in 1999, a distraught Maculano had written the follow-

ing words to Francesco Barberini: "Last night Galileo was afflicted with pains which assaulted him, and he cried out again this morning. I have visited him twice, and he had received more medicine. This makes me think that his case should be expedited very quickly." Recently, he said, the three theological consultors' reports had been filed, all of which concluded that Galileo "defends and teaches the opinion which is rejected and condemned by the Church, and that the author also makes himself suspect of holding it. That being so, the case could immediately be brought to a prompt settlement." In another letter to Francesco, Maculano reported on April 23 that he had proposed a plan for a settlement to the Holy Office, requesting that the cardinals "grant me the authority to deal extrajudicially with Galileo, in order to make him understand his error and, once having recognized it, to bring him to confess it. The proposal seemed too bold at first . . . however, after I mentioned the basis on which I proposed this, they gave me the authority . . . In order not to lose time," Maculano continued, which suggests that as a clergyman with a conscience he was extremely worried about Galileo's health and mindful of canon-law restrictions about trying defendants *in extremis*, "yesterday afternoon I had a discussion with Galileo, and, after exchanging innumerable arguments and answers, by the grace of the Lord I accomplished my purpose: I made him grasp his error, so that he clearly recognized that he had erred and gone too far in his book; he expressed everything with heartfelt words, as if he were relieved by the knowledge of his error; and he was ready for a judicial confession. However, he asked me for a little time to think about the way to render his confession honest. [Thus] the Tribunal will maintain its reputation; the culprit

can be treated with benignity; and, whatever, the final outcome, he will know the favor done to him, with all the conse-
quent satisfaction one wants in this."

One cannot resist picturing the two Tuscans arguing in the
waning afternoon light, the humane and practical-minded
priest sitting erect in his crimson vestments, now and then
bending to help the half-reclining scientist ease his pain. The
high point of the drama had been reached: one sees Galileo
falling into his irrepressible habit of talking mathematics and
biblical hermeneutic, and Maculano as a priest (and an engineer, with mathematical training) tactfully warning him off
this course. One can only speculate about Maculano's reasons
for wishing to strike a deal with Galileo, but certain possibilities (in addition to the latter's ill health) come to mind: the
pope's own hesitancy and former friendship with Galileo; the
bad publicity generated by the trial in Tuscany, Italy, and
Europe at large; the obvious weaknesses in the prosecution;
and the probability that, given Galileo's reputation and the
widespread interest in Copernicanism among Italian intellectuals and even many churchmen, his massive condemnation
and punishment would constitute very poor public policy.
Finally, one cannot exclude a truant sympathy for the man
and his ideas: Benedetto Castelli, now mathematician to the
pope, had written Galileo on October 2, 1632, that Maculano
himself (who as a fortifications expert was probably the only
member of the Congregation capable of following the demonstrations in the *Dialogue*) "was of the same [heliocentric]
opinion, that the question should not be concluded with the
authority of the Sacred Letters [i.e., Scripture]; and he even
told me that he wanted to write about it." Aside from the foregoing, there must have been, as always in the secrecy-bound

Rome of 1633, a number of hidden political reasons for the plea bargain that we cannot remotely imagine. It would be naive to think that Maculano had no personal or familial motive for ending the trial posthaste.

As for Galileo, in this face-to-face meeting with a prelate of Maculano's standing, he must suddenly have realized that his soul was in mortal danger. On one level, he could not rid his mind of a world picture that he knew to be true. On another, he was doubtless saddened, though perhaps for the first time, to have so displeased his Church. Caught between two contradictory emotions, and doubtless ashamed of his recent perjury, he surely needed the interval he sought to arrive at what he called an "honest confession." Eventually, under the impress of fear, genuine piety, and a desire to espouse Copernicanism while simultaneously rejecting it, Galileo's duplicity probably turned into self-delusion.

At this point, Latin Christendom hung in the balance. Let us for a moment play a parlor game, the game of counterfactual fantasy. Had Maculano's scheme been realized, Galileo would probably have confessed his error. He would have been convicted of some lesser crime, such as "rashness," and been obliged, most likely, to do penance for several months. He would have been permitted to return to his research with full freedom of movement as long as he left Copernicanism and the Bible alone. But by this time Maculano, and perhaps Niccolini too, must have grown aware that not all the inquisitors of the Holy Congregation were of their opinion. Let us continue our parlor game and speculate about the two camps into which they conceivably had separated. In the first, inclining to leniency, were surely Maculano and Francesco Barberini; and, we might guess, Gaspare Borgia and Laudivio Zacchia

(because they, like Francesco, would decline to present themselves at Galileo's eventual recantation); also, just possibly, Antonio Barberini, the pope's brother, who had been shoehorned into the curia without qualifications nine years earlier and who, according to Niccolini, liked to be wooed, since it made him feel important: Niccolini told Cioli that Antonio Barberini might well be "helping" Galileo more than anybody else, and what could this mean but badgering the pope? So we might suppose (without, I repeat, any evidence) that these inquisitors were "soft," and the others presumably less so or——unhappily—not at all so.

Two days after Maculano's fateful meeting with Galileo, the accused gave his second deposition. His story had completely changed. No longer did it seem to him that his book refuted Copernicanism. It had now "dawned" on him, he said, to review the book, and he had "started to read it with the greatest concentration and to examine it in the most detailed manner. Not having seen it for so long, I found it almost a new book by another author." He felt that the Aristotelian side in the dialogue had not been presented fairly, and that in trying to show off his polemical abilities in defending the weaker, Copernican side, he had exaggerated. "I resorted," he said, ". . . to the natural gratification everyone feels for his own subtleties and for showing himself to be cleverer than the average man, by finding ingenious and apparent considerations even in favor of false propositions. . . . My error then was, and I confess it, one of vain ambition, pure ignorance, and inadvertence."

In essence, Galileo was being so bold as to tell the tribunal to convict him of vanity, ignorance, and carelessness, none of which approached heresy. The court adjourned briefly, after which he returned to confirm his denial of the heliocentric

thesis; he also offered to add, by the tribunal's leave, two more literary "days" to the *Dialogue*'s four, in which his characters would demolish the arguments already presented. The cardinals of the Holy Congregation did not seem thrilled by this suggestion.

SO GALILEO HAD made his confession. He had bowed his head. Niccolini, Cardinal Barberini, Maculano, and anyone else who had discovered what was going forward awaited a rapid resolution of the trial. Yet nothing had actually been determined. There seemed to be some mysterious holdup.

Why had Galileo not been convicted, scolded, given a penance, and released? Unknown to him, after the trial's second session a large amount of material from the Inquisition's 1615–16 investigation of his beliefs, itself notably sloppy, had been entered into evidence as an attachment to a remarkably mendacious summary—not a transcript—of the proceedings to date; this joined the negative reports on the *Dialogue* filed on April 27 by the three theological consultors, none of whom was proficient in astronomy but all of whom concluded that Galileo supported heliocentrism. It is impossible to say who compiled the falsified summary, bloated with the 1615–16 allegations, or to what degree it may have misled the magistrates; after all, they had read Galileo's own account and taken note of his statement of contrition. The three consultors' reports were distinctly damaging, but Maculano had told Niccolini that they strengthened, not weakened, his case for a plea deal. Unfortunately, however, both the summary and one of the reports, a vitriolic attack by the Hungarian-born Jesuit named Melchior Inchofer, who mentioned Galileo's dispute over the sunspots with his still-aggrieved fellow Jesuit

Christopher Scheiner, cited a certain passionate letter that
Galileo had written in 1614 to Benedetto Castelli, then a
mathematics professor at Pisa. The letter was a private com-
munication to a Benedictine priest; it did not publicly teach or
advocate anything; moreover, the Inquisition's copy had been
shamelessly doctored. But unfortunately it revealed its
author's genuine belief that the Old Testament passages
appearing to contradict science should not be taken literally.
This was news to nobody, but it may have muddied the case
for a plea deal, because the Index had established the contrary
in 1616 and because Galileo as a layman had no right to pronounce on the Bible.

It is impossible to deduce from the trial records (which,
spotty as they are, may have been typical for that period and
that institution) whether Cardinal Maculano's extrajudicial
settlement was brutally sabotaged or simply delayed, watered
down, denatured through bureaucratic manipulation. In the
end, it never went through, and there is no acknowledgment
in the documents that it didn't or an explanation as to why it
didn't. We do not know what Maculano or Francesco Barberini or their allies made of this shipwreck, but it is not
unreasonable to suppose that they grew aware that their
antagonists had outfoxed them. They were men of the cloth,
however, and they left no lament.

Galileo had been escorted back to Villa Medici on April 30
and instructed to wait. Never had his perception of other people, indistinct at best, been as fuzzy as it was now. To his
friends in Tuscany he wrote buoyant letters suggesting that he
would be released in short order, and, indeed, exculpated.
Niccolini for his part seemed fairly heartbroken. The grand
duke, proprietor of the enormous villas of Palazzo Pitti,

Careggi, Cafaggiolo, Poggio a Caiano, Poggio Imperiale, L'Ambrogiana, Serravezza—one could go on and on—was still fussing about the defraying of Galileo's expenses, a detail over which the ambassador scarcely had the will to bicker. Ever since autumn, even before Galileo's arrival, he had been trying to swing an expedited trial, enlisting all his political savvy, all his knowledge of the papal court, and now it was slipping through his fingers. In late May, when the situation had drastically deteriorated, he was still keeping the truth from the old fellow, who was often bedridden. "I haven't told him everything yet," he wrote Cioli, "because I intend, in order to avoid causing him pain, to go at it little by little."

On May 10, Maculano summoned Galileo before the Congregation and offered him eight days to prepare a defense. He coolly replied that he would hand in a written defense forthwith, together with the original of Bellarmine's certificate. Galileo's defense, what we would call a "sworn statement," stated that the Copernicanism of the *Dialogue* was inadvertent and that he was prepared to make amends for it, and he begged the judges to consider his ill health and the slander he had suffered. The Italians have a charming term for an insincere confession of this sort, that it is *figlia della convenienza processuale*, a "daughter of trial strategy," yet it only reflected the plea that Galileo had contrived under Maculano's supervision. Unknown to him, however, the deal was off, and the plea fell on deaf ears.

Someone, or some people, somewhere, had decided to make an example of Galileo. Pope Urban VIII had ultimate authority over the Congregation, and we shall come to this preeminent fact; certain Jesuits may have thrown their weight against the Tuscan physicist; and he had plenty of other ene-

mies. What should be noted, however, is that the conflict between heliocentrism and religion had made only the most garbled appearance at the trial; no scientific knowledge or reasoning had been put to use. At all events, a brief and tendentious report was submitted to the pope, and a pontifical order was issued prohibiting further publication of the *Dialogue* and ordering that Galileo be interrogated concerning his real or secret intentions, under threat of torture if necessary (*interrogandum esse super intentione, etiam comminata ei tortura*), the assumption being that he might be acting with malice or deception.

Canon law barred the torture of Galileo, and there is no evidence that he was tortured. As Maurice Finocchiaro has emphasized, however, the salient fact is that Galileo's final interrogation, on June 21, 1633, was conducted under the threat of torture, and this gives it a peculiar psychological dimension. One must distinguish carefully between the institutional justification for torture and the real, psychic reason why it might be resorted to or threatened. Torture is widely practiced in most countries, but its adoption and abandonment (at least nominal) in the postwar period by such Western nations as France, Great Britain, Greece, Argentina, and Israel, and its current use by the United States, have led to systematic study. It is very difficult for a person or group possessing physical power over a defenseless antagonist to resist using it, or threatening to use it, and that this explains the practice of torture much more convincingly than does the flimsy hope of extorting secret information of any practical value. The Roman Inquisition's relatively restrained use of torture, compared to that of the civil authorities, has been cited in mitigation of its horrors. Yet this argument cuts two ways, as stated

earlier, for its appearance in a religious context is precisely what excites our repugnance. If, ultimately, the pope resorted only symbolically to the threat of force to resolve his dispute with Galileo, the symbol—the *corda*—was absolutely central, as the popular imagination has long and justly observed.

At the end of the final session the interrogator asked, "*An aliquid ei occurrat ex se dicendum.*" Has anything occurred to him that he wishes to say?

"I have nothing whatever to say," Galileo replied, usually so voluble.

And what does he hold, concerning the sun being at the center of the world?

"Assured by the prudence of my superiors, all ambiguity within me has ceased," he said, adding that he accepted the stability of the earth.

And with respect to the opinion expressed in his book?

"I conclude that within me I do not hold nor have I held, according to the determination of my superiors, the damned opinion."

At this point he was formally warned that he was nonetheless under suspicion of still harboring the Copernican opinion, and that unless he "resolved to make the truth known," the "remedies of the law" and "opportune measures" would be used against him. This again referred to the "rigorous examination" of torture.

"I do not hold nor have held this opinion of Copernicus," he said, "after I was notified by the injunction that I was to abandon it. Besides, I am here in your hands, you may do as you please."

This last phrase, which can also be translated "I am here in their hands, let them do what they want," gives away Galileo's

true state of mind. Read either way, it is a rude, peremptory way of addressing a tribunal of cardinals, especially in 1633. At last he has realized that the plea bargain has evaporated: the rebellious, frightened prisoner who for weeks has been hiding inside the penitent culprit devised by Maculano has, with these exasperated words, climbed out. Everything in this interrogation discloses Galileo's terminal acknowledgment of the brute fact that this trial is really about nothing but discipline and obedience, strength and weakness, menace and fear. Close scrutiny of every one of his answers at the final interrogatory shows that he embraces the doctrine of the Church only because he has been told to, and for no other reason. *I have nothing to say . . . the prudence of my superiors . . . I am in your hands, do as you please.*

"And he was told to tell the truth," reads the transcript, "otherwise recourse would be had to torture."

"I am here to show obedience," said Galileo, "and I have not held this opinion after the determination was made, as I have said."

IN THE INTERESTS of following the story of the trial as directly as possible, we have not paused over one circumstance that gravely complicated Galileo's case. This was the appearance, among the documents entered into evidence, of the letter he had written early in 1614 to Benedetto Castelli, in which he had trespassed into the field of biblical exegesis. The letter, as already noted, was private, but once it became public Galileo faced what Richard J. Blackwell has called a kind of "double jeopardy," one point of which related "to the content of the interpretation, the other to assuming the role of being an interpreter. No matter what the merits of the former, the

individual was always in jeopardy on the latter." In other words, whether or not the biblical analysis found clerical acceptance, and in this case it did not, Galileo as a layman had no dispensation to interpret the Scriptures. This prohibition had figured among the decrees of the Fourth Council of Trent, published in April of 1546.

What was in this letter, and why did Galileo write it? At the end of 1613, Dowager Duchess Cristina of Lorraine, mother of Cosimo de' Medici, had raised the issue of the compatibility of Catholicism and Copernicanism at a breakfast that Benedetto Castelli had attended. Informed of this, Galileo resolved to draft a short essay expressing his views on the matter, which became known as the "Letter to Benedetto Castelli." He must have intended for this letter to reach some influential people beyond Castelli and so perhaps to forfeit its private character, but he could hardly have suspected that it would end up as evidence for a heresy charge against him. His aim was simply to show that the heliocentric world view did not contradict Scripture and thus posed no threat to the Church. With reference to the best-known scriptural passages that seemed to support the geocentric cosmology, such as Ecclesiastes 1:5 ("The sun rises, and sets, and returns to its place, from which, reborn, it revolves through the meridian, and is curved toward the North") and Joshua 10:12 ("Sun stand thou still . . ."), he argued that the writers of the Bible had accommodated their language to the grasp of ordinary people, offering them the historical chronicles and ethical precepts needed for their salvation; it was perfectly obvious that the Bible had no concern with astronomy. Since God had created both nature and the Scriptures, both necessarily had to be true: if an apparent conflict came about, the Bible should take second

place, not because it was any less valid, but because its meaning was often obscure or ambiguous, whereas mathematical
demonstrations were straightforward and final. Above all,
Galileo warned of how unwise it would be to fix a biblical
meaning in advance if it were unrelated to the spiritual cause
of salvation, for in time such an interpretation might be
shown to be false. From the theological standpoint there were
snags in this reasoning, and when they emerged, Galileo sat
down and wrote another, more fulsome "Letter to the Grand
Duchess Cristina" in which he tried to untangle them, noting,
in addition—probably at Castelli's prodding—that Saint
Augustine, in his long treatise *On the Literal Interpretation of
Genesis*, had admonished the faithful not to read Genesis too
narrowly, lest a mulish adherence to the letter of this difficult
text should cause a "scandal" to the Church. For the Inquisition, this layman's citation of Augustine must have seemed a
piece of effrontery. But the cardinals at the 1633 trial did not
need, as lawyers say, to "reach" that damning fact, since the
"Letter to Benedetto Castelli" answered to their purpose, and
they left the letter to Cristina alone, probably out of respect for
the Tuscan ducal family.

We do not know what Galileo or anybody really believed at
this period, since religious belief was prescribed by an autocracy and heresy was an actionable offense. If one had misgiv-
ings, one kept them to oneself, so it would be naive to take
religious ruminations penned in the papal realm or its client
territories at face value. The Inquisition's own records confirm
that many people harbored reservations and heretical beliefs:
before the Counter-Reformation, they had been much more
candid about them. Giorgio Vasari, author of the *Lives* of the
Italian artists, of 1550, tells us that Perugino, painter of innu-

merable sweetly pious altarpieces, "was a person of very little religion, and no one could ever make him believe in the immortality of the soul"; Perugino can hardly have been alone. The biblical text had also lost credence. If it was still overwhelmingly felt to be in some sense the word of God, the Reformation had led to numerous books being expunged from the Protestant biblical canon, and the Vulgate, the Roman Catholic Bible in Latin, had also been called into serious question. This text had been assembled from a collection of manuscripts and largely translated into Latin by Saint Jerome in AD 390–405. Yet Hebrew joined Latin and Greek as a widely studied language in sixteenth-century Italy, and by the end of the Council of Trent, which lasted from 1545 to 1563, Jerome's Vulgate was thought to be inaccurate and too often obscure: though declared canonical in 1546, an emended edition was called for. It took three pontifical commissions to produce a corrected text in 1588 and yet another in 1592, and the two editions differed at almost five thousand points, leading to heated criticism and some derision. Meanwhile, the whole idea of a sacrosanct text had fared rather poorly in European humanist circles, where the basic concepts of modern philology had struck root. That Moses, the supposed scribe of the Pentateuch, could certainly not have written some of it—he dies toward the end of Deuteronomy—had already been noted in the Babylonian Talmud, a fact further discussed by Abn Ezra, a twelfth-century rabbinical commentator much consulted by Christian exegetes, and this and similar problems had been quietly reviewed. Within a few years of Galileo's trial, a Catholic, Isaac La Peyrère, two Protestants, Thomas Hobbes and Samuel Fisher, and, later, a Jew, Benedict Spinoza, were to deny that humanity has an accurate

copy of the true Bible, or that it was written by any but a number of authors.

So the Counter-Reformation was fighting a losing battle against a freer, and often less literal, reading of the Scriptures. Yet if there was one thing that had concerned the Council of Trent, it was the possibility that laymen would decide for themselves what passages in the Bible could be interpreted other than literally. In fact, the issue of the earth traveling about the sun had little if any bearing on the Catholic faith. But the notion that persons without theological training could decide for themselves to read this or that biblical passage in a nonliteral sense constituted a mortal danger for Catholicism in the early seventeenth century. The Protestant churches had split from Rome in denial of the literal meaning of the Sacrament of the Eucharist, reaffirmed in Session VIII of the Council of Trent, in 1551, as "truly, actually, and substantially the body and blood" of Christ. It places no great demand upon our sympathetic imagination to understand that the Church, in an age before the ideal of free expression had attained currency anywhere, was not about to allow laymen to decide whether phrases like "This is my body" (Mathew 26:26) were to be taken literally. A magnificent civilization had been rent asunder over just these issues. Consequently, though what Galileo said about Scripture was perfectly reasonable (indeed, a wise commentary on 2 Corinthians 3:6, "For the letter killeth, but the spirit giveth life"), it was also most impractical, reflecting a devout but overconfident sense of mission.

If the Inquisition's charges against Galileo were that he had championed Copernicanism conclusively rather than counterfactually, as a hypothetical argument, and that he had interpreted Scripture according to his own lights, then he was

basically guilty. The serious mitigating factors—Bellarmine's certificate and the Church's ample opportunity to review the *Dialogue* before its printing—were not taken to heart. The entire trial focused on the issue of insubordination, with only three short paragraphs being devoted to the substance of Galileo's science; and these paragraphs, which occur in the report from the consultor Zaccaria Pasqualigo and concern the causes of the tides, merely show that Pasqualigo did not understand what he had read. Certain Galilean scholars have asserted that the cognitive issues raised by the conflict between Galileo's Copernicanism and the Church's views about astronomy and the Bible had already been taken up in the period between 1610 and 1616, and especially during the Inquisition's investigation of Galileo at the end of that period. With the best will in the world, I cannot agree: there are no records to support such a view. No Church document from 1615 to 1616 carefully examines either the scientific plausibility of heliocentrism or the complex issue of its compatibility with the Roman faith. The "Consultants' Report on Copernicanism" of February 24, 1616, in large measure a response to Galileo, states only that the heliocentric proposition is "foolish and absurd in philosophy" and "explicitly contradicts in many places the sense of Holy Scripture." Bellarmine's already mentioned letter to Father Foscarini of April 12, 1615, also intended for Galileo, represents the Vatican's broadest discussion of this topic, yet basically it dismisses the question out of hand. When, at the outset, Bellarmine says, "It appears to me that Your Reverence and Sig. Galileo have acted prudently in being satisfied with speaking in terms of assumptions," denoting with this term a merely suppositional Copernicanism, he is not complimenting Foscarini and Galileo on their intellec-

tual tact. No: he is issuing a veiled, Jesuitical warning. Already the dynamic of intimidation has begun.

Did Galileo and the Church ever square off intellectually? The answer must be negative. Though initially acclaimed by the Collegio Romano, Galileo ran afoul of the Jesuits around 1624 with the "Reply to Ingoli" and *The Assayer*, which was to some degree an attack on Father Orazio Grassi. Both Ingoli and Grassi were churchmen who disputed various aspects of Copernicanism, Ingoli seriously and consistently and Grassi confusedly and tangentially. Neither represented the Vatican, and neither accused Galileo of heresy. Moreover, it is likely that Maffeo Barberini, as newly elected pope, supported Galileo against the Jesuits at the time; one recalls that *The Assayer* was reprinted with a dedication to Urban VIII. (Neither the "Reply" nor *The Assayer* was cited at the trial.) In fact, at various periods Galileo had devoted supporters inside the Church—not to mention Maffeo Barberini himself, one thinks of Sarpi, Castelli, Dini, and Ciampoli—and he had many ruthless antagonists in the secular academic community. It is only toward 1633, rather suddenly and sadly, that science and religion as social forces began identifiably to collide. What often goes unremembered is that the Church was then a temporal power with a police force (the papal *sbirri*) and a court system at its command. Thus the confrontation with Galileo, when it came, was not a genuine discussion but an act of coercion. The coercion was mild by modern standards— those, let us say, of the Moscow show trials of 1938—but it was coercion nevertheless. If there is any lesson to be derived from the Galileo affair, it is probably that the dialogue between science and religion, which is valuable on many, especially ethical, points, should not be contaminated by the use or

threatened use of state power. The obedience to a hierarchy, the reliance on formalistic or ritualistic structures, the reference to sanctified interpretations of scriptural texts whose very spirituality derives from their fluidity and continual expansion of meaning—these aspects of the authoritarian posture destroy any hope of a positive exchange. Religion is often demagogically enlisted to rally a threatened social order, and such gambits, too, frustrate the intellectual probity of the debate.

There is another dimension to this trial that is seldom if ever mentioned, though it may be the most critical one. Historically the Italians have had little use for ideologies, which have tended merely to mask their mundane needs and wishes. In Galileo's period, as now, people's ambitions were tied to family allegiances, to the desire to further the fortunes and powers of one's *casa*, one's dynastic house. Little else mattered. The great political writing of the preceding decades— Machiavelli, Castiglione, Guicciardini—had assumed a clear-eyed cynicism on the part of the more capable princes and bishops, dismissing as hypocrisy all noble sentiment. One suspects, therefore, that a tissue of underhanded pressure, linked not to ecclesiastical but to family interests, underlay this important trial. Indeed the dominant presence of the Barberini, Medici, and Borgia clans suggests a mind-boggling maze of *clientelismo*, and who knows whether the famous plea bargain was not the consequence of somebody's "having the goods" on somebody else—for a few blessed weeks at least. Calling to mind this world of hushed antechambers and darkened sacristies, of signifying nods and whispered Latin words, we should be on our guard against the tendency to indulge in high-minded philosophical interpretations of the skimpy trial

records that have survived from that spring of 1633. In the way of documents, however, they are, along with a few letters, all we have (unless more of the Barberini correspondence should come to light). If the usual Roman machinations took place, we don't know who may have been importuning or cajoling whom.

We have seen that between 1615 and 1632, the Vatican's position toward Galileo was exquisitely ambivalent, tending now toward favor, now toward hostility. This lay in the nature of autocracy, which exercises power most effectively if its subjects do not know exactly where they stand—their very uncertainty enforces awe and timidity. But what if a subject comes along whose ideas are both monumentally brilliant and potentially transgressive? For the prince to accept such ideas would diminish him by comparison, while their suppression would make him look stupid and cruel. This was the dilemma the pope faced in 1632–33.

Historians have two chief theories as to why Maculano's extrajudicial settlement collapsed, leading to a long and punitive trial. One was that a Jesuit cabal was launched against Galileo. The other was that a vindictive Maffeo Barberini intervened and axed the plea deal. Both may be true at once, and much of interest has been written outlining complex possibilities. One thing, however, cannot be denied: nothing could be done without the pope's consent.

In order to understand Maffeo, a cultivated and intelligent if vain and temperamental man, it is important to note that he had several different *kinds* of reasons to be angry at Galileo. First, he had personal reasons: this fellow Tuscan whom he had encouraged, whom he had addressed as "son" and "brother," and whom he admired rather as he admired

Bernini, had betrayed him in the *Dialogue* by assigning his theology to a weak-minded debater with the unfortunate monicker of Simplicio; the pope may also have thought that Galileo had enlisted Ciampoli to hoodwink him into sanctioning the book. Second, Maffeo had doctrinal reasons, relating to his pontifical role: Galileo had espoused heliocentrism and trespassed into biblical territory, which were the crimes for which the Inquisition had indicted him. Third, Maffeo may have had political reasons, resulting from his decisions as a temporal ruler. This is not the place to explain Pope Urban VIII's claims to theocratic absolutism after 1627; his disastrous meddling in the Mantuan succession; his support for a powerful Protestant leader, Gustavus Adolphus of Sweden; his befuddled attempts to arbitrate among the great Catholic powers; or his provocation of the Spanish Hapsburgs, who, it may be remembered, ruled the Kingdom of Naples, just south of the Papal States. In a secret consistory of March 1632, Cardinal Gaspare Borgia, the Spanish ambassador, had attacked the pope so vehemently that guards had had to be summoned: Maffeo was an inveterate francophile, and Rome was rife with rumors that the Spanish wanted him deposed. Ciampoli had been implicated in the disturbance, and the pope, as we have seen, always made much of Ciampoli's connection with Galileo (though certainly Galileo did not). It is hard not to surmise that, all things considered, Maffeo felt obliged to set in motion an imposing institutional machine that would demonstrate his control over the Galileo affair. A quick fix did not serve his publicity purposes.

But Galileo's ordeal and ultimate condemnation were the reverse of everything Maffeo's papacy, with its yearning to fuse Renaissance humanism and post-Tridentine piety, had stood

for. It was a tragedy for him as well as for Galileo. Long ago he had written to the philosopher Tommaso Campanella about the 1616 decree: "It was never our intention; had it been up to us, that decree would not have been passed." But once again, as in the days of Pope Paul V and Cardinal Bellarmine, the Vatican had refused to see, refused to look up and perceive the world as it is, the heavens as they are made. It was a kind of blindness, as Kepler had so insightfully put it, though in the long run this blindness had little or nothing to do with optics, or astronomical representation, or any objection to Galileo's comfortable reliance on vision, on sense data. It had to do with the sheer difficulty of looking fixedly at what seems to be distressing or upsetting. Maffeo may have had lingering regrets about this. Years later, when Benedetto Castelli implored the pope, through his brother Antonio, to accept Galileo's assurance that he had never intended to hurt his feelings, Maffeo answered—rather sorrowfully, I think—"We believe him, we believe him."

THE SENTENCE OF the Holy Congregation, handed down on June 22, convicted Galileo of holding and teaching Copernicanism, of interpreting Scripture according to his own meaning, and of deceitfully gaining permission to publish the *Dialogue*. Bellarmine's certificate was discounted. Galileo's crime was that of being "vehemently suspected of heresy," a religious offense situated between full-fledged heresy and temerity, or rashness. He was to be imprisoned indefinitely, though this part of the sentence was very soon commuted. He spent the rest of his life under house arrest at Arcetri, in the province of Florence.

His abjuration of Copernicanism, at the convent of Santa

Maria sopra Minerva, paved the way for his penitence and—in the Vatican's view—his benign treatment. "I abjure, curse, and detest the above-mentioned errors and heresies," he said on his knees. Cardinal Francesco Barberini was not present. Also absent were Cardinals Borgia and Zacchia.

Invidia

There is a moment in Galileo's *Dialogue* in which one of the participants says something that must, at the time, have sounded incredible, perhaps even profoundly disturbing. Reading this passage, you feel that you're standing at one of the turning points in the story of human thought. In this part of Galileo's fictional discussion, Salviati, the champion of the heliocentric theory, is talking about gravity and the apparent fact that bodies fall straight downward along a plumb line. "We observe the earth to be spherical," he says, "and therefore we are certain that it has a center, toward which we see that all its parts move. We are compelled to speak in this way, since their motions are all perpendicular to the surface of the earth, and we understand that as they move toward the center of the earth, they move toward their whole, their universal mother. Now let us have the grace to abandon the argument that their natural instinct is to go not toward the center of the earth, but toward the center of the universe; for we do not know where that may be, or whether it exists at all.

Even if it exists, it is but an imaginary point; a nothing, without any quality."

In rapid succession, then, Salviati has asserted that the center of the earth is not the center of the universe; that we don't know where that center is, or indeed if it exists; and that even if it does exist, it is just a point of no particular interest. This arresting assertion, which places the earth in the middle of nowhere, contradicts not only two thousand years of educated assumptions concerning its position, but also the teachings of the Catholic Church's most authoritative theologian, Saint Thomas Aquinas, and the vision of Italy's greatest poet, Dante Alighieri, whose *Inferno* (like much Christian folklore) situates Hell at the center of the earth, at the farthest distance from Heaven.

Salviati's remark is worth examining. Following on the heels of the expression "their universal mother," used with reference to the earth, it sounds at first like a metaphor. Generally he is a hard-headed customer, given to offering geometrical proofs for his assertions, but he proposes none for this one—it is as if he wants to shock his contemporaries out of their cosmological complacency, to float an image of a universe radically different from the one they conveniently imagine. Yet a moment's reflection is enough to convince us that this is not a metaphor. It has the magic of a metaphor, but it is a statement of fact.

Galileo, who was not only a scientist but also a trained musician, a prose stylist, and an occasional poet, always recognized the distinction between metaphors and factual assertions. As we have seen, one of his objections to Aristotle was leveled against the notion of the incorruptible firmament, an array of perfect heavenly bodies circling about the earth, as

opposed to our corruptible, sublunary world. For Galileo, this lingo of corruptibility did not belong in "mathematical philosophy"—what we call science: it smacked of ethics or mythology. Despite his admiration for Ovid, Dante, and Ariosto, Galileo warned on several occasions against the contagion of science by thought processes proper to poetry or metaphysical speculation. Yet when he wrote about science, as opposed to doing it, and especially when he defended his ideas against potential attacks from theologians, he used many phrases with metaphorical undertones, others that might well be construed as metaphors, and still others that simply were metaphors.

Galileo had a keen comic sense. Elsewhere in the *Dialogue* Salviati talks about a man walking around the globe, noting that after a while his head will have traveled farther than his feet. For a moment we wonder whether we're reading a work by Edward Lear, and not a seventeenth-century scientific genius, until we realize that the statement is literally true. And Galileo can range farther still. In a famous letter to a friend, the former archbishop Piero Dini, of March 23, 1615 (that is, about two years after beginning to study the sunspots), Galileo wrote, "It seems to me that there is to be found in nature a most spirited, tenuous, and fast-moving substance which is diffused throughout the universe, which penetrates everything without resistance, and warms up, vivifies, and fecundates all living creatures. It seems to me that the senses themselves show that the main recipient of this spirit is the body of the sun, from which light is radiated throughout the universe, accompanied by that caloric spirit which penetrates all vegetable bodies, making them alive and fruitful." This brief meditation, with its undertone of mystical sun-worship and its suggestion that the universe is organically constituted, was not intended for public

consumption; it was, perhaps, a flight of fancy. But Galileo's most famous trope, the idea of "the book of nature," was unquestionably launched as a promotional tool in his campaign to secure the Church's acceptance for his brand of astronomy. "Philosophy," he wrote in *The Assayer*, of 1623, "is written in this grand book, the universe, which stands continually open to our gaze. But it cannot be understood unless one first learns to comprehend the language and recognize the letters in which it is composed. It is written in the language of mathematics, and its characters are triangles, circles, and other geometric figures, without which it is humanly impossible to understand a single word of it. Without these, one wanders about in a dark labyrinth." Here Galileo champions a mathematical science that can produce results that are quantifiable at least in the sense that Euclidean geometry can entertain quantities; but he also goes farther than that. He suggests, enlisting a metaphor, that nature is a book, analogous to Scripture and created by God, who has given us minds in order that we may decipher and understand it. Like the Vulgate, which is written with the twenty-four characters of the Latin alphabet, this book has its language, which is that of mathematics, and so far the trope seems to work; but as we think about it, we might wonder what Galileo is really talking about, for typically this brand of Galilean rhetoric wavers between the metaphorical and the simply factual. After all, the universe is not "composed" in any language—such a metaphor is technically inapt. What is composed in the language of mathematics is the ongoing project, the "book," if you will, of scientific research, and this "book"—the ideal library of all we will ever know about nature—Galileo has conflated with the image (rather hard to summon up mentally) of the universe as an infinite text. So the

promotional tag "the book of nature" (already a time-honored phrase in 1623, and which he sometimes also refers to as "the book of philosophy") can refer to the universe, or to science, or to both at once.

The Italian humanists of the cinquecento and early seicento were much more at home with metaphorical thought than we are. As dab hands at Latin they all knew the works of Cicero, Longinus, and Quintilian, which exhaustively catalogue the forms and uses of the classic tropes. A lot of Mannerist and Baroque literature, especially of the minor sort, actually amounts to a maze of hypertrophied metaphors, rather like Vicino Orsini's garden of stone monsters at Bomarzo, near Viterbo, which we know means something though we are hard pressed to figure out what. And Galileo, of course, was part of this culture: his own contribution to poetic literature contains, among other oddments, two lectures on Dante's *Inferno*, which are mathematical rather than literary, and a slender sheaf of poems known as *Le Rime*.

Galileo's "lessons" on the geometrical layout of the *Inferno*, which he presented at two consecutive meetings of the Florentine Academy in late 1587 or early 1588, were a young man's geometrical divertimento. They have no scientific interest, but they remind us of the poetic content of his life. By poetic content I mean the scenery of a man's or woman's yearnings, which may be connected to a lost love, a house under construction, a city from which one has been exiled—whatever gives meaning to one's deepest reveries. Of course some people's lives have no poetic content, for not everyone is given to self-communion; but Galileo was much preoccupied with the story of his life, which, in his mind, was the romance of a great natural philosopher, the tale of his glorious discoveries and of

the defeat of his envious, slanderous enemies. And, in a curious way, it must be said that one could viably build a Baroque romance around the biography of Galileo. That the Pisan mathematician as a very young man should have given talks on an imaginary world, and later gone on to describe the solar system and to defend his views against the contention that they were imaginary—there is a poetic logic in this, as if he himself were living a metaphor. In his Dante lectures Galileo explicated a treatise "On the Site, Form, and Measure of the Inferno and of the Giants and Lucifer" by Antonio Manetti, a Florentine architect of the late fifteenth century, and in the figures he drew for the audience he made repeated use of conic sections. His demonstrations have not survived, but they must perforce

Botticelli's Drawing of Dante's Inferno

Galileo analyzed Dante's *Inferno* as a series of conic sections in two lectures to the Florentine Academy in 1587 or 1588.

have resembled regularized, geometrical versions of Botticelli's famous illustration of the Inferno.

Galileo's poetry, which was of the "occasional" variety, shows the influence of Francesco Berni, the gifted satirist and wit from Pistoia who died, in his mid-thirties, in 1535. The mathematician's most interesting effort, written sometime between 1589 and 1592, concerns the compulsory wearing of the academic gown, in which a simile, far-fetched but not atypical of the period, is drawn between certain sexual practices and the Aristotelian manner of reasoning. Reading these verses, one suddenly perceives Galileo as an amphibian, equally at home on the terra firma of science and in the fluvial rapids of the metaphorical. Mario Biagioli has discussed what he calls Galileo's "remarkable skills in emblematics," by which he means his ability to harness the symbolic aspects of his discoveries and ideas for their promotional value at the Medici court. This was no minor aspect of his social role. During this period there were no endowed chairs for great scientists, no multimillion dollar research grants, no positions on the boards of cutting-edge engineering firms, no protection for one's scientific discoveries, no Nobel prizes. Beset by financial problems even after his appointment as court philosopher in 1610, Galileo had to defend himself against a horde of petty detractors, intellectual-property bandits, and serious astronomical and theological opponents. That he was continually assailed by a flock of spiteful, pecking "pigeons"—a term he derived from the name of one of his chief Aristotelian opponents, Ludovico delle Colombe (*colombo* means "pigeon" in Italian)—was almost literally true. But he parlayed this truth into something else, something more gripping, more emotionally laden—the myth of himself, the myth of Galileo.

That the myth of Galileo was founded on truth is borne out by the many warnings in the letters of Galileo's close friends concerning the envy and enmity of his rivals, which it would be tedious to enumerate here. Of particular interest, though, is the correspondence that he conducted with the Tuscan painter Ludovico Cigoli, who, it will be remembered, was employed in Rome in 1610–12, together with Domenico Passignano, on the decoration of the cupola of the Pauline Chapel in Santa Maria Maggiore. Cigoli and Galileo had a high regard and much affection for each other, though the former, as a sort of elder brother, did not hesitate to criticize the latter's writing style when it grew too pompous. Cigoli's letters reveal that he was embittered by the envy of other painters over whom he had been preferred for the cupola commission: "all my pleasure," he confessed, "is accompanied by so much bitterness." Only two of Galileo's letters have survived, to twenty-nine of Cigoli's, but much of the exchange clearly concerns the backbiting that both men faced: the painter, as the wiser and calmer of the pair, makes use of his vicissitudes to steady Galileo's nerves against the disparagement he has to endure, and to encourage him to persevere in his research and in his will to publish. Cigoli in 1607 had already painted a strange *Deposition* showing both the sun and the moon, which, as Eileen Reeves has persuasively argued, had distinct pro-Copernican implications, even hinting through pictorial means that to oppose heliocentrism meant turning one's face away from the world as God had created it.

In October of 1610, we find Cigoli reporting to Galileo in disgust that Father Clavius, the elderly chief astronomer of the Jesuit Collegio Romano, has told a friend of his that he laughs at the Jovian moons, and that to convince anyone of their exis-

tence it would be necessary to make a spyglass that would first
create them, hey presto, and then show them, and that Galileo
was welcome to his opinion and he to his. Cigoli also warns
Galileo about the possible negative outcome of publishing the
Starry Messenger in Italian, so that anyone can read and attack
it. "And it also irritates them and they make a great fuss over
[somebody else] having invented the spyglass . . . and I tell all
this to you dear sir so that you may gird yourself and that your
enemies may not find you ill-armed for your defense." In
November he tells Galileo that he is proudly showing off his
letters and not to lose heart, because "every beginning creates a
difficulty for those who are hardened and have grown sclerotic
in an opinion. Yet in the end the truth shall have its way." By
January of 1611 he can report that Father Clavius has, despite
himself, confirmed the existence of the Jovian moons, but by
summer he warns him of the intrigues of his Aristotelian
opponents (probably Ludovico delle Colombe's followers),
who "plant mines behind your back," and suggests that he
"publicly dispose of their opinions." A little later he gets wind
of Clavius's insistence on the moon's absolute smoothness. "I've
thought and thought about it," he says, "and I can find no other
fallback for his defense than this, that a mathematician, however grand he may be, if he happens to have no [ideas], is not
only half a mathematician, but also a man without eyes."

Toward autumn he announces that Passignano has received
a telescope from Venice and is beginning to look at the
sunspots, which "seem to be wandering within the body of the
sun." From this point forward, for as long as he remains in
Rome, he will forward considerable information to Galileo
about the sunspots, including a diagram of the rudimentary
camera obscura that Passignano has contrived, along with a

concave green lens, to observe the sun. Touchingly, he has elected himself Galileo's research assistant. There are, however, more warnings about Ludovico delle Colombe, about *invidia*, or envy, and about the *maledicenti*, or "slanderers." A certain Archbishop Marzimedici, he says, is trying to get a preacher to draft a broadside on parchment against Galileo with reference to the movement of the earth, but Cigoli will warn his friend if anything untoward actually happens. For much of 1612 he forwards sunspot observations, including careful drawings, noting on one occasion that he has recently made twenty-six separate viewings. He also tells Galileo with barely concealed amusement that Passignano, by nature solemn and opinionated, now seems to be claiming that he has observed stellar parallax. In the autumn of 1612, however, his disgusted tone returns: he speaks of the Aristotelians seeking not truth but the preservation of entrenched social position, and of having had wind of Galileo's controversy with Scheiner. He returns again to the theme of *invidia*, and the ever-conspiring mob of evildoers and charlatans; some defamers accuse him of secretly painting in oil, not fresco; and he has noticed Galileo's hesitation to let Prince Cesi publish certain eloquent missives (which would become *Letters on the Sunspots*). On no account should Galileo let anyone talk him out of this. "Do it, do it, do it," he says, "and do not let your own cause down, any more than you have done in the past. Write the truth, without overdoing it and without currying adulation or yielding the field to fortune's whims; and do not slow your course for them whether they be pigeons or geese—laugh them off, Signor Galileo."

But Cigoli was sometimes dejected himself, and then he liked to say (with an exasperated puff, one supposes), "Cigoli

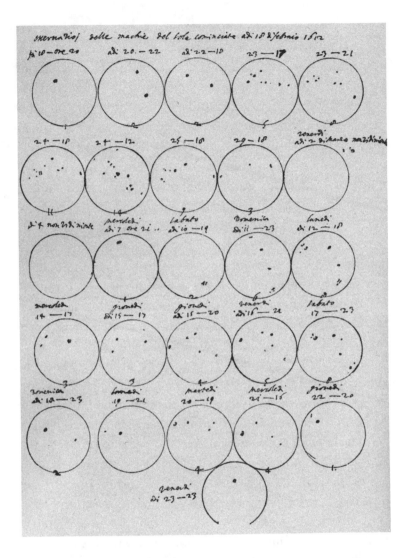

Cigoli's Sunspot Drawings

A page of sunspot drawings sent by Cigoli to Galileo early in 1612.

blows away with wind"—*Cigoli per il vento va via*. It was a wry pun that played his name off one of the most famous lines in Italian poetry,

> *E cigola per vento che va via*
> And hisses in the air that rushes away

which comes from the story of Pier della Vigna, in the *Inferno* (XIII, 42). This passage, which most educated Italians knew (and still know) by heart, tells the story of how Dante the protagonist, walking with Virgil through the Seventh Circle, comes upon the terrible Wood of the Suicides, and there, breaking a twig off a bush, hears the cry "Why do you tear me?" come hissing out of it, like air escaping from a burning branch. Dante's tale was based on sad reality. When alive, at the court of Frederick II of Naples, Pier della Vigna had been the victim of calumny, itself the product of the "whore" Invidia; envy had made him "unjust against my own just self," and he had taken his life in despair. Later, in *Purgatory* XIII, Dante describes the penance of the envious, who must endure having their eyes sewn tight with *fil di ferro*, "threads of iron"; and this may partly explain Cigoli's remark to Galileo about Father Clavius being "a man without eyes." In the Italian speech of this period, the term *invidia* often carries Dantesque connotations. The invidious refuse to see, or, when they do see, it is only to covet what is not theirs.

There is in the Uffizi a pair of drawings by Cigoli that Miles Chappell identified in 1975 as studies for an allegory of Invidia intended for Galileo, with reference presumably to what the

Cigoli's "Invidia"

A drawing by Cigoli showing Virtue or Truth menaced by Envy (Invidia). Cigoli saw both Galileo and himself as the targets of relentless social envy: hence this work, created with the mathematician in mind.

scientist was enduring at hands of his enemies and slanderers. The drawings show Virtue, in the form of a maiden of ideal aspect, overcoming Invidia, a hideous hag with Medusa-like hair who cowers beneath a boulder. In the drawings Virtue is half-metamorphosed into a tree or bush, which led Chappell

to connect her, tentatively, with the nymph Daphne, who was changed into a laurel tree to evade violation by Apollo. But Daphne had no connection to virtue, nor Apollo to envy, and it is far more likely that Cigoli was thinking of the hissing sound of his own name and of poor Pier della Vigna. Most probably, then, these drawings emblematically depict Truth in the guise of a Pier della Vigna–like figure triumphing over the wicked persecution of Invidia. As images, they are the perfect embodiment of the dark, conspiratorial side of the Galileo myth.

Envy is the feeling of spite that others have what you do not have. In the catalogue of the sins it has several variations, chiefly jealousy, which is the fear that another will take or has already taken what is yours. Another variation, one might say, is pretense, or the claim that things are as you see them, and the refusal or incapacity to see them in any other, usually less flattering, way. Envy, jealousy, and pretense are all inspired by self-love. When Galileo and his supporters talk about Invidia, they are to some degree talking about the pretense of those who will not look at the world as God has made it, but only at the world as they, for their own vain purposes, would prefer it to be. In subtly warning Christians to avoid such contortions, Galileo was surely not thinking altruistically of the welfare of the Church: mostly he was aiming to protect his own scientific pursuits. But his argument was very sound advice.

NOTES

Prologue: The Summons

page
22 *The Assayer*: See Galileo Galilei, *Discoveries and Opinions of Galileo* (ed. and trans. Stillman Drake), Doubleday Anchor Books, Garden City, N.Y., 1957, pp. 256–258.

23 "failed to diminish": Ibid.

24 Erwin Panofsky and Eileen Reeves: See Erwin Panofsky, *Galileo as a Critic of the Arts*, M. Nihoff, The Hague, 1954; and Eileen Reeves, *Painting the Heavens: Art and Science in the Age of Galileo*, Princeton University Press, Princeton, N.J., 1997.

25 "the Tuscan artist": Noted by Reeves, *Painting the Heavens*, p. 13.

25 "encoded" or "indexed" in a subject's memory: I have culled these two examples from p. 486 of Richard Boyd's "Metaphor and Theory Change: What Is a 'Metaphor' a Metaphor for?" in Andrew Ortony, ed., *Metaphor and Thought*, Cambridge University Press, Cambridge, U.K., 1979, pp. 480–532. See also Ricardo Nirenberg, "Metaphor: The Color of Being," in Louis Armand, ed., *Contemporary Poetics*, Northwestern University Press, Evanston, Ill., 2007, pp. 153–174.

Galileo Galilei and Maffeo Barberini

28 Viviani's biography: Vincenzo Viviani's "Racconto istorico" is in vol. 19, pp. 597–632, of Galileo Galilei, *Le opere di Galileo Galilei*

(ed. Antonio Favaro), Barbera, Florence, 1929–39. Hereafter abbreviated *OGG*.

29 Aristotelian dynamics: For a discussion of the lack of an idea of force in Aristotle, see Paul Tannery, "Galileo and the Principles of Dynamics," in Ernan McMullin, ed., *Galileo Man of Science*, Basic Books, New York, 1968, pp. 163–177.

30 entomological illustration: For an example, see Stillman Drake, *Galileo at Work: His Scientific Biography*, University of Chicago Press, Chicago, 1978, p. 290.

30 "theory of the concave spherical mirror": Galileo's hypothetical ray diagram is particularly well illustrated in Edward R. Tufte, *Beautiful Evidence*, Graphic Press, Cheshire, Conn., 2006, pp. 80–81.

30 regarded Ariosto: All of Galileo's "Scritti letterari" are in *OGG*, vol. 9.

31 critical essay on Tasso's *Jerusalem Delivered*: This is the "Considerazioni al Tasso," in *OGG*, vol. 9, pp. 63–148.

31 anamorphosis: See the "Considerazioni," ibid., pp. 129–130.

31 taught perspective: Perspective, a drafting technique derived from geometrical optics, was part of the mathematics curriculum at many Italian universities in the late sixteenth century. There is to my knowledge no proof that Galileo taught it, but he referred to it in his letters to Ludovico Cigoli and he would have mastered it with ease.

31 elaborate treatises on the subject [of perspective]: Among others, *Comendarius di F. Commandinus*, Venice, 1558; Lorenzo Sirigatti, *Pratico della prospettiva*, 1596; Giordano Nunonario and Guidobaldo Dal Monte, *Perspectivae Libri Sex*, Pesaro, 1600.

31 It has been noted: Panofsky, *Galileo as a Critic of the Arts*, and Reeves, *Painting the Heavens*.

33 As William R. Shea has pointed out: See William R. Shea, *Galileo's Intellectual Revolution*, Neale Watson Academic, New York, 1972, pp. 39–40.

33 Aristotle's cosmos: As elaborated in the *De Caelo*. See *On the Heavens* [Greek and English] (trans. W. K. C. Guthrie), Heinemann, London, 1939, esp. Book 2, pp. 130–255.

36 the mathematical concept of integration: See Galileo Galilei, *Dialogue Concerning the Two Chief World Systems—Ptolemaic &*

Copernican (trans. Stillman Drake), University of California Press, Berkeley, 1967, pp. 228–229. For the comparison of infinite sets, see Galileo Galilei, *Dialogue on the Two New Sciences* (1638) (trans. Henry Crew and Alfonso de Salvio), Macmillan, New York, 1914, p. 30ff.

40   Maffeo told Paul V: For this incident, I have noted: Herrera, "Memorie intorno la vita d'Urbano VIII," Barb. 4901, Biblioteca Apostolica Vaticana, pp. 48–50, which is cited in Ludwig [Freiherr] von Pastor, *The History of the Popes* (trans. Ernest Graf), Kegan, Paul, French, Trubner & Co., London, 1937, vol. 26, p. 387, n. 4; Herrera is cited again in Paolo Portoghesi, *Roma barocca: storia di una civiltà architettonica*, C. Bestetti, Roma, 1966, p. 54.

42   Maffeo's poetry: See *Maphaei S.R.E., Card. Barberini nunc Urbani PP VIII poemata*, ex typographia R. Cam. Apost., Romae, 1631. See also Lucia Franciosi, "Immagini e poesia alla corte di Urbano VIII," in Marcello Fagiolo, ed., *Gian Lorenzo Bernini e le arti visive*, Istituto della Enciclopedia italiana, Roma, 1987, pp. 85–90. Maffeo Barberini's poetry consists of paraphrases of the Psalms, sacred odes, meditations on the fugitive nature of life, and occasional and congratulatory verses. I confess that I haven't read much of it, though I have studied Franciosi's analysis. Of the third category, here is an example, slightly edited by me ("Poemata," p. 113):

> *Serio desiderium fugacis*
> *Vitae fascinat! Ut trahit voluptas!*
> *Ut cor abripit aura blanda plausus,*
> *Implicat laqueis opum cupido,*
> *Fallit Ambitio, tenetq[ue] luxus!*
> *Stulti quid sequimur [c]aduca? Fulgens*
> *Caeli Regia nos vocat; sed armis*
> *Obniti Pietatis est necesse*
> *Contra nequitiae dolos.*

45   Maffeo was extraordinarily well disposed: For a detailed chronicle of this early affection, including the examples offered here, see Antonio Favaro, "Gli oppositori di Galileo: VI, Maffeo Barberini," *Atti del Reale Istituto Veneto di Scienze, Lettere ed Arti*, vol. 80, 1920–21, pp. 1–16, and vol. 81, pp. 17–46.

The Telescope; or, Seeing

53   Kepler, writing to Galileo on March 28, 1611: Letter 611 from M. Caspar et al., eds., *Johannes Kepler Gesammelte Werke*, C. H. Becksche Verlag, Munich, 1937, vol. 16, p. 372.

53   Stillman Drake has claimed: Stillman Drake, "Galileo's Steps to Full Copernicanism and Back," *Studies in the History and Philosophy of Science*, vol. 18, no. 1, 1987, pp. 93–105.

54   letter of 1597 to Jacopo Mazzoni: For a discussion, see Stillman Drake, *Galileo at Work: His Scientific Biography*, University of Chicago Press, Chicago, 1978, p. 40.

54   *ex suppositione*: For a full examination of the many forms of Aristotelian suppositional reasoning (most not used by Galileo), see William A. Wallace, "Aristotle and Galileo: The Use of Hypothesis (*Suppositio*) in Scientific Reasoning," *Studies in Aristotle*, vol. 9, 1981, pp. 47–77.

54   Galileo wrote Kepler: On August 4, see *OGG*, vol. 10, pp. 67–68.

55   three lectures at the University of Padua: Only tiny fragments of these lectures have survived. See Drake, *Galileo at Work*, pp. 104–106.

56   Ronchi suggested a reason: Vasco Ronchi elaborated this thesis in his *Nature of Light: An Historical Survey* (trans. V. Barocas), William Heinemann, London, 1970.

57   vigorously disputed in 1972 by David C. Lindberg: David C. Lindberg and Nicholas H. Steneck, "The Sense of Vision and the Origin of Modern Science," in Allen G. Debus, ed., *Science, Medicine, and Society in the Renaissance: Essays to Honor Walter Pagel*, Heinemann, London, 1972, vol. 1, pp. 29–45.

58   a focal length of 12 to 20 inches: I am citing Albert Van Helden's figures. See Albert Van Helden, "The Invention of the Telescope," *Transactions of the American Philosophical Society*, vol. 67, no. 4, 1977, p. 11.

58   the first primitive spyglasses: See Engel Sluiter, "The Telescope before Galileo," *Journal for the History of Astronomy*, vol. 28, 1997, pp. 223–234, and Colin A. Ronan, G. L'E. Turner, et al., "Was There an Elizabethan Telescope?" *Bulletin of the Scientific Instrument Society*, vol. 37, 1993, pp. 2–10.

58   The first telescope: See Van Helden, "Invention of the Telescope," p. 42.

60   "a report reached my ears": Galilei, *Discoveries and Opinions*, pp. 28–29.

61   "News came": From *OGG*, vol. 6, pp. 258–259. Translated by Stillman Drake and C. D. O'Malley, *The Controversy on the Comets of 1618: Galileo Galilei, Horatio Grassi, Mario Guiducci, Johann Kepler*, University of Philadelphia Press, Philadelphia, 1960, pp. 212–213, quoted in Van Helden, "Invention of the Telescope."

61   "My reasoning": From *OGG*, vol. 4, pp. 258–259. Translated by Albert Van Helden, "Galileo and the Telescope," in Paolo Galluzzo, ed., *Novità celesti e crisi del sapere*, Giunti, Florence, 1984, p. 152.

62   Joining a bitter race: Recounted in Galileo to Beneditto Landucci, August 29, 1609, in *OGG*, vol. 10, p. 253.

62   Galileo wrote to the doge: See *OGG*, vol. 10, p. 250.

66   This description of the Galilean telescope is largely derived from the extraordinarily clear presentation by Tom Pope and Jim Mosher, "Galilean Telescope Homepage," available at www.pacifier.com/~tpope/Galilean_Optics_Page.htm. Last accessed July 27, 2008.

67   a letter from Arcetri to Fortunio Liceti: From *OGG*, vol. 18, p. 233, quoted in Vasco Ronchi, *Il Cannocchiale di Galileo e la scienza del Seicento*, Edizioni Scientifiche Einaudi, Torino, 1958, p. 139.

68   Galileo discussed magnification very warily: See "The Starry Messenger," in Galilei, *Discoveries and Opinions*, p. 30. In the passage beginning "Now in order to determine . . . the magnifying power of an instrument . . .," he is forthcoming enough to offer the reader the following information: "Let [the user] draw two circles or two squares on paper, one of which is four hundred times larger than the other . . . [that is, has twenty times its width]. He will then observe from afar both sheets fixed to the same wall, the smaller one with one eye applied to the glass and the larger one with the other, naked eye. This can easily be done with both eyes open at the same time. Both figures will then appear of the same size if the instrument multiplies objects according to the desired proportion [x20]." This is all very well and good, but it reveals nothing of how to make a telescope! It is so obvious that it seems addressed to readers who have never even possessed a pair of reading glasses.

68   His competitors: See Mario Biagoli, *Galileo's Instruments of Credit: Telescopes, Images, Secrecy*, University of Chicago Press, Chicago, 2006, p. 93ff.

69   "And I shall describe it": Letter from Giovanni Battista Della Porta to Federico Cesi, August 28, 1609, from *OGG*, vol. 10. Translated by Van Helden, "Invention of the Telescope," p. 44.

71   Ippolito Francini: See Albert Van Helden, *Catalogue of Early Telescopes*, Istituto e Museo di Storia delle Scienze/Giunti, Florence, 1999, p. 30.

71   The composition of Galileo's lenses: The information offered here is based on my interview with Dr. Giorgio Strano of the Institute and Museum of the History of Science, Florence.

74   The very large dusky patches: From "The Starry Messenger," in Galilei, *Discoveries and Opinions*, p. 31.

75   "The surface of the Moon": Ibid., p. 31.

75   to Antonio de' Medici: From *OGG*, vol. 10, pp. 273–278.

76   "Now on Earth": From "The Starry Messenger," in Galilei, *Discoveries and Opinions*, p. 33. For the geometrical proof of the minimal height of a lunar mountain cited here, see ibid., pp. 31–42.

80   C. W. Adams found: C. W. Adams, "A Note on Galileo's Determination of the Height of Lunar Mountains," *Isis*, vol. 17, 1932, pp. 427–429.

81   his long letter: From *OGG*, vol. 10, pp. 273–278.

82   he wrote his friend Belisario Vinta: From *OGG*, vol. 10, p. 280.

83   Galileo soon concluded that they were moons: Galilei, *Discoveries and Opinions*, p. 57.

83   he manufactured many hundreds of telescopes: See Biagoli, *Galileo's Instruments of Credit*, pp. 90–94.

89   At last, in October of 1610: For dating, see letter of December 30 to Benedetto Castelli, 1610, in *OGG*, vol. 10, pp. 502–504.

89   If his earlier observations: For the *Letters on Sunspots*, see Galilei, *Discoveries and Opinions*, pp. 59–85.

90   "I confess to your Excellency": Ibid., p. 113.

91   Carrington: To picture what Richard Carrington discovered, remember that the earth rotates from west to east. Then imagine a cloud that while also rotating west to east has a proper motion, a motion of

its own, much slower than the rotation of the earth. With respect to longitude, its proper motion is east to west.

92   "I seem to have observed": Ibid., p. 130.

93   axis of the sun's tilt: Ibid., p. 125.

93   "to verify the rest of the [Copernican] system": Ibid., p. 144.

93   "perhaps this planet also": Ibid., p. 144.

93   Cesare Cremonini: See Biagioli, *Galileo's Instruments of Credit*, p. 113.

94   Giulio Libri: See Drake, *Galileo at Work*, p. 162.

94   Martin Horky wrote a letter to Kepler: From *OGG*, vol. 10, pp. 142–143.

94   the classical literary etymology of the Italian word for envy, *invidia*: With respect to envy, Charles S. Singleton writes (in *Dante Alighieri: The Divine Comedy: Inferno: 2. Commentary*, Bolligen Series LXXX, Princeton University Press, Princeton, NJ, 1970, p. 213: "The sin of envy is thought of as movement of the eyes, first of all. Thus, in *Purg.* XIII, souls are purged of that sin by having their eyelids sewed shut. Pietro di Dante, commenting on *Purg.* XIII, says: 'Invidia facit, quod non videatur, quod expedit videre; et ideo dicitur *invidia*, quasi *non visio*.' ('Envy causes that which should be seen not to be seen. And therefore is called *invidia*, almost as if to say, nonvision.') [Thus also] the *Magnae derivationes* of Uguccione da Pisa: 'Invideo tibi, idest non video tibi, idest non fero videre te bene agentem.' ('I envy you—that is to say, I do not see you; that is, I cannot bear to see you doing so well.')" This etymology goes back to the Latin grammarian Priscian (ca. AD 500) and beyond him to Cicero.

94   "I must write of a harsh objection": Quoted in Ronchi, *Cannocchiale di Galileo*, p. 139. For *cronicatori* I read "star chroniclers," that is, those who note the exact times of the rising and setting of the stars.

95   the problem of confirmation: Biagioli, *Galileo's Instruments of Credit*, pp. 27–44, 132.

96   a primitive camera obscura: This whole issue is best described in ibid., p. 192, n. 141.

97   public sessions: See ibid., pp. 86–90.

97   Galileo's wash drawings: See *OGG*, vol. 10, p. 274ff.

99   which Galileo apparently had a large hand in: See Horst Bredekamp, *Galilei der Kunstler: Der Mond. Die Sonne. Die Hand*, Akademie Verlag, Berlin, 2007, p. 189ff, under the heading "Galilei als Stecher?" Bredekamp, having examined the hatching employed in the shading of the lunar craters and compared it with Galileo's drawings, writes, "One is allowed to suspect that Galileo, pressed by time as the book neared production, enjoyed a phase as a graphic artist" (my translation).

99   Gugliemo Righini: For a discussion, see M. L. Righini Bonelli and William R. Shea, eds., *Reason, Experiment, and Mysticism in the Scientific Revolution*, Macmillan, New York, 1975, pp. 59–76.

103   Father Christopher Clavius: See Reeves, *Painting the Heavens*, p. 151.

103   The other tradition: Of the many works discussing the doctrine of the immaculate conception, see especially Jaroslav Pelikan, *Mary through the Centuries: Her Place in the History of Culture*, Yale University Press, New Haven, Conn., 1996, pp. 177–200.

105   Gallanzone Gallanzoni: A *cavaliere*, or knight, from Rimini, he was at this time Cardinal Joyeuse's secretary. See Galileo's letter to Gallanzone of July 16, 1611, in *OGG*, vol. 11, p. 143, and the commentary in Reeves, *Painting the Heavens*, pp. 17–18, 216–220, who reads it as an "implicit criticism of the emerging doctrine of the Immaculate Conception, the basis of many of the Marian associations with the moon."

106   "Having crossed that fiery sphere": This as yet unpublished translation of Astolfo's moon voyage is by David Slavitt. See Ludovico Ariosto, *Orlando furioso* [Canto Trentesimoquarto], Successori Le Monnier, Firenze, 1888, pp. 254–256.

108   "They have, as it were": Johannes Kepler, *Conversations with the Sidereal Messenger* (trans. and ed. Edward Rosen), Johnson Reprint Corporation, New York, 1965, p. 28.

109   painted dome: See John Shearman, "The Chigi Chapel in S. Maria del Popolo," *Journal of the Warburg and Courtauld Institute*, vol. 24, 1961, p. 138ff, and Philippe Morel, "Morfologia delle cupole dipinte da Correggio a Lanfranco," *Bolletino d'arte*, ser. 6, vol. 69, no. 23, 1984, pp. 1–34.

111 *invidia* implied: See Reeves, *Painting the Heavens*, p. 17.

111 Cigoli, then living in Rome, wrote Galileo: From OGG, vol. 10, pp. 290–291.

112 Passignano: See again Biagioli, *Galileo's Instruments of Credit*, p. 192, n. 141; also Reeves, *Painting the Heavens*, p. 5.

117 Samuel Edgerton: Samuel Y. Edgerton, *The Renaissance Rediscovery of Linear Perspective*, Basic Books, New York, 1975, p. 162. Giotto's acquaintance with Alhazen's optics, long suspected, has received further confirmation by Giuliano Pisani in the January 2008 issue of *Bolletino del museo civico di Padova*, widely reviewed in the Italian press. I have not, however, been able to obtain a copy.

119 decree condemning Copernicanism: Maurice A. Finocchiaro, ed. and trans., *The Galileo Affair: A Documentary History*, University of California Press, Berkeley, 1989, pp. 146–150.

119 Bellarmine warned him: Ibid., pp. 147–148.

121 Bellarmine stated: Translated by Richard J. Blackwell in *Galileo, Bellarmine, and the Bible: Including a Translation of Foscarini's Letter on the Motion of the Earth*, University of Notre Dame Press, Notre Dame, Ind., 1991, p. 266.

121 "Considerations on the Copernican Opinion": This can be found in Finocchiaro, *Galileo Affair*, pp. 70–86.

122 These are circular sophisms: See Guido Morpurgo-Tagliabue, *I Processi di Galileo e l'epistemologia*, Armando, Rome, 1981, pp. 51–59.

123 Francesco Ingoli: See OGG, vol. 5, pp. 403–412. Ingoli was made a consultor to the Holy Office and recognized as a quasi-official anti-Copernican (and by extension anti-Galilean) critic only *after* writing this notably weak polemic. See Annibale Fantoli, *Galileo, for Copernicanism and for the Church* (trans. George V. Coyne), Vatican Observatory Publications, Vatican City, 1996, p. 255, n. 50.

123 Galileo politely refuted: See "Galileo's Reply to Ingoli (1624)" in Finocchiaro, *Galileo Affair*, pp. 154–197.

124 Stillman Drake's phrase: See Galilei, *Discoveries and Opinions*, p. 264, n. 14.

128 as William R. Shea has written: Shea, *Galileo's Intellectual Revolution*, p. 163.

129 "Now when we see this beautiful order": Galilei, *Dialogue*, p. 367.

130   Shea has termed: Shea, *Galileo's Intellectual Revolution*, p. 163.

130   a different graphic representation: See Galilei, *Dialogue*, pp. 342–345.

130   Mount a crossbow: Ibid., p. 168ff.

131   the church of San Petronio: Ibid., p. 463. I have seen this discussed in detail only by J. L. Heilbron in his fascinating *The Sun in the Church: Cathedrals as Solar Observatories*, Harvard University Press, Cambridge, 1999, pp. 176–180. I thank my friend Norman Derby, professor of physics, for his help in understanding Professor Heilbron's enlightening but somewhat confusing diagram.

132   as Shea has observed: Shea, *Galileo's Intellectual Revolution*, p. 181.

The Trial; or Not Seeing

136   In 1575, as the Counter-Reformation: All this and more information concerning Tuscany's increasing pliability with respect to the Papal States can be found in Furio Diaz, *Il Granducato di Toscana: I Medici*, UTET Libreria, Turin, 1976, Part III, "La Toscana nell'età della Controriforma," pp. 274–278, 287–288, 321–323, and 323–326.

138   Ciampoli's verse: See Giovanni Ciampoli, *Poesie sacre*, Carlo Zenero, Bologna, 1648. This collection consists of verses on the utility of sacred poetry; meditations based on the Psalms; and songs of praise, notably for the Santa Casa di Loreto. There is also a "Cantico delle Benedittioni" for the coronation of Pope Urban VIII. For an interpretation, see Franciosi, "Immagini e poesia alla corte di Urbano VIII."

139   "calumny of ": From *OGG*, vol. 14, pp. 383–385.

139   "I began to think, as you so rightly say": Ibid., pp. 388–399.

140   "In such affairs of the Holy Office": Ibid.

142   "*Basta, basta!*": Letter of September 18, from ibid., pp. 391–393.

142   "speak cautiously": Ibid.

142   "my great esteem for you": From *OGG*, vol. 14, pp. 118–119.

144   The missive: Ibid., pp. 406–410.

146   he wrote Galileo a shrewd note: Letter of October 23, from ibid., pp. 418–419.

147   "I tried to awaken in him": Letter of November 13, from ibid., pp. 427–428.

148   harvesting hundreds of thousands of lives: Lodovico Antonio Mura-
       tori, in his classic *Annali d'Italia: Dal principio dell'era volgare sino
       all'anno MDCCXLIX*, Classici Italiani Contrada del Cappuccio,
       Milan, 1820, vol. 15, p. 117, gives the figure 560,000, including
       500,000 for the terra firma of the Veneto and 60,000 for the other
       northern and central Italian regions. He does not offer
       documentation.

150   Benedetto Castelli: Letter of November 20, from *OGG*, vol. 14, pp.
       430–431.

152   to his old friend Elia Diodati: From *OGG*, vol. 10, January 15, pp.
       23–26.

153   "Though the affairs of this tribunal": February 16, 1633, from *OGG*,
       vol. 15, p. 41.

153   He wrote Cioli on February 19: From *OGG*, vol. 10, pp. 43–45.

154   "in the guise of a visitor": February 19, 1633, from *OGG*, vol. 15, pp.
       43–45.

154   "I think Serristori": Ibid.

155   He decided to write two cardinals: From *OGG*, vol. 15, pp. 46, 49.

155   "From what I gather": Niccolini to Cioli, February 27, from *OGG*,
       vol. 10, pp. 54–55.

157   "At the Palace": I have used James J. Langford's translation (from
       Galileo, Science, and the Church, Desclee, New York, 1966), quoted in
       Richard J. Blackwell, *Behind the Scenes at Galileo's Trial: Including the
       First English Translation of Melchior Inchofer's Tractatus sylleticus*,
       University of Notre Dame Press, Notre Dame, Ind., 2006, p. 5.

157   Galileo wrote to a friend: To Geri Bocchineri, February 25, from
       OGG, vol. 10, p. 50.

158   *amorevolezza*: Niccolini to Cioli, February 27, from ibid., p. 55: "*non
       mancano chi dubiti che difficilmente [Galileo] habbia a scansar d'esser
       ritenuto al S. Offizio, bensí si proceda seco sin adesso con molta
       amorevolezza e placidità.*"

159   And indeed Francesco conceded: Niccolini to Cioli, February 27,
       from ibid., pp. 55–56.

159   "may God forgive": Niccolini to Cioli, March 13, from ibid., pp.
       67–68.

160   As a temporal prince: On foreign ambassadors visiting Urban VIII, see

Leopold von Ranke, *History of the Popes: Their Church and State* (trans. E. Fowler), Colonial Press, New York, 1901, vol. 2, pp. 371–374.

161   Maffeo's response: Letter of March 13, from *OGG*, vol. 15, pp. 67–68.

161   On April 9, only three days before the trial: Niccolini to Cioli, April 9, from *OGG*, vol. 10, p. 84.

164   The Roman Inquisition has hardly been studied: The principal source used here on the legal practices of the Roman Inquisition is John Tedeschi, *The Prosecution of Heresy: Collected Studies on the Inquisition in Early Modern Italy*, Medieval & Renaissance Texts & Studies, Binghamton, N.Y., 1991. For early Tuscan resistance to the Inquisition, see p. 92; see also p. 126ff; for the influence of the Roman legist Ulpian, see p. 143.

167   the *corda*: In the concluding chapter of his magisterial two-volume work, *La Tortura giudiziaria nel diritto comune* (Giuffré, Rome, 1954), Piero Fiorelli notes (pp. 231ff) that Cicero, Quintilian, Saint Augustine, Ulpian, Boccaccio, and Montaigne had all inveighed against the the use of judiciary torture. By 1633, it was regarded by many theologians and philosophers as an indefensible practice, and two important works by Jesuit fathers had bitterly criticized it: *Universa teologia scholastica* (1627) of Adam Tanner, a Tyrolese, and *Cautio criminalis* (1631) of Friedrich von Spee, a German.

167   "Torture," went a canon-law maxim: Tedeschi, *Prosecution of Heresy*, p. 144.

168   canon-law guidelines: For these, see L. Garzend, "Si Galilée pouvait être juridiquement torturé," *Revue des questions historiques*, vol. 90, 1911–12, pp. 353–389, and vol. 91, 1911–12, pp. 36–67.

168   the trial transcript for June 16, 1633: Sergio M. Pagano, ed., with Antonio G. Lucciani, *I Documenti del processo di Galileo Galilei* [Contro Galileo Galilei], Pontificiae Academiae Scientiarum, Vatican City, 1984, p. 154. Hereafter abbreviated *DPGG*.

170   Galileo at his first deposition: Ibid., pp. 124–130.

171   "I was notified": My translation. Ibid., p. 127.

172   "We, Robert Cardinal Bellarmine": Translated by Blackwell, *Behind the Scenes at Galileo's Trial*, p. 9.

172   Galileo described: Galileo's first deposition is found in *DPGG*, pp. 124–130.

176   On April 16, Niccolini wrote Cioli: From *OGG*, vol. 10, pp. 94–95.

176   He wrote Cioli a jubilant note on April 23: From ibid., p. 103.

176   The day before, in a letter discovered only in 1999: Translated by Blackwell, *Behind the Scenes at Galileo's Trial*, p. 14.

177   In another letter: Ibid., p. 14.

178   "was of the same [heliocentric] opinion": This letter of October 2, 1632, is cited in Fantoli, *Galileo, for Copernicanism and for the Church*, p. 407. Here Fantoli convincingly cites it as evidence of the "disparity of views that existed among the Church authorities themselves."

180   Two days after Maculano's fateful meeting with Galileo, the accused gave his second deposition: *DPGG*, pp. 130–132.

183   "I haven't told him everything yet": Letter of May 22, from *OGG*, vol. 10, p. 132.

183   On May 10, Maculano: *DPGG*, pp. 135–137.

184   As Maurice Finocchiaro has emphasized: Maurice A. Finocchiaro, *Retrying Galileo*, University of California Press, Berkeley, 2005, p. 11.

185   At the end of the final session: From *DPGG*, pp. 154–155.

185   "I am here in your hands": *del resto, son qua nelle loro mani, faccino [sic] quello gli piace [sic]*.

186   what Richard J. Blackwell has called: See Blackwell, *Galileo, Bellarmine, and the Bible*, p. 37. See chapter 3 for a full analysis.

189   Perugino: See Giorgio Vasari, *The Lives of the Artists* (trans. Julia Conway Bondanella and Peter Bondanella), Oxford University Press, New York, 1991, p. 266.

189   concepts of modern philology: For the beginnings of modern biblical criticism, see Richard H. Popkin, "Spinoza and Bible Scholarship," in Don Garrett, ed., *The Cambridge Companion to Spinoza*, Cambridge University Press, Cambridge, U.K., 1986, pp. 383–407.

191   The "Consultants' Report on Copernicanism": A translation can be found in Finocchiaro, *Galileo Affair*, p. 146.

191   "It appears to me": Translated by Blackwell in *Galileo, Bellarmine, and the Bible*, p. 266.

196   "It was never our intention": Quoted in Favaro, "Gli oppositori di Galileo," p. 18.

196   "We believe him": Ibid., p. 39.

Epilogue: Invidia

198   a moment in Galileo's *Dialogue*: Galilei, *Dialogue*, p. 36–37.

200   a famous letter: Translated by Blackwell in *Galileo, Bellarmine, and the Bible*, p. 206.

201   "Philosophy . . . is written in this grand book": Galilei, *Discoveries and Opinions*, pp. 237–238.

202   "lessons" on the geometrical layout of the *Inferno*: From *OGG*, vol. 9, pp. 31–57.

204   The mathematician's most interesting effort: From ibid., pp. 213–223.

204   Mario Biagioli has discussed: See Mario Biagioli, *Galileo Courtier: The Practice of Science in the Culture of Absolutism*, Chicago University Press, Chicago, 1993, pp. 107–159.

205   Cigoli: This account of the Galileo-Cigoli correspondence is based on the letters as they appear in *OGG*, vol. 10, pp. 241, 243, 290, 441, 456, 475, 478; and vol. 11, pp. 36, 132, 167, 175, 208, 212, 228, 241, 268, 286, 290, 318, 347, 361, 369, 386, 410, 418, 424, 475, 484, 501.

205   Eileen Reeves: Reeves, *Painting the Heavens*, pp. 138–183.

209   a pair of drawings by Cigoli: See Miles Chappell, "Cigoli, Galileo, and *Invidia*," *Art Bulletin*, vol. 57, no. 1, March 1975, pp. 91–98.

Selected Bibliography

Original Sources

Galilei, Galileo. *Le opere di Galileo Galilei* (ed. Antonio Favaro), 20 vols. Barbera, Florence, 1929–39. Abbreviated *OGG*.

Pagano, Sergio M., ed., with Antonio G. Lucciani, *I Documenti del processo di Galileo Galilei* [Contro Galileo Galilei]. Pontificiae Academiae Scientiarum, Vatican City, 1984. Abbreviated *DPGG*.

Books

Ariosto, Ludovico. *Orlando furioso* (trans. David Slavitt). Harvard University Press, Cambridge, in press.

Aristotle. *On the Heavens* (trans. W. K. C. Guthrie). Heinemann, London, 1939.

Banfi, Antonio. *Vita di Galileo Galilei*. Cultura, Milan, 1930.

Barberini, Maffeo. *Maphaei SRE, Card. Barberini nunc Urbani PP VIII poemata*. R. Cam. Apost., Rome, 1631.

Biagioli, Mario. *Galileo Courtier: The Practice of Science in the Culture of Absolutism*. Chicago University Press, Chicago, 1993.

———. *Galileo's Instruments of Credit: Telescopes, Images, Secrecy*. University of Chicago Press, Chicago, 2006.

Blackwell, Richard J. *Galileo, Bellarmine, and the Bible: Including a Transla-tion of Foscarini's Letter on the Motion of the Earth.* University of Notre Dame Press, Notre Dame, Ind., 1991.

——. *Behind the Scenes at Galileo's Trial: Including the First English Trans-lation of Melchior Inchofer's Tractatus syllepticus.* University of Notre Dame Press, Notre Dame, Ind., 2006.

Bredekamp, Horst. *Galileo der Kunstler: Der Mond. Die Sonne. Die Hand.* Akademie Verlag, Berlin, 2007.

Chappell, Miles L., ed. *Disegni di Ludovico Cigoli.* Exhibition catalogue. Olschki, Florence, 1992.

Ciampoli, Giovanni. *Poesie sacre.* C. Zenero, Bologna, 1648.

Deo Feo, Vittorio, and Vittorio Martinelli, eds. *Andrea Pozzo.* Electa, Milan, 1996.

Diaz, Furio. *Il Granducato di Toscana: I Medici.* UTET Libreria, Turin, 1976.

Drake, Stillman. *Galileo at Work: His Scientific Biography.* University of Chicago Press, Chicago, 1978.

Edgerton, Samuel Y. *The Renaissance Rediscovery of Linear Perspective.* Basic Books, New York, 1975.

——. *The Heritage of Giotto's Geometry: Art and Science on the Eve of the Scientific Revolution.* Cornell University Press, Ithaca, N.Y., 1991.

Evans, Robin. *The Projective Cast: Architecture and Its Three Geometries.* MIT Press, Cambridge, 1995.

Fagiolo, Marcello, ed. *Gian Lorenzo Bernini e le arti visive.* Istituto della Enciclopedia italiana, Rome, 1987.

Fantoli, Annibale. *Galileo, for Copernicanism and for the Church.* Vatican Observatory Publications, Vatican City, 1996.

Finocchiaro, Maurice A., ed. and trans. *The Galileo Affair: A Documentary History.* University of California Press, Berkeley, 1989.

Finocchiaro, Maurice A. *Retrying Galileo.* University of California Press, Berkeley, 2005.

Fiorelli, Piero. *La Tortura giudiziaria nel diritto comune,* 2 vols. Giuffré, Rome, 1954.

Galilei, Galileo. *Discoveries and Opinions of Galileo* (trans. and ed. Stillman Drake). Doubleday Anchor Books, Garden City, N.Y., 1957.

Galilei, Galileo. *Dialogue Concerning the Two Chief World Systems—Ptolemaic & Copernican* (trans. Stillman Drake). University of California Press, Berkeley, 1967.

———. *Sidereus nuncius, or, The Sidereal Messenger*. Chicago University Press, Chicago, 1989.

———. *Le Rime* (ed. Antonio Marzo). Salerno, Rome, 2001.

Gingerich, Owen. *The Eye of Heaven: Ptolemy, Copernicus, Kepler*. American Institute of Physics, New York, 1993.

Haskell, Francis. *Patrons and Painters: A Study in the Relations between Italian Art and Society in the Age of the Baroque*. Knopf, New York, 1963.

Heilbron, J. L. *The Sun in the Church: Cathedrals as Solar Observatories*. Harvard University Press, Cambridge, 1999.

Kemp, Thomas. *The Science of Art: Optical Themes in Western Art from Brunelleschi to Seurat*. Yale University Press, New Haven, Conn., 1990.

Kepler, Johannes. *Conversations with the Sidereal Messenger* (trans. and ed. Edward Rosen). Johnson Reprint Corporation, New York, 1965.

Kirwin, William Chandler. *Powers Matchless: The Pontificate of Urban VIII, the Baldachin, and Gian Lorenzo Bernini*. P. Lang, New York, 1997.

Lindberg, David C., and Robert S. Westman. *Reappraisals of the Scientific Revolution*. Cambridge University Press, Cambridge, U.K., 1990.

McMullin, Ernan, ed. *The Church and Galileo*. University of Notre Dame Press, Notre Dame, Ind., 2005.

Morpurgo-Tagliabue, Guido. *I Processi di Galileo e l'epistemologia*. Armando, Rome, 1981.

Muratori, Lodovico Antonio. *Annali d'Italia: Dal principio dell'era volgare sino all'anno MDCCXLIX*, 14 vols. Classici Italiani Contrada del Cappuccio, Milan, 1820.

Nussdorfer, Laurie. *Civic Politics in the Rome of Urban VIII*. Princeton University Press, Princeton, N.J., 1992.

Onori, Lorenza Mochi, Sebastian Schütze, and Francesco Solinas, eds. *I Barberini e la cultura europea del Seicento*. De Luca Editori d'Arte, Rome, 2007.

Panofsky, Erwin. *Galileo as a Critic of the Arts*. M. Nihoff, The Hague, 1954.

Pasquali, Giorgio. *Storia della tradizione e critica del testo*. F. Le Monnier, Florence, 1962.

Pastor, [Freiherr] Ludwig von. *The History of the Popes* (trans. Ernest Graf), Kegan Paul, French, Trubner & Co., London, 1937.

Pedrotti, Frank L., Leno M. Pedrotti, and Leno S. Pedrotti. *Introduction to Optics*. Pearson/Prentice Hall, Upper Saddle River, N.J., 2007.

Pérez-Gómez, Alberto, and Louise Pelletier. *Architectural Representation and the Perspective Hinge*. MIT Press, Cambridge, 1997.

Prosperi, Adriano. *L'Inquisizione romana: letture e ricerche*. Edizioni di storia e letteratura, Rome, 2003.

Ranke, Leopold von. *History of the Popes: Their Church and State* (trans. E. Fowler), 3 vols. Colonial Press, New York, 1901.

Redondi, Pietro. *Galileo Eretico*. Einaudi, Turin, 1983.

Reeves, Eileen. *Painting the Heavens: Art and Science in the Age of Galileo*. Princeton University Press, Princeton, N.J., 1997.

Ronchi, Vasco. *Il Cannocchiale di Galileo e la scienza del Seicento*, Edizioni Scientifiche Einaudi, Turin, 1958.

———. *The Nature of Light: An Historical Survey* (trans. V. Barocas). William Heinemann, London, 1970.

Shea, William R. *Galileo's Intellectual Revolution*. Neale Watson Academic, New York, 1972.

Shea, William R., and Mariano Artigas. *Galileo in Rome: The Rise and Fall of a Troublesome Genius*. Oxford University Press, New York, 2004.

Taton, R., and C. Wilson, eds. *Planetary Astronomy from the Renaissance to the Rise of Astrophysics: Part A. Tycho Brahe to Newton* [*The General History of Astronomy*]. Cambridge University Press, Cambridge, U.K., 1989.

Van Helden, Albert. *Catalogue of Early Telescopes*. Istituto e Museo di Storia della Scienza/Giunti, Florence, 1999.

Wlassics, Tibor. *Galilei critico letterario*. Longo Editore, Ravenna, 1974.

Articles and Chapters

Adams, C. W. "A Note on Galileo's Determination of the Height of Lunar Mountains." *Isis*, vol. 17, 1932, pp. 427–429.

Beretta, Francesco. "Le Procès de Galilée et les Archives du Saint-Office." *Revue des sciences philosophiques et théologiques*, vol. 83, 1999, pp. 441–490.

Cajori, Florian. "History of the Determination of the Heights of Mountains." *Isis*, vol. 12, 1929, pp. 482–512.

Campanella, Tommaso. "The Defense of Galileo" (trans. Grant McColley). *Smith College Studies in History*, vol. 22, nos. 3–4, April–July 1937.

Chappell, Miles. "Cigoli, Galileo, and *Invidia*." *Art Bulletin*, vol. 57, no. 1, March 1975, pp. 91–98.

"Cigoli, Ludovico [Ludovico Cardi detto il Cigoli]," in *Bollettino della Accademia degli Euteleti della città di San Miniato*, S. Miniato, Florence, n.d.

Drake, Stillman. "Galileo's Steps to Full Copernicanism and Back." *Studies in the History and Philosophy of Science*, vol. 18, no. 1, 1987, pp. 93–105.

Favaro, Antonio. "Gli oppositori di Galileo, VI; Maffeo Barberini." *Atti del Reale Istituto Veneto di Scienze, Lettere ed Arti*, vol. 80, 1920–21, Part 2, pp. 1–46.

Garzend, L. "Si Galilée pouvait être juridiquement torturé." *Revue des questions historiques*, vols. 90 and 91, 1911–12, pp. 353–389 and pp. 36–67.

Giacchi, Orio. "Considerazioni giuridiche sui due processi contro Galileo," in *Nel Terzo Centenario della morte di Galileo: saggi e conferenze*. Università Cattolica del Sacro Cuore, Vita e Pensiero, Milan, pp. 383–406.

Lindberg, David C., with Nicholas H. Steneck. "The Sense of Vision and the Origins of Modern Science," in Allen G. Debus, ed., *Science, Medicine and Society in the Renaissance: Essays in Honor of Walter Pagel*. Heineman, London, 1972, vol. 1, pp. 29–46.

Morel, Philippe. "Morfologia delle cupole dipinte da Correggio a Lanfranco." *Bolletino d'arte*, ser. 6., vol. 69, no. 23, 1984.

Popkin, Richard H. "Spinoza and Bible Scholarship," in Don Garrett, ed., *The Cambridge Companion to Spinoza*. Cambridge University Press, Cambridge, U.K., 1996, pp. 383–407.

Ronan, Colin A., G. I.'E Turner, et al. "Was There an Elizabethan Telescope?" *Bulletin of the Scientific Instrument Society*, vol. 37, 1993, pp. 2–10.

Sluiter, Engel. "The Telescope before Galileo." *Journal for the History of Astronomy*, vol. 28, 1997, pp. 223–234.

Van Helden, Albert. "The Invention of the Telescope." *Transactions of the American Philosophical Society*, vol. 67, no. 4, 1977, pp. 5–64.

———. "Galileo and the Telescope," in Paolo Galluzzo, ed., *Novità celesti e crisi del sapere*. Giunti, Florence, 1984, pp. 149–158.

———. "The Telescope and Authority from Galileo to Cassini." *Osiris*, 2nd ser., vol. 9, "Instruments," 1994, pp. 8–29.

Van Helden, Albert, and Mary Winkler. "Representing the Heavens: Galileo and Visual Astronomy." *Isis*, vol. 82, no. 2, June 1992, pp. 195–217.

Illustration Credits

Page 34: Dan Hofstadter and Nino Mendolia.

Page 41: Courtesy Casa Buonarroti, Florence.

Page 65: Courtesy Tom Pope and Jim Mosher.

Page 69: From *Le opere di Galileo Galilei*, Barbera, Florence, 1929–39.

Page 77: Courtesy Houghton Library of the Harvard College Library, Typ525.96.791(A)F.

Page 81: Dan Hofstadter and Nino Mendolia.

Page 85: From *Le opere di Galileo Galileo*, Barbera, Florence, 1929–39.

Page 87: Dan Hofstadter and Nino Mendolia.

Page 88: Dan Hofstadter and Nino Mendolia.

Page 97: Dan Hofstadter and Nino Mendolia.

Page 98: From *Le opere di Galileo Galileo*, Barbera, Florence, 1929–39.

Page 111: Early photograph restored by Nino Mendolia.

Page 113: Drawing by Dan Hofstadter.

Page 116: *Perspectiva pictorum et architectorum Andreae Putei*, Rome, 1673–1700.

Page 117: Nino Mendolia.

Page 203: Courtesy of Biblioteca Apostolica Vaticana.

Page 208: From *Le opere di Galileo Galileo, 1929–39*, published by Barbera, Florence.

Page 210: Courtesy of Gabinetto Fotografico Uffizi Museum, Florence.

Index

Page numbers in *italics* refer to illustrations.